Foreword

You have in your hand the second *Official Guide* designed to help students prepare for examinations produced under the auspices of the Division of Chemical Education (DivCHED) of the American Chemical Society (ACS). The first guide, published in 1998 for general chemistry students, has proved to be a very popular—some would say absolutely essential—resource for students preparing to sit for ACS general chemistry examinations. The printer's ink from the first *Official Guide* was still wet when requests started pouring in for an organic chemistry study guide.

As we have done so often in the past, we called on colleagues in the chemistry education community to help us put the guide together. A distinguished group of active, and respected, organic faculty members from all over the country accepted the challenge, rolled up their sleeves, and went to work. (A complete list of who did what appears on the *Acknowledgements* page.) This *Official Guide*, more than two years in the making, is the culmination of their wonderful *volunteer* efforts.

As a discipline, chemistry is surely unique in the extent to which its practitioners provide beneficial volunteer service to the teaching community. ACS exams have been produced by volunteer teacher-experts for more than seventy years. Other projects of the Examinations Institute invariably benefit from the abundance of donated time, effort, and talent. The result is that invariably high quality chemistry assessment materials are made available to the teaching (and learning) community at a fraction of their real value.

The two *Official Guides* that have been released so far are intended to be ancillary student materials, particularly in courses that use ACS exams. As we noted in the Foreword to the general chemistry study guide, the care that goes into producing ACS exams may be lost on students who view the exams as foreign and unfamiliar. The purpose of this series of guides is to remove any barriers that might stand in the way of students demonstrating their knowledge of chemistry. The extent to which this goal is achieved in the organic study guide will become known only as future generations of chemistry students sit for an ACS exam in organic chemistry.

We wish them the best.

I. Dwaine Eubanks
Lucy T. Eubanks

Clemson, South Carolina
February, 2002

Acknowledgements

As far as we are able to ascertain, this *Official Guide* is the largest volunteer undertaking in the history of the Examinations Institute. The individuals who agreed to serve on the committee to deliver the technical content must have had many second thoughts during the two years the guide was in the making. Nonetheless, they came through with an excellent piece of work—an in-depth guide that will benefit students studying organic chemistry for many years to come. The frequently felt, but often-unsaid, *"Thank you!"* from our students is the real reward for this kind of endeavor. A *"Nice job, guys!"* from colleagues is also important. Well, here's our *"Nice job, guys, and thanks a million!"* from the staff of the Examinations Institute.

Organic *Official Guide* Committee	
Marshall W. Logue, Chair	Michigan Tech University
John Michael Ferguson	University of Central Oklahoma
James G. MacMillan	University of Northern Iowa
H. Mark Perks	University of Maryland, Baltimore County
Ron Wikholm	University of Connecticut

This *Official Guide* also benefited from the careful proofreading by several colleagues. We extend our special thanks to these faculty members.

Carmen M. Simone	Casper College
Richard W. Morrison	University of Georgia
Tamera S. Jahnke	Southwest Missouri State University
Michael A. McKinney	Marquette University
Stephen E. Branz	San Jose State University
James O. Schreck	University of Northern Colorado
Susan M. Schelble	University of Colorado at Denver
Amina K. El-Ashmawy	Collin County Community College
Michelle M. Dose	Hardin–Simmons University

The personnel of the ACS Division of Chemical Education Examinations Institute played a central role in helping us to produce ***Preparing for Your ACS Examination in Organic Chemistry: The Official Guide.*** A very special thank you for all of the work involved is owed to our staff members.

Brenda A. Rathz	Clemson University
Sherri P. Morrison	Clemson University
W. Sam Burroughs	Clemson University

While all of these reviewers have been very helpful in finding problems large and small, any remaining errors are solely our responsibility. You can assist us in the preparation of an even better product by notifying the Exams Institute of any errors you may find.

I. Dwaine Eubanks
Lucy T. Eubanks

Clemson, South Carolina
February, 2002

How to Use This Book

Every year, students tell us that they really know more than they demonstrate on their final examination in organic chemistry. Sometimes the questions are described as "tricky"; sometimes insufficient time is thought to be the problem; and sometimes the material is judged to differ from what was covered in class. These problems most often result from chemistry having been learned as a set of formulas and techniques, rather than as a coherent set of conceptual models that enables comprehension of the submicroscopic world. We urge you, as you are learning chemistry, to strive for genuine understanding of the concepts and models. For example, rote learning of named organic reactions that are meaningless to you is *not* learning chemistry; and that knowledge will not serve you well at exam time.

The major divisions of this book correspond to the common groupings of topics covered by ACS exams for organic chemistry. If you are taking an exam that covers the full year of organic chemistry, all of the topic groups can be expected on the exam.

Each topic group is introduced with a short discussion of the important ideas, concepts, and knowledge that are most frequently stressed in organic chemistry courses. *These discussions are not a substitute for studying your textbook, working the problems there, and discussing the challenging ideas with your teachers and fellow students.* Rather, they are reminders of what you have studied and how it fits into a larger understanding of that part of the natural world that we call chemistry.

Next, questions are presented that address those ideas. These questions have been drawn from past ACS exams, and they should give you a good idea of the depth and range of understanding that is expected. Each question is dissected, and you will see how chemists think through each of the questions to reach the intended response. You will also see how choosing various wrong responses reveal misconceptions, careless computation, misapplication of principles, or misunderstandings of the material. Knowing how each incorrect answer is generated will assist you in diagnosing problems with your grasp of the principle being examined.

The most effective way to use this book is to answer each **Study Question** before looking at the discussion of the item. Jot down a note of how you arrived at the answer you chose. Next, look at the analysis of the question. Compare your approach with that of the experts. If you missed the item, do you understand why? If you chose the correct response, was it based on understanding or chance? After you have spent time with the **Study Questions**, treat the **Practice Questions** as if they were an actual exam. Allow yourself 50 minutes, and write down your response to each question. Finally, score yourself. Go over the practice questions again. Write down what you needed to know before you could answer the question; and write down how you should think the problem through to reach the intended answer.

This book is designed to help you demonstrate your *real* knowledge of chemistry. When you take a organic chemistry examination prepared by the American Chemical Society Examinations Institute, you should be permitted to concentrate on demonstrating your knowledge of chemistry, and not on the structure of the examination. We sincerely hope that *The Official Guide* will enrich your study of chemistry, and minimize the trauma of effectively demonstrating what you have learned.

Sample Instructions

You will find that the front cover of an ACS Exam will have a set of instructions very similar to this. This initial set of instructions is meant for both the faculty member who administers the exam and the student taking the exam. You will be well advised to read the entire set of instructions while waiting for the exam to begin. This sample is from the organic chemistry exam released in 2002.

TO THE EXAMINER: This test is designed to be taken with a special answer sheet on which the student records his or her responses. All answers are to be marked on this answer sheet, not in the test booklet. Each student should be provided with a test booklet, one answer sheet, and scratch paper; all of which must be turned in at the end of the examination period. The test is to be available to the students **only** during the examination period. For complete instructions refer to the *Directions for Administering Examinations.* Norms are based on:

Score = Number of right answers
70 items — 120 minutes

TO THE STUDENT: DO NOT WRITE ANYTHING IN THIS BOOKLET! Do not turn this page until your instructor gives the signal to begin. When you are told to begin work, open the booklet and read the directions on page 2.

Note the **restriction** on the **time** for administering the exam. This restriction applies to allow your results to be compared to national norms, ensuring that all students have had the same tools and time to display their knowledge. Your instructor may choose not to follow the time restriction, particularly if they do not plan to submit your data as part of the national process for calculating norms.

Be sure to notice that scoring is based **only** on the number of right answers. There is no penalty, therefore, for making a reasonable guess even if you are not completely sure of the correct answer. Often you will be able to narrow the choice to two possibilities, improving your odds at success. You will need to keep moving throughout the examination period, for it is to your advantage to attempt every question. Do not assume that the questions become harder as you progress through an ACS Exam. Questions are not grouped by difficulty, but by topic.

Next, here is a sample of the directions you will find at the beginning of an ACS exam.

DIRECTIONS

- When you have selected your answer, blacken the corresponding space on the answer sheet with a soft, black #2 pencil. Make a heavy, full mark, but no stray marks. If you decide to change an answer, erase the unwanted mark very carefully.
- Make no marks in the test booklet. Do all calculations on scratch paper provided by your instructor.
- There is only one correct answer to each question. Any questions for which more than one response has been blackened **will not be counted.**
- Your score is based solely on the number of questions you answer correctly. **It is to your advantage to answer every question.**
- The best strategy is to arrive at your own answer to a question before looking at the choices. Otherwise, you may be misled by plausible, but incorrect, responses.

Pay close attention to the mechanical aspects of these directions. Marking your answers without erasures helps to create a very clean answer sheet that can be read without error. As you look at your Scantron® sheet before the end of the exam period, be sure that you check that every question has been attempted, and that only one choice has been made per question. As was the case with the cover instructions, note that your attention is again directed to the fact that the score is based on the total number of questions that you answer correctly. You also can expect a reasonable distribution of **A**, **B**, **C**, and **D** responses, something that is not necessarily true for the distribution of questions in *The Official Guide*.

Table of Contents

Nomenclature

Nomenclature is an important part of chemistry. Being able to determine the proper name for a structure or writing out the structure corresponding to the name of a compound is essential to communicate chemical information. Nomenclature is the vocabulary of chemistry. It is impossible to communicate effectively about chemistry without knowing the proper names of compounds. Fortunately, organic nomenclature is quite well organized and systematic.

The most widespread system of nomenclature is that devised by the International Union on Pure and Applied Chemistry (IUPAC). It is more useful to know how to name or identify a few basic structural units than it is to memorize every nuance of IUPAC nomenclature. A solid knowledge of identifying and naming chains, aliphatic and aromatic rings, alcohols, alkenes, alkynes, carboxylic acids (and derivatives), aldehydes and ketones is essential. Practicing a wide variety of nomenclature problems is the best way to become proficient in organic nomenclature.

A long, detailed guide on how to name every kind of organic compound is not very useful, because there is simply too much to remember. What you need for most nomenclature questions is encapsulated in this brief list of necessary skills and helpful hints:

(1) It is *essential* to know the names and structures of the major functional groups. Also, you must know how to incorporate each functional group into the name of a compound. Nomenclature questions often give nonexistent functional groups as incorrect choices. If you can recognize the bogus groups, you can eliminate those choices from consideration.

(2) Nomenclature questions often involve distinguishing among various types of isomers. It is important to know the basic definitions of isomers (constitutional isomers, stereoisomers, enantiomers, diastereomers, and so on). You should be able to determine the (*R*)/(*S*) configuration of stereo centers. Other important naming conventions to know are when (and how) to use *cis* and *trans* or (*Z*) and (*E*) when describing alkenes; *cis* and *trans* to describe the relationship of substituents on cyclic compounds; and *ortho*, *meta* and *para* when dealing with aromatic compounds.

(3) Most nomenclature problems should take no more than 2 minutes to complete. Problems with more than three functional groups, or with very unusual functional groups, are seldom included on exams.

Study Questions

NM-1. In the structure shown, the functional groups in the triangle, rectangle, pentagon and circle (in this same order) are

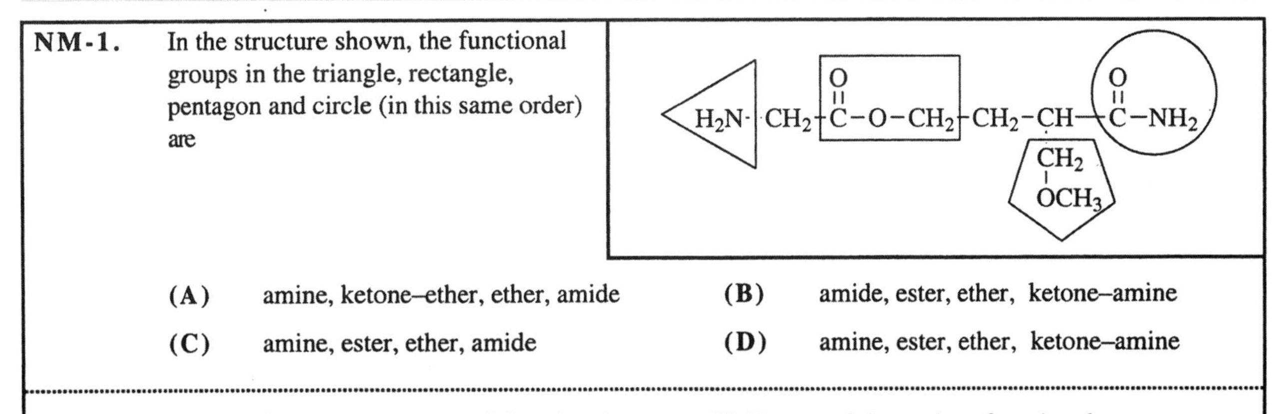

(A) amine, ketone–ether, ether, amide

(B) amide, ester, ether, ketone–amine

(C) amine, ester, ether, amide

(D) amine, ester, ether, ketone–amine

Knowledge Required: (1) Recognition of functional groups. (2) Names of the various functional groups.

Thinking it Through: The functional group inside the triangle is an amine; an amide would have a nitrogen atom attached to a carbonyl group. Therefore, choice **(B)** is eliminated. The functional group inside the rectangle is an ester, not a falsely named ketone–ether group. Thus, choice **(A)** can be eliminated. This leaves choices **(C)** and **(D)**, both of which have the pentagon containing an ether functional group. The group inside the circle has an NH_2 attached to a carbonyl group and is an amide. Thus, choice **(C)** is the correct answer. Choice **(D)** has a falsely named ketone–amine functional group.

NM-2. What is a correct IUPAC name for this compound?

$$
\begin{array}{l}
\qquad\qquad\qquad\qquad CH_3-CH-CH_3 \\
\qquad\qquad\qquad\qquad\qquad | \\
\qquad CH_2-CH_2-CH_2-CH_2 \\
\qquad | \\
CH_3-C-CH_3 \\
\qquad | \\
\qquad CH_3
\end{array}
$$

(A) 1,1,1,6-tetramethylheptane

(B) 1-*tert*-butyl-4-isopropylbutane

(C) 2,2,7-trimethyloctane

(D) 1-*tert*-butyl-4-isopropyloctane

Knowledge Required: (1) Procedure for naming alkanes. (2) Identification of alkyl groups.

Thinking it Through: The number of carbon atoms in the longest continuous chain is eight, and the parent hydrocarbon is an octane. Thus choices **(A)** and **(B)** cannot be correct. Although choice **(B)** does contain an isopropyl group and a *tert*-butyl group, some of the carbon atoms that make up these groups are actually part of the main chain. Naming the compound as a substituted butane does not indicate that the longest chain has eight carbon atoms. Choice **(D)** can be rejected because the so-called isopropyl group and a *tert*-butyl group are counted both as substituents and as part of the main chain. Such 'double counting' is not allowed. Choice **(C)** is correct because the eight-carbon atom chain and the three methyl groups are clearly identified with the lowest set of locants (numbers). In this case, carbon atom number 1 is at the lower left.

Note that the name "2,7,7-trimethyloctane" (resulting from numbering from the opposite end of the molecule) is not acceptable because the two methyl groups on the carbon atom next to the left end of the chain give this end priority in numbering.

NM-3. What is the IUPAC name for this compound?

$$
\begin{array}{l}
\qquad\qquad\quad OH \qquad\quad O \\
\qquad\qquad\quad | \qquad\qquad \| \\
CH_3-CH-CH-CH_2-C-CH_3 \\
\qquad\quad | \\
\qquad\quad CH_2-CH_3
\end{array}
$$

(A) 4-hydroxy-5-methyl-2-heptanone

(B) 5-ethyl-4-hydroxy-2-hexanone

(C) 2-ethyl-3-hydroxy-5-hexanone

(D) 4-hydroxy-3-methyl-6-heptanone

Knowledge Required: (1) Procedure for naming and numbering ketones. (2) Setting priorities of functional groups when assigning numbers.

Thinking it Through: Because ketones (and aldehydes) have higher priority than alcohols, the compound should be named as a ketone rather than as an alcohol. The alcohol should be named as a substituent on the ketone molecule. (Related note: carboxylic acids and their derivatives have higher priority than ketones, so a carboxylic acid carbon atom is *always* number one.) When assigning names to ketones, find the longest continuous carbon atom chain containing the carbonyl group. Here, the longest chain has seven carbon atoms; therefore the compound name is based upon heptanone. Next, identify the substituents, which are a hydroxyl group and a methyl group. Number the carbon atoms starting from the end that will give the carbonyl carbon atom the lowest possible number.

Choice **(B)** is incorrect because it designates the horizontal carbon atom chain as the main chain. This chain has only six carbon atoms and is not the longest present in the molecule. This is a common error; the longest chain is not necessarily the horizontal one. Choice **(C)** is incorrect for the same reason. Moreover, the hydroxyl group is given higher priority in numbering than the carbonyl group, which is also incorrect. Choice **(D)** is incorrect because the hydroxyl group is incorrectly given higher priority than the carbonyl group. Choice **(A)** is the correct answer.

NM-4. What is the IUPAC name for this compound?

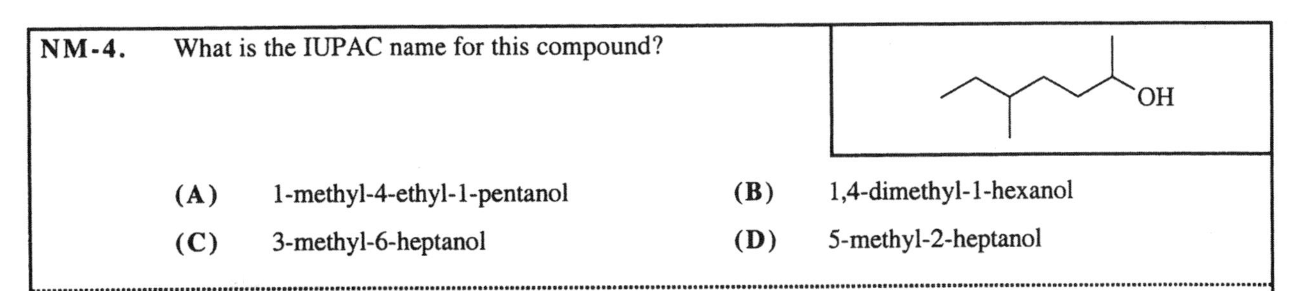

(A) 1-methyl-4-ethyl-1-pentanol (B) 1,4-dimethyl-1-hexanol

(C) 3-methyl-6-heptanol (D) 5-methyl-2-heptanol

Knowledge Required: (1) Procedure for naming alcohols. (2) Identification of alkyl groups.

Thinking it Through: The first step in naming alcohols is to identify the longest continuous carbon atom chain containing the hydroxyl group. Here, the longest chain containing the –OH has seven carbon atoms, so the compound is a heptane derivative. Second, identify the substituents that are present. Here, a methyl group is present. Third, number the chain (or ring) to give the hydroxyl group the lowest number possible.

Choice **(A)** is incorrect because its name indicates the compound is a pentane derivative. Choice **(B)** is also incorrect because it does not correctly indicate the longest chain present. Choice **(C)** is wrong because even though the compound is correctly identified as a heptane derivative, the molecule is numbered starting at the wrong end of the chain. Choice **(D)** is the correct answer because the seven-carbon atom chain is correctly identified and the hydroxyl group is correctly assigned the lowest possible number.

NM-5. What is the IUPAC name for this compound?

$CH_3CH_2CH_2CH_2CHCH_2CH_2CH_3$, with $C(=O)O^- \ Na^+$ attached to the CH carbon

(A) sodium α-propylcaproate (B) sodium 2-propylhexanoate

(C) sodium 4-octanoate (D) sodium 1-pentanecarboxylate

Knowledge Required: Procedure for naming carboxylic acids.

Thinking it Through: When assigning names to carboxylic acids, first find the longest continuous carbon atom chain containing the carboxylic acid group. The carboxyl group counts as a carbon atom when computing the length of the chain. In this case, the longest chain has six carbon atoms, so the compound is a derivative of hexanoic acid. As before, the second step is to locate and identify the substituents. A propyl group is present. When numbering a carboxylic acid, the carboxylate carbon atom is always number one, so the propyl group is on C–2. The compound is a sodium salt of 2-propylhexanoic acid, so the correct answer is **(B)**.

Choice **(A)** is incorrect because it uses non-IUPAC terms. Moreover, caproic acid is a trivial name for hexanoic acid. Choice **(C)** can be rejected because the name indicates that the carboxylic acid is a substituent on an eight-carbon atom main chain. This is not part of IUPAC naming procedure. Choice **(D)** is incorrect because it states that the main chain has five carbon atoms and that the carboxyl group is a substituent on it. Remember that the carboxyl group counts when determining how many carbon atoms are in a chain.

NM-6. What is the IUPAC name for this compound?

$$CH_3-\overset{\overset{\displaystyle O}{||}}{C}-OCH_2CH_2CH_3$$

(A) methyl butanoate **(B)** methyl propanoate

(C) propyl methanoate **(D)** propyl ethanoate

Knowledge Required: Procedure for naming esters.

Thinking it Through: Esters are derivatives of carboxylic acids and alcohols. To name an ester, identify the alkyl group that is in the "alcohol" portion of the molecule and name it, then identify and name the "acid" part of the molecule. The ester is named with the alkyl part first, and the acid part is named as a carboxylate. For example, the ester of benzoic acid and methanol would be named as methyl benzoate.

$$R-\overset{\overset{\displaystyle O}{||}}{C}-OH + H-O-R' \longrightarrow \underbrace{R-\overset{\overset{\displaystyle O}{||}}{C}}_{\text{"acid part"}}-\underbrace{O-R'}_{\text{"alcohol part"}} + H_2O\ .$$

In the problem, the alcohol portion is derived from 1-propanol. Therefore, the alkyl group is a propyl group. The carboxylic acid portion of the molecule is derived from ethanoic acid. Thus the answer is propyl ethanoate, which is **(D)**.

A helpful hint: Use "carbon atom counts" to eliminate choices which do not correspond to that of the given structure. For example, the ester $CH_3CH_2CH_2COOCH_2CH_3$ contains a total of six carbon atoms. Therefore any suggested name whose components would contain other than six carbon atoms must be wrong and that choice can be eliminated. If the name "propyl ethanoate" were a possible answer, it could be rejected immediately because the propyl group contains three carbon atoms and ethanoate contains two carbon atoms, for a total of five. Similarly, if the question asked which of four structures correspond with a given name, it should be possible to count up the number of carbon atoms suggested in the name and reject any possible structures that differ from this number. This approach will often narrow the choices, but it is usually necessary to examine the alkyl and acid portions of the ester closely to be absolutely sure.

Using this "carbon atoms counts" approach, choice **(B)** can be rejected because methyl propanoate is derived from methanol (1 carbon atom) and propanoic acid (3 carbon atoms). The molecule shown in the question has five carbon atoms, so choice **(B)** can be eliminated from consideration. Similarly, choice **(C)** is eliminated because the name describes an ester derived from a 3-carbon-atom alcohol and a 1-carbon atom carboxylic acid. If two or more choices have the correct number of carbon atoms, then look at the structures of the alcohol's alkyl group and at the carboxylic acid portion to determine which choice is correct. For example, choice **(A)** cannot be the correct answer because methyl butanoate would have a methyl group as the alcohol part of the ester and be derived from butanoic acid. A glance at the given molecule shows that it is not derived from butanoic acid, so this choice must be incorrect.

NM-7. What is the IUPAC name for this compound?

$$CH_3-CH{=}CH-CH(CH_3)-C(=O)-OH$$

(A) 4-methyl-2-pentenoic acid

(B) 2-methyl-3-pentenoic acid

(C) 5-hydroxy-4-methyl-2-penten-5-one

(D) 1-hydroxy-2-methyl-3-penten-1-one

Knowledge Required: (1) Recognition of functional groups. (2) Procedure for naming carboxylic acids. (3) Procedure for naming alkenes. (4) Setting priorities of functional groups when assigning numbers.

Thinking it Through: When solving nomenclature problems, immediately reject any choice containing a nonexistent functional group. For example, if a molecule contains the $CONH_2$ group, one of the choices may contain the phrases such as "amino ketone" or "ketoamine." Note that these two terms are *not* true functional groups. Correct terms for this group are "amide" or "carboxamide." Similarly, take care not to mistake one real functional group for another, such as amide for amine or vice versa.

The procedure for determining the proper name for this compound is similar to the previous problems. The two functional groups present are a carboxylic acid and an alkene. First number the main chain containing the carboxylic acid. This chain has five carbon atoms, so the name will be derived from pentanoic acid. Choices **(C)** and **(D)** can be eliminated because they call the carboxylic acid group a hydroxyketone—a nonexistent functional group. (Note that there are ketones containing hydroxyl groups, but they *never* have the hydroxyl group attached to the carbonyl carbon atom. A hydroxyl attached to a carbonyl group is a carboxylic acid, period.)

Next, the chain has to be numbered. Carboxylic acid groups have absolute priority in numbering, so the carboxylic acid carbon atom is always C–1. Therefore, the methyl group is on C–2, and the double bond is between C–3 and C–4. Choice **(A)** can be eliminated because the chain is numbered from the end opposite the carboxylic acid. Choice **(B)** is the correct answer.

NM-8. Which is the structure of *N*-benzylethanamide?

(A) $CH_3-C(=O)-NH-CH_2-C_6H_5$

(B) $CH_3-CH_2-NH-CH_2-C_6H_5$

(C) $CH_3-CH_2-NH-C_6H_5$

(D) $CH_3-C(=O)-NH-C_6H_5$

Knowledge Required: (1) Procedure for naming amides. (2) Distinction between the phenyl and benzyl groups.

Thinking it Through: Consider the name of the compound: *N*-benzylethanamide. This name says it is based on ethanamide (the primary amide derived from ethanoic acid) so it is based on the general structure CH_3CONH_2. The name also says that there is a benzyl group attached to the nitrogen atom. The benzyl group is C_6H_5–CH_2–. Therefore the correct answer is choice **(A)**. Choice **(C)** can be eliminated because it does not contain an amide group. This compound is *N*-ethylaniline. Choice **(B)** is also incorrect because there is no amide group present. This compound is benzylethylamine. Choice **(D)** contains an amide group and is indeed derived from ethanamide. However, note that the group attached to the nitrogen atom is a phenyl group, not a benzyl group. It is all too easy to confuse the phenyl and benzyl groups, so care must be taken when looking at names or structures.

NM-9. What is the IUPAC name for this compound?

OH
H
Br
H

(A) *ortho*-bromocyclohexanol **(B)** *endo*-2-bromocyclohexanol

(C) *cis*-2-bromocyclohexanol **(D)** *trans*-2-bromocyclohexanol

Knowledge Required: (1) Procedure for naming alcohols. (2) Procedure for naming cyclic compounds. (3) Designating stereochemistry in cyclic compounds, especially cyclohexane derivatives.

Thinking it Through: Each of the choices includes a term describing the relative locations of two substituents. Only one of these is correct. Choice **(A)** can be rejected because the term "*ortho*" is used in conjunction with naming benzene derivatives (The term "*ortho*" is used to describe two substituents which are on adjacent carbon atoms on a benzene ring.) Choice **(B)** is incorrect because the term "*endo*" is not used when naming cyclic compounds. (The term "*endo*" is often used to describe the position of substituents on a bridged bicyclic ring system.) To distinguish between the remaining two choices, examine the relative positions of the bromo and hydroxyl groups in the structure. The bromine atom is on the top face of the ring, as is the hydroxyl group. Therefore the two substituents are *cis* in configuration. Therefore the correct answer is choice **(C)**. If the two substituents were on opposite faces, they would be *trans*.

NM-10. What is the IUPAC name for this compound?

CH_3 CH_3
C=C
H $CH_2CH(CH_3)_2$

(A) (*E*)-3,5-dimethyl-2-hexene **(B)** (*Z*)-3,5-dimethyl-2-hexene

(C) (*Z*)-1,2-dimethyl-2-isobutylethylene **(D)** (*E*)-3,5,5-trimethyl-2-pentene

Knowledge Required: (1) Procedure for naming alkenes. (2) Procedure for determining *Z*/*E* relationships.

Thinking it Through: When naming alkenes, the first step is to locate the longest chain that contains the double bond(s). In this case, the longest chain contains six carbon atoms, so the compound is a hexene derivative. This eliminates choices **(C)** and **(D)** from consideration. Next, the substituents are identified and their locations determined. In this case, choices **(A)** and **(B)** both correctly indicate that there are two methyl groups present, on C–3 and C–5.

The third step when naming alkenes is to determine whether *Z*/*E* isomers can exist. Here, the two groups attached to C–2 are different, as are the two attached to C–3. Therefore *Z*/*E* isomers are possible. To determine whether the alkene is *Z* or *E*, use the Cahn–Ingold–Prelog convention to assign priorities to the groups at each end of the double bond. (The Cahn–Ingold–Prelog convention is also used to assign (*R*) or (*S*) to chiral compounds.) On C–2, the methyl group has higher priority than the hydrogen atom, and on C–3, the isobutyl group has higher priority than the methyl group. Because the two higher-ranking alkyl groups are on opposite sides of the double bond (*trans*), it is the *E* isomer. Choice **(A)** is the correct answer. If the two higher-ranking groups were on the same side of the double bond (*cis*), then it would be the *Z* isomer.

Helpful hints: (1) The *Z* comes from the German word *zusammen*, "together," whereas *E* comes from *entgegen*, "opposite." (2) To help keep *Z* and *E* straight, one can remember the saying "*Z* means the higher ranking groups are on the 'Zame Zide' (same side)."

NM-11. What is the proper name for this compound?

(A) 4-bromo-2-ethyl-1-nitrobenzene **(B)** 4-bromo-2-ethylaniline

(C) *p*-bromo-*o*-ethylnitrobenzene **(D)** *p*-bromo-*o*-ethylaniline

Knowledge Required: (1) Procedure for naming aromatic compounds. (2) Proper use of *ortho*, *meta* and *para* .

Thinking it Through: The $-NH_2$ group means that instead of being named as a substituted benzene, it is named as a substituted aniline. Therefore, choices **(A)** and **(C)** can be eliminated. A "nitro" group would be NO_2. When numbering aromatic rings, it is generally done to give the lowest possible set of numbers. When a substituent causes the parent name to be something other than benzene (such as the $-NH_2$ changing the name to "aniline"), the carbon atom with that substituent is designated C–1. The terms *ortho*, *meta* and *para* are used when benzene derivatives possess exactly *two* substituents; therefore, choice **(D)** can be rejected. Choice **(B)** is the correct answer.

NM-12. Which functional groups does a peptide bond contain?

(A) ester **(B)** aromatic ring

(C) acetal **(D)** amide

Knowledge Required: (1) Identification of functional groups. (2) Functional groups in biological molecules.

Thinking it Through: Peptide bonds are formed when amino acids join to form peptides and proteins. Every amino acid contains carboxylic acid and amine functional groups, and most also contain functional groups in their side chains. To form a peptide bond, the amine group on one amino acid reacts with the carboxylic acid portion of another amino acid molecule, forming an amide. Thus, choice **(D)** must be the correct answer. Choice **(A)** can be rejected because esters are found in lipids such as triacylglycerols (fats and oils) and waxes. Aromatic rings are found in several amino acids, but they are part of the side chains, rather than part of the peptide bonds. Therefore, choice **(B)** can be rejected. Choice **(C)** cannot be the correct answer because an acetal has two alkoxy groups bonded to the same carbon atom.

Practice Questions

1. What is the IUPAC name for this compound?

$$CH_3-\overset{\displaystyle CH_3}{\underset{\displaystyle CH_3}{C}}-CH_2-\overset{\displaystyle OH}{CH}-CH_2-CH_3$$

(A) 1-*tert*-butyl-2-butanol

(B) 5,5-dimethyl-3-hexanol

(C) 2,2-dimethyl-4-hexanol

(D) 1,1,1-trimethyl-3-pentanol

2. What is the proper name for this compound?

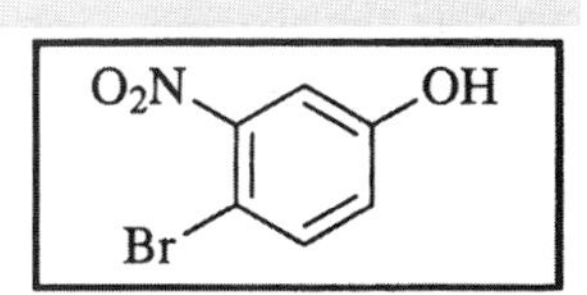

(A) 4-bromo-3-nitrophenyl

(B) *p*-bromo-*m*-nitrophenol

(C) 4-bromo-3-nitrophenol

(D) 1-bromo-2-nitrophenol

3. Which compound is (*E*)-1,2-dichloro-2-pentene?

(A) $CH_3CH_2(H)C=C(CH_2Cl)(Cl)$

(B) $CH_3CH_2(H)C=C(Cl)(CH_2Cl)$

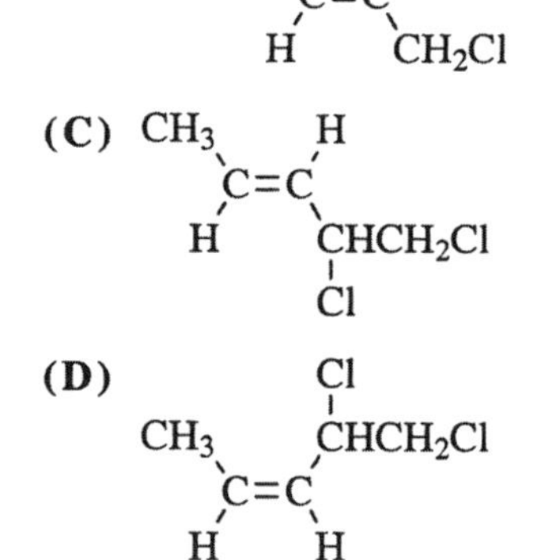

(C) $CH_3(H)C=C(H)(CHClCH_2Cl)$

(D) $CH_3(H)C=C(H)(CHClCH_2Cl)$

4. What is the IUPAC name for this structure?

Br O

(A) 3-bromo-4-methylheptanone

(B) 5-bromo-4-methylheptanone

(C) 5-bromo-4-methyl-3-heptanone

(D) 3-bromo-4-methyl-5-heptanone

5. Which is the structure of (*Z*)-3,5-dimethyl-3-nonene?

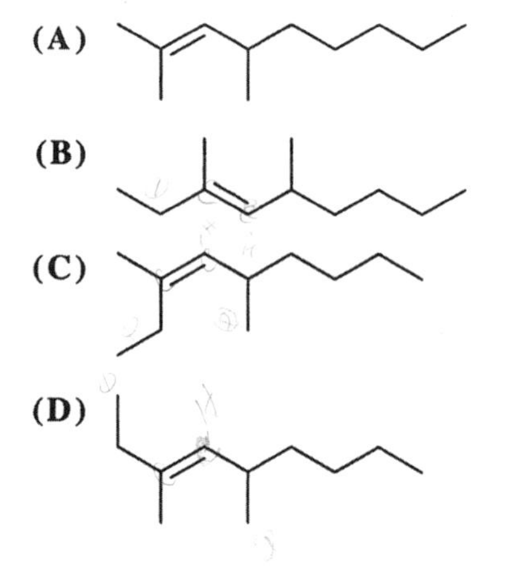

(A)

(B)

(C)

(D)

6. What is the IUPAC name for this compound?

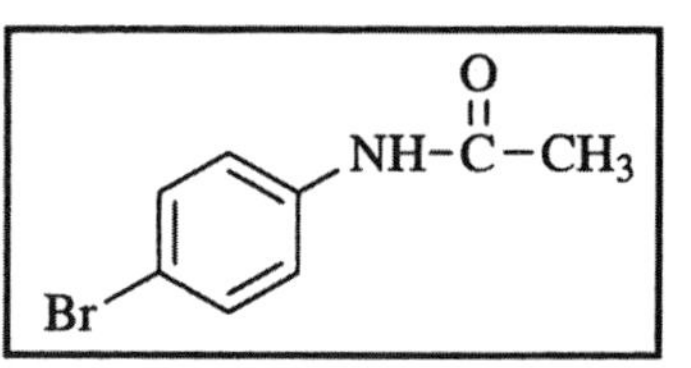

(A) *N*-(4-bromophenyl)ethanamide

(B) *p*-bromobenzylethanamide

(C) 4-bromobenzamide

(D) *N*-(4-bromobenzyl)ethanamide

7. What is the IUPAC name for this compound?

$CH_3CH_2(CH_3)C=C(H)CH_2-CH(OH)-CH_3$

(A) (*Z*)-5-ethyl-4-hexen-2-ol

(B) (*Z*)-3-methyl-3-hepten-6-ol

(C) (*E*)-5-methyl-4-hepten-2-ol

(D) (*Z*)-5-methyl-4-hepten-2-ol

8. Which is the structure of *N*-phenyl-4-nitrobenzamide?

(A)

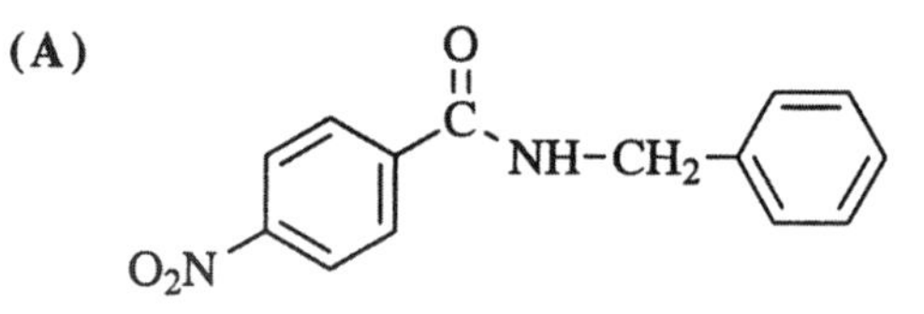

(B) $C_6H_5-C(=O)-NH-CH_2-C_6H_4-NO_2$

(C) $O_2N-C_6H_4-C(=O)-NH-C_6H_5$

(D)

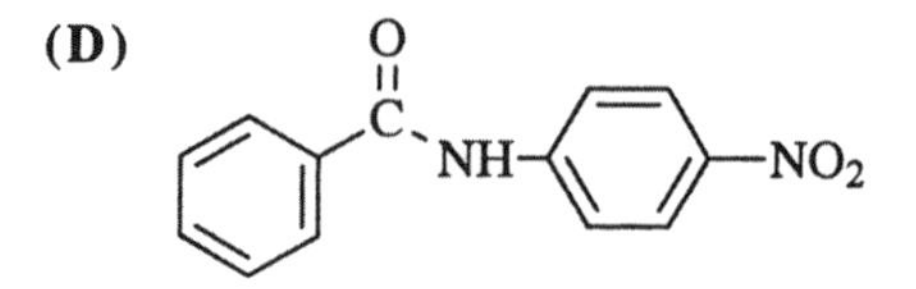

9. What is the IUPAC name of this compound?

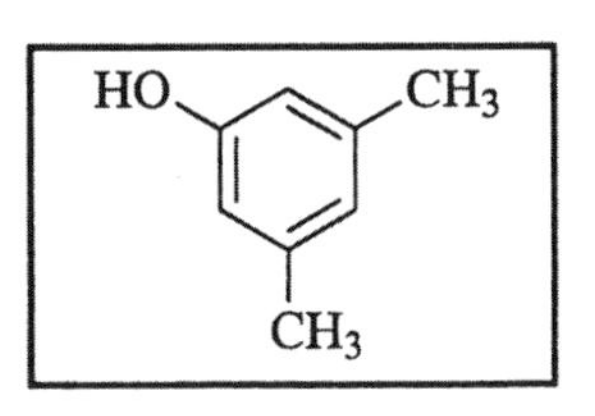

(A) *m*-hydroxy-*m*-xylene

(B) 3,5-dimethylphenol

(C) 2,4-dimethyl-6-hydroxybenzene

(D) 3-hydroxy-5-methyltoluene

10. What is the IUPAC name of this compound?

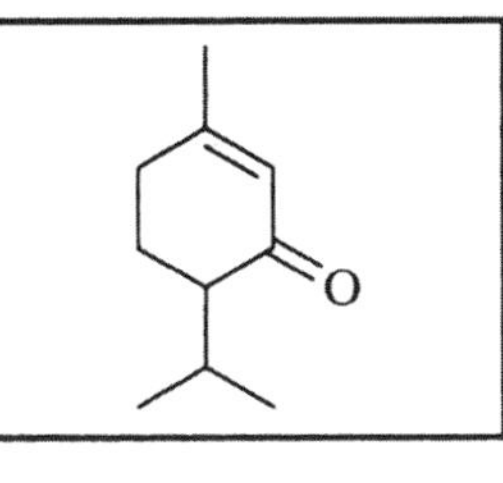

(A) 1-methyl-4-isopropyl-1-cyclohexen-3-one

(B) 1-methyl-4-isopropyl-3-ketocyclohexane

(C) 5-methyl-2-methylethyl-5-cyclohexenone

(D) 3-methyl-6-methylethyl-2-cyclohexenone

11. Which of these molecules contains **amide, ketone, ester**, and **ether** functional groups?

(A) $CH_3\overset{\overset{O}{\|}}{C}CH_2O\overset{\overset{O}{\|}}{C}CH_2CH_2OCH_2CH_2NH_2$

(B) $CH_3\overset{\overset{O}{\|}}{C}CH_2O\overset{\overset{O}{\|}}{C}CH_2CH_2OCH_2\overset{\overset{O}{\|}}{C}NH_2$

(C) $CH_3O\overset{\overset{O}{\|}}{C}CH_2\overset{\overset{O}{\|}}{C}CH_2\overset{\overset{O}{\|}}{C}NH_2$

(D) $CH_3O\overset{\overset{O}{\|}}{C}\underset{\underset{NH_2}{|}}{C}HCH_2CH_2OCH_2CH_2CH_2\overset{\overset{O}{\|}}{C}H$

Answers to Study Questions

1. C
2. C
3. A
4. D
5. B
6. D
7. B
8. A
9. C
10. A
11. B
12. D

Answers to Practice Questions

1. B
2. C
3. A
4. C
5. C
6. A
7. D
8. C
9. B
10. D
11. B

Structure, Hybridization, Resonance, Aromaticity

A simple way to understand the bonding between two or more atoms is to use the observation of G. N. Lewis that atoms are usually in a low energy arrangement if they have or share 8 valence electrons. Lewis structures should show all valence electrons and, where possible, the atoms should have octets. To determine the number of valence electrons, sum the number of valence electrons for each neutral atom and add an electron for each negative charge or subtract one electron for each positive charge.

Formal charge on an atom in a molecule can be calculated by counting the non-bonded electrons associated with an atom and one-half of all bonding electrons. The formal charge on the atom is the difference between the count from above and the number of valence electrons associated with the neutral atom. The sum of the formal charges on the atoms of a molecule should equal the overall charge on the molecule or ion. The lowest energy structure is usually the Lewis structure where all atoms have octets (if possible) and there is the least amount of charge separation.

A more sophisticated approach to bonding considers the atomic orbitals of atoms that combine to form molecular orbitals. With carbon atoms, there are four valence electrons, $2s^2$, $2p^2$. The *s* electrons are paired and the two *p* electrons are unpaired. To explain the four bonds that carbon atoms form, each carbon atom begins with four unpaired electrons. To get four orbitals the hybridization method is used. The four orbitals, one *s* orbital and three *p* orbitals, combine to form four new orbitals called sp^3 hybrid orbitals. These four orbitals orient toward the corners of a tetrahedron. The angle between each is 109.5°. Similar hybridization is found in nitrogen and oxygen atoms.

When carbon atoms form multiple bonds, the hybridization is different and, consequently, the bond angles are different. For carbon–carbon double bonds, the carbon atoms are sp^2-hybridized with 120° bond angles. The remaining *p* orbital is used to make a π bond. For carbon–carbon triple bonds, each carbon atom is *sp*-hybridized with a bond angle of 180°. The two remaining *p* orbitals are used to make two π bonds.

With some molecules or ions there is more than one way to draw the Lewis structures where all the atoms have octets. With these structures the charge (if the structure is an ion) may be written on different atoms. These two or more structures represent what are called "resonance structures." These Lewis structures represent delocalization of the π electrons of the molecule or ion. The "Real" molecule or ion is a composite of all the resonance structures. If more than one resonance structure can be written for a molecule this usually demonstrates that the molecule will be more stable than the individual Lewis structures would indicate.

With ring systems resonance is often dramatic. Molecules are unusually stable if they are cyclic, planar, and have a continuous π system with $4n+2$ electrons where $n = 0, 1, 2, \ldots$ These molecules are said to be "aromatic".

Study Questions

ST-1. Which is an acceptable Lewis structure of CO (where formal charges are not shown)?

(A) :C≡O: **(B)** C≡Ö: **(C)** :C̤=Ö: **(D)** :C̤̈–Ö̤:

Knowledge Required: (1) Rules for drawing Lewis Structures. (2) The number of valence electrons for carbon and oxygen atoms. (3) Rules for determining formal charge.

Thinking it Through: Carbon atoms have four (4) valence electrons, oxygen atoms have six (6) valence electrons, and CO does not have a charge. The Lewis structure must show 10 valence electrons. Choice **(C)** with 12 valence electrons can be eliminated as can choice **(D)** with 14 valence electrons. The lowest energy Lewis structures for second row element atoms such as carbon and oxygen will have octets. This eliminates choice **(B)** where the oxygen atom has 10 valence electrons. Choice **(A)** is the correct answer. Both oxygen and carbon atoms have 8 valence electrons.

ST-2. In which of these Lewis structures would the iodine be assigned a formal charge of +2?

(A) $CH_3—\ddot{I}—\ddot{O}:$ (with lone pairs below I and O)

(B) $CH_3—I(=\ddot{O}:)—\ddot{O}:$ (with :Ö: above I)

(C) $CH_3\ddot{O}—\ddot{I}—\ddot{O}:$

(D) $CH_3—\ddot{I}=\ddot{O}$

Knowledge Required: (1) Rules for drawing Lewis Structures. (2) The number of valence electrons for carbon, oxygen, and iodine atoms. (3) Rules for determining formal charge.

Thinking it Through: The formal charge on an atom is the difference between the number of electrons assigned to an atom and the number of valence electrons on the neutral atom. The assignment of electrons to an atom is all the nonbonded electrons plus one half of the bonding electrons. In choice **(A)** the iodine would be assigned 6 electrons (2 electrons from 1/2 of the 4 bonding electrons, plus 4 nonbonding electrons). A neutral iodine would have 7 valence electrons, so the formal charge in choice **(A)** would be +1. The same argument would be used for choice **(C)**. In choice **(D)** 7 electrons would be assigned to the iodine (3 electrons from 1/2 of the 6 bonding electrons, plus 4 nonbonded electrons). A neutral iodine should have 7 valence electrons, so the charge on the iodine in choice **(D)** would be 0. Choice **(B)** must be correct. There would be 5 electrons assigned to iodine (3 electrons form 1/2 of the 6 bonding electrons, plus 2 nonbonding electrons). Since a neutral iodine would have 7 valence electrons, the formal charge must be +2 (7 – 5).

ST-3. What are the respective hybridizations of the atoms numbered **1** to **4** in this compound?

$H_2\ddot{N}—CH_2—C(=\ddot{O}:)—\ddot{N}H—CH_2—C\equiv N:$

	1 (N)	2 (C)	3 (C)	4 (C)
(A)	sp^2	sp	sp^2	sp^3
(B)	sp^2	sp^2	sp^3	sp
(C)	sp^3	sp^2	sp^2	sp
(D)	sp^3	sp^2	sp^3	sp

Knowledge Required: (1) The definition of hybridization as a combination of atomic orbitals. (2) The connection between hybridization and the number of attached groups. (3) The fact that nitrogen atoms have a nonbonded pair of electrons that must exist in a hybrid orbital.

Thinking it Through: Atom (1) is a nitrogen atom that has three bonds and a nonbonded electron pair. It must have four orbitals and must be sp^3-hybridized. This eliminates choices **(A)** and **(B)**. Atom (2) is a carbonyl group that must have three bonds to other atoms. The carbon atom must be sp^2-hybridized. Both choices **(C)** and **(D)** have atom (2) as sp^2 so this does not help. Atom (3) is a carbon atom with four bonds and must be sp^3-hybridized. This eliminates choice **(C)**. The answer must be **(D)**. As a check, atom (4) is a carbon atom only bonded to two other atoms and must be sp-hybridized which is what is indicated in choice **(D)**.

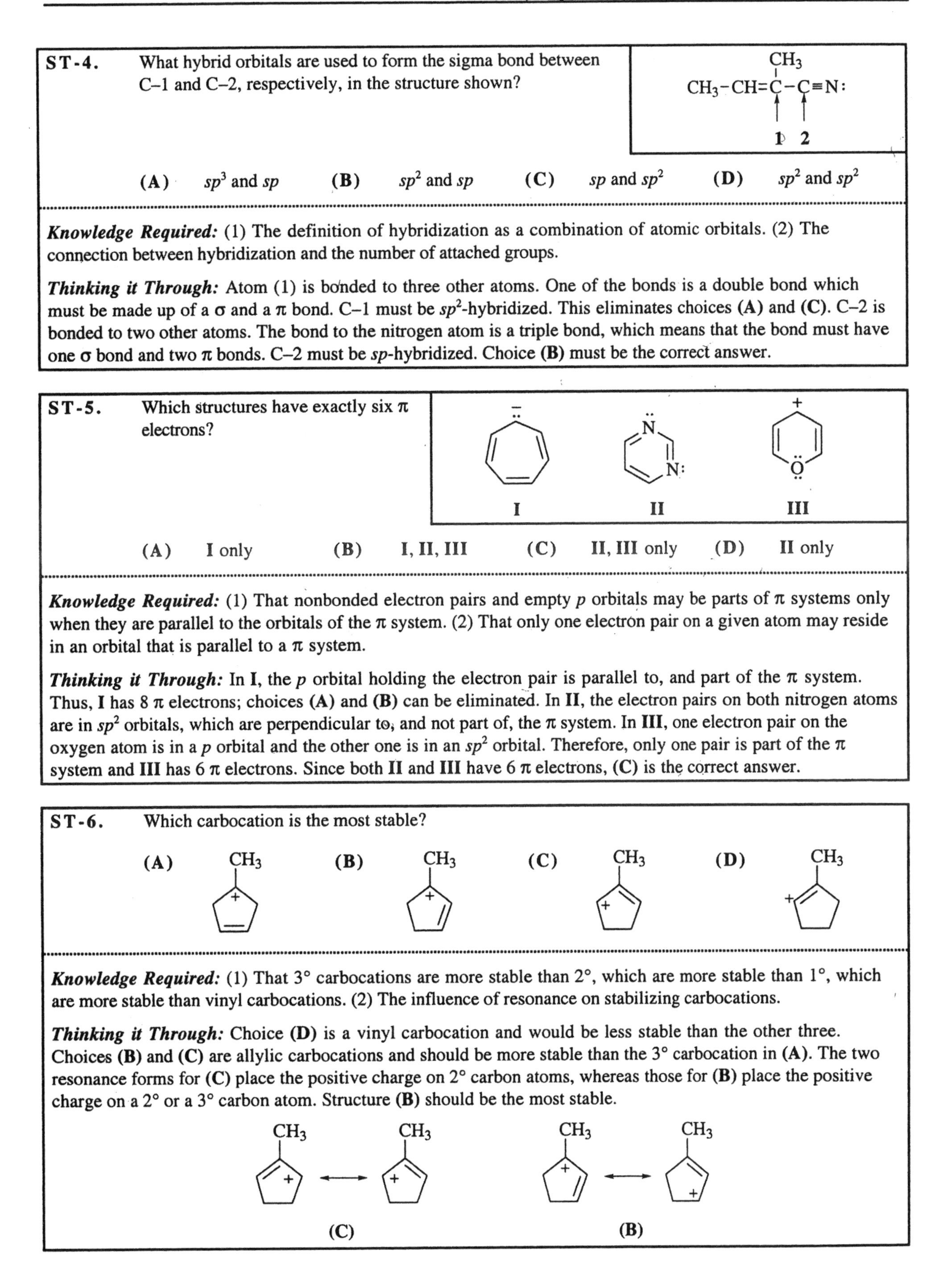

ST-4. What hybrid orbitals are used to form the sigma bond between C–1 and C–2, respectively, in the structure shown?

$CH_3-CH=\underset{1}{C}(CH_3)-\underset{2}{C}\equiv N:$

(A) sp^3 and sp (B) sp^2 and sp (C) sp and sp^2 (D) sp^2 and sp^2

Knowledge Required: (1) The definition of hybridization as a combination of atomic orbitals. (2) The connection between hybridization and the number of attached groups.

Thinking it Through: Atom (1) is bonded to three other atoms. One of the bonds is a double bond which must be made up of a σ and a π bond. C–1 must be sp^2-hybridized. This eliminates choices **(A)** and **(C)**. C–2 is bonded to two other atoms. The bond to the nitrogen atom is a triple bond, which means that the bond must have one σ bond and two π bonds. C–2 must be sp-hybridized. Choice **(B)** must be the correct answer.

ST-5. Which structures have exactly six π electrons?

I II III

(A) I only (B) I, II, III (C) II, III only (D) II only

Knowledge Required: (1) That nonbonded electron pairs and empty *p* orbitals may be parts of π systems only when they are parallel to the orbitals of the π system. (2) That only one electron pair on a given atom may reside in an orbital that is parallel to a π system.

Thinking it Through: In **I**, the *p* orbital holding the electron pair is parallel to, and part of the π system. Thus, **I** has 8 π electrons; choices **(A)** and **(B)** can be eliminated. In **II**, the electron pairs on both nitrogen atoms are in sp^2 orbitals, which are perpendicular to, and not part of, the π system. In **III**, one electron pair on the oxygen atom is in a *p* orbital and the other one is in an sp^2 orbital. Therefore, only one pair is part of the π system and **III** has 6 π electrons. Since both **II** and **III** have 6 π electrons, **(C)** is the correct answer.

ST-6. Which carbocation is the most stable?

(A) (B) (C) (D)

Knowledge Required: (1) That 3° carbocations are more stable than 2°, which are more stable than 1°, which are more stable than vinyl carbocations. (2) The influence of resonance on stabilizing carbocations.

Thinking it Through: Choice **(D)** is a vinyl carbocation and would be less stable than the other three. Choices **(B)** and **(C)** are allylic carbocations and should be more stable than the 3° carbocation in **(A)**. The two resonance forms for **(C)** place the positive charge on 2° carbon atoms, whereas those for **(B)** place the positive charge on a 2° or a 3° carbon atom. Structure **(B)** should be the most stable.

(C) (B)

ST-7. Which structure is equivalent to the condensed formula $(CH_3)_2CH(CH_2)_3Br$?

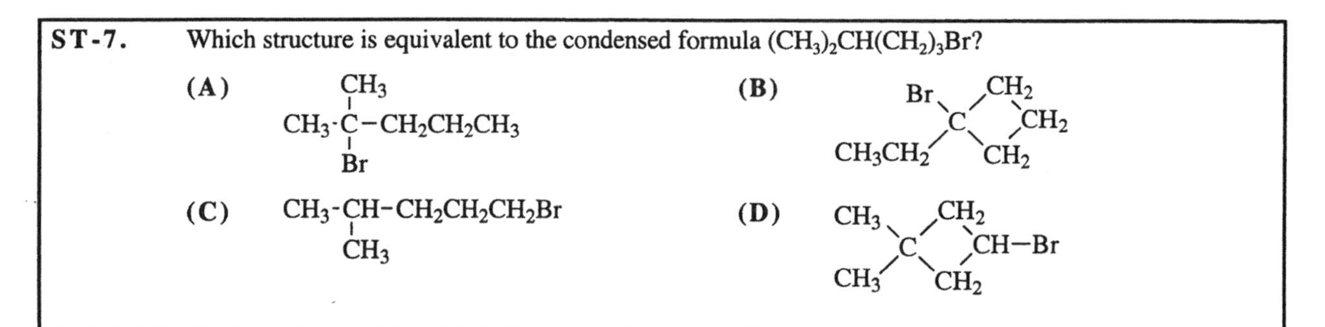

Knowledge Required: (1) That carbon atoms form four bonds and hydrogen and bromine atoms each form single bonds. (2) That numerical subscripts indicate the number of atoms or groups of atoms considered. (3) That condensed formulas are read from left to right, with each atom being bonded to the atom to the right until all the bonds are filled.

Thinking it Through: The condensed formula has parentheses around the CH_3 and a subscript of 2. This means that the CH to the right must have two CH_3's attached to it. This eliminates choices **(B)** and **(D)**. The CH has one more bond and the next atom to the right is the $(CH_2)_3$ which represents a string of three CH_2 groups; $CH_2CH_2CH_2$. The first of the CH_2's is attached to the CH that has the two CH_3's attached and the last CH_2 of the string is attached to the next atom to the right that is the Br. This makes choice **(C)** correct and eliminates choice **(A)** since the Br is attached to the wrong carbon atom.

ST-8. What factor is responsible for a greater heat of combustion per CH_2 for cyclopropane than the heat of combustion per CH_2 for cyclohexane?

(A) Cyclohexane has a different hydrogen-to-carbon atom ratio than cyclopropane.

(B) Cyclohexane is a strained ring relative to cyclopropane.

(C) Cyclopropane is a strained ring relative to cyclohexane.

(D) Cyclohexane has more carbon atoms than cyclopropane.

Knowledge Required: (1) That the unstrained bond angle for a carbon atom is 109.5°. (2) That the heat of combustion is the energy released when a hydrocarbon is converted into CO_2 and H_2O. (3) That cyclopropane is planar, with bond angles of 60°; and that cyclohexane is non-planar with bond angles very close to 109.5°.

Thinking it Through: The formula of cyclopropane is C_3H_6 and the formula of cyclohexane is C_6H_{12}. In each molecule the C to H ratio is 1 to 2. Choice **(A)** can be eliminated. The question says that cyclopropane has a greater heat of combustion per CH_2, which means that the heat of combustions has been normalized for the number of CH_2 groups. This eliminates choice **(D)**. For more energy to be given off per CH_2 there must be more strain in the molecule. Cyclopropane has C–C–C bond angles of 60°, each of which is far from the unstrained bond angle of 109.5°. Cyclohexane has bond angles close to 109.5° so it would be unstrained. This would indicate that choice **(C)** is correct and choice **(D)** is incorrect.

ST-9. The heat of combustion of neopentane is 3250 kJ/mol; that of pentane is 3269 kJ/mol. Which statement best describes the relative stability of these two compounds?

(A) Pentane and neopentane are isomers and therefore exhibit the same stability.

(B) Pentane is more stable than neopentane by 19 kJ/mol.

(C) Neopentane is more stable than pentane by 19 kJ/mol.

(D) Neopentane and pentane are not isomers so no statement can be made regarding relative stability.

Knowledge Required: 1) That the term "neo" would indicate an isomer of pentane, so that neopentane and pentane have the same formula. (2) That heats of combustion represent the energy released when a molecule is burned to CO_2 and H_2O. (3) That the isomer that releases the most energy is the least stable.

Thinking it Through: Neopentane and pentane are isomers so choice **(D)** must be false. Both compounds would burn to 5 CO_2 and 6 H_2O molecules. There is a difference in the energy given off so there must be a difference in stability so choice **(A)** cannot be true. The heat of combustion is negative meaning that energy is given off or that the alkanes are less stable than 5 CO_2 and 6 H_2O molecules. The molecule that gives off the most energy, pentane, is the least stable. Choice **(B)** can be eliminated. Choice **(C)** is correct.

ST-10. Which represents a pair of resonance structures?

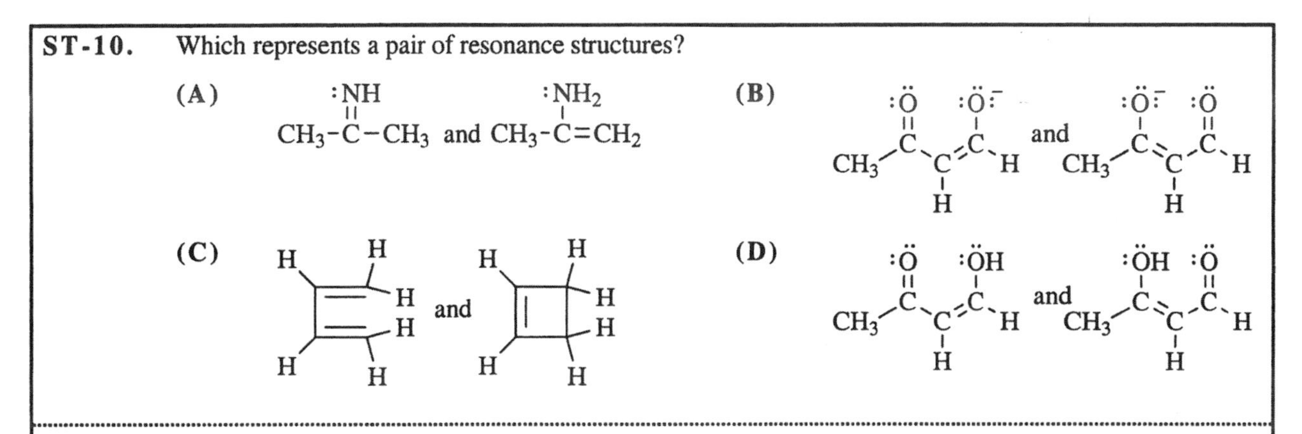

Knowledge Required: (1) That resonance structures differ only in the way *p* and π electrons are illustrated as being distributed among atoms. (2) That the positions of atoms (nuclei) do not change relative to each other.

Thinking it Through: In **(A)** one of the hydrogen atoms on the methyl group moves from the carbon atom to the nitrogen atom. This change involves two σ bonds. In other words, the nuclei have moved, and this is not allowed in resonance structures. Choice **(A)** is eliminated. Choice **(D)** can be eliminated for the same reason. The hydrogen atom is σ bonded to one oxygen atom in one structure and to a different oxygen atom in the other structure. Choice **(C)** looks good for a moment but the hybridization of the terminal carbon atoms is changing from sp^2 to sp^3, which means that bond angles are changing and atoms must be moving. Choice **(C)** cannot be correct. In choice **(B)**, only the π electrons are repositioned when moving the charge from one oxygen atom to the other. Thus, choice **(B)** must be correct. For choice **(B)**, there is a third resonance structure not shown that places the negative charge on the central carbon atom.

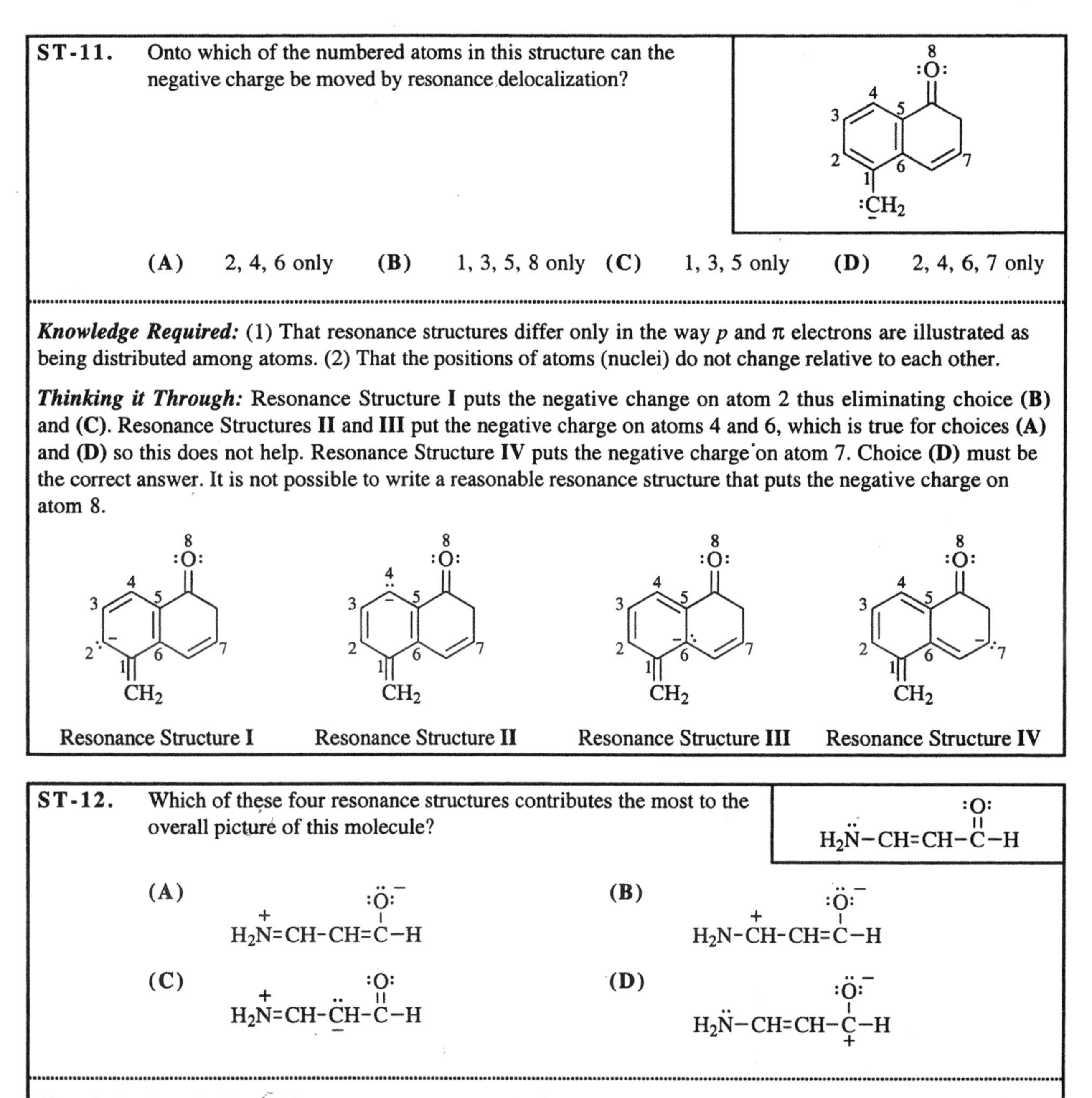

ST-11. Onto which of the numbered atoms in this structure can the negative charge be moved by resonance delocalization?

(A) 2, 4, 6 only **(B)** 1, 3, 5, 8 only **(C)** 1, 3, 5 only **(D)** 2, 4, 6, 7 only

Knowledge Required: (1) That resonance structures differ only in the way *p* and π electrons are illustrated as being distributed among atoms. (2) That the positions of atoms (nuclei) do not change relative to each other.

Thinking it Through: Resonance Structure **I** puts the negative change on atom 2 thus eliminating choice **(B)** and **(C)**. Resonance Structures **II** and **III** put the negative charge on atoms 4 and 6, which is true for choices **(A)** and **(D)** so this does not help. Resonance Structure **IV** puts the negative charge on atom 7. Choice **(D)** must be the correct answer. It is not possible to write a reasonable resonance structure that puts the negative charge on atom 8.

Resonance Structure **I** Resonance Structure **II** Resonance Structure **III** Resonance Structure **IV**

ST-12. Which of these four resonance structures contributes the most to the overall picture of this molecule?

$H_2\ddot{N}-CH=CH-C(=O)-H$

(A) $H_2\overset{+}{N}=CH-CH=C(-\ddot{O}:^-)-H$

(B) $H_2\overset{+}{N}-CH-CH=C(-\ddot{O}:^-)-H$

(C) $H_2\overset{+}{N}=CH-\ddot{C}H^--C(=O)-H$

(D) $H_2\ddot{N}-CH=CH-\overset{+}{C}(-\ddot{O}:^-)-H$

Knowledge Required: (1) That resonance structures differ only in the way *p* and π electrons are illustrated as being distributed among atoms. (2) That the positions of atoms (nuclei) do not change relative to each other. (3) That all first row elements have octets in low energy contributing structures. (4) That positive charge is most stable on the least electronegative element and negative change is most stable on the most electronegative element.

Thinking it Through: In structure **(B)**, the carbon atom with the positive charge does not have an octet. This is also true for the carbon atom with the positive charge in choice **(D)**. These two choices can be eliminated. The structures in choices **(A)** and **(C)** all have octets; however, in choice **(C)** the negative charge is on a carbon atom rather than on the oxygen atom as in choice **(A)**. Choice **(A)** would be the more stable contributor and would make a greater contribution to the overall structure.

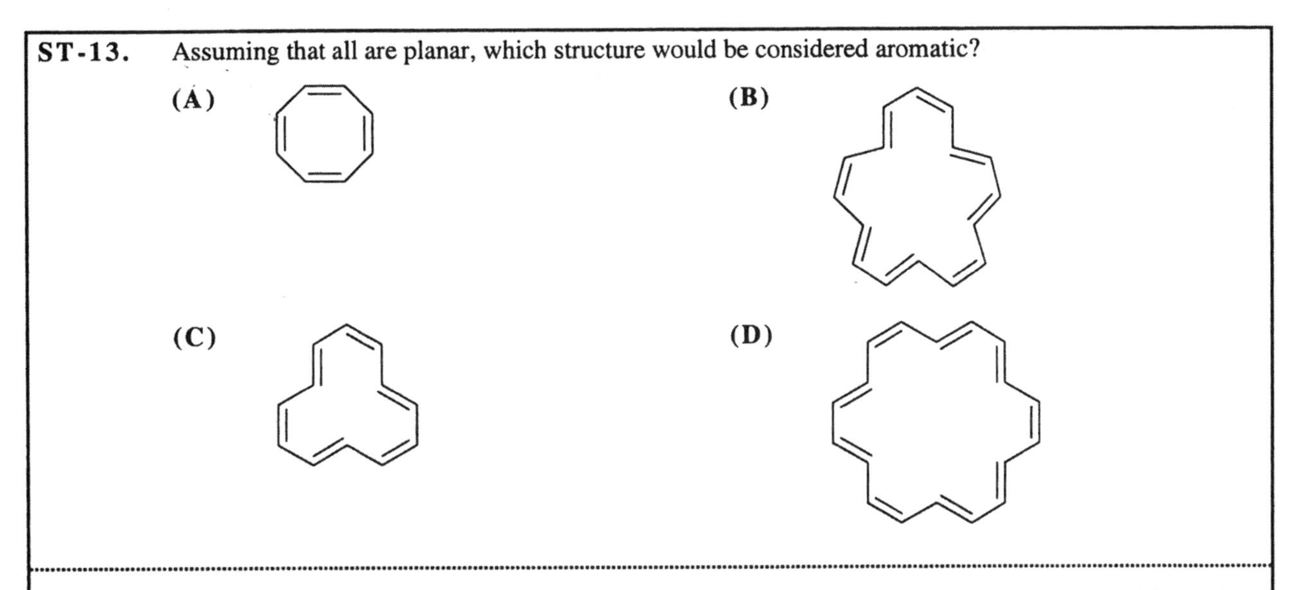

ST-13. Assuming that all are planar, which structure would be considered aromatic?

(A) (B) (C) (D)

Knowledge Required:. That to be aromatic a system must be a cyclic, planar, conjugated system with $4n+2$ π electrons, where n is an integer (Huckel's Rule).

Thinking it Through: All systems are planar (assumed), cyclic and conjugated systems. All could potentially be aromatic. Thus, all we need to determine is whether or not each one contains $4n+2$ π electrons. Structure **(A)** has 8 π electrons, **(B)** has 16 π electrons, **(C)** has 12 π electrons and **(D)** has 18 π electrons. Thus, only **(D)** obeys Huckel's $4n+2$ rule and is aromatic.

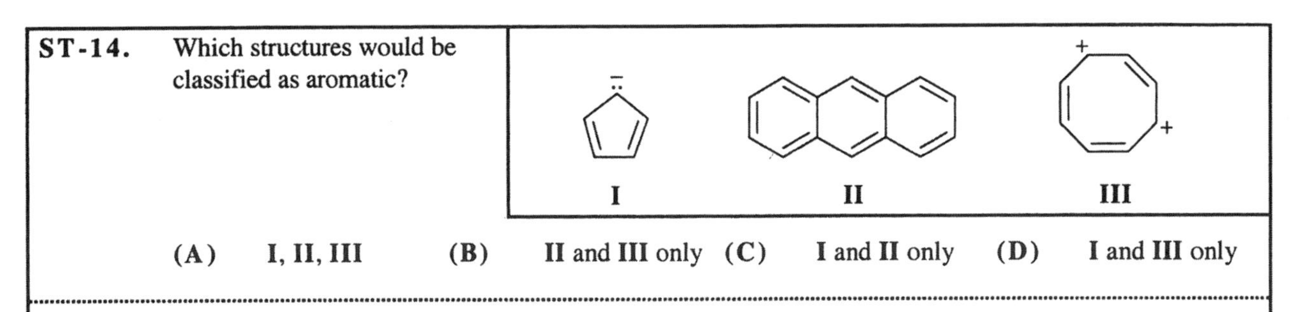

ST-14. Which structures would be classified as aromatic?

I II III

(A) I, II, III (B) II and III only (C) I and II only (D) I and III only

Knowledge Required: That to be aromatic a system must be a cyclic, planar, conjugated system with $4n+2$ π electrons, where n is an integer (Huckel's Rule).

Thinking it Through: All three systems could be planar and all are cyclic, conjugated systems. The p orbital with the electron pair in **I** and the empty p orbitals in **III** complete the conjugation of the respective π systems. All we need to determine is whether or not **I**, **II** and **III** contain $4n+2$ π electrons. Both **I** and **III** have 6 π electrons and **II** has 14 π electrons. These all fit the $4n+2$ rule where n = 1 or 3, and all three are aromatic. Thus, choice **(A)** is correct

Practice Questions

1. Identify the lowest-energy Lewis structure for nitrogen oxide. (Formal charges not shown.)

(A) $:\ddot{N}=\ddot{N}-\ddot{O}:$ (B) $:\ddot{N}=N=\ddot{O}:$

(C) $:N\equiv N-\ddot{\underset{..}{O}}:$ (D) $:\ddot{\underset{..}{N}}-N\equiv O:$

2. Which value is closest to the internal C–C–C bond angle in cyclohexane?

(A) 90° (B) 100°

(C) 110° (D) 120°

3. Which is an acceptable Lewis structure for diazomethane?

(A) 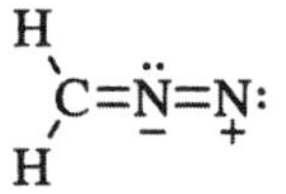 $H_2C=\ddot{N}^{-}=N^{+}:$ (B) 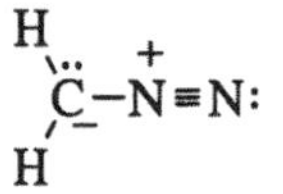 $H_2\ddot{C}^{-}-\overset{+}{N}\equiv N:$

(C) $H_2C=N=N:$ (D) 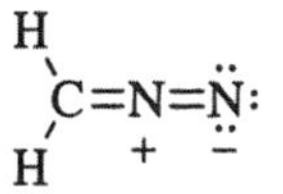 $H_2C=N^{+}=\ddot{N}:^{-}$

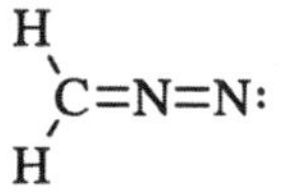

4. In which of these Lewis structures would the sulfur be assigned a formal charge of +1?

(A) 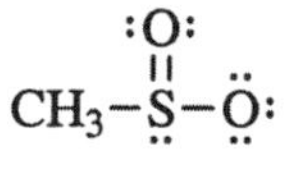 $CH_3-\ddot{S}(=\ddot{O}:)-\ddot{O}:$ (B) 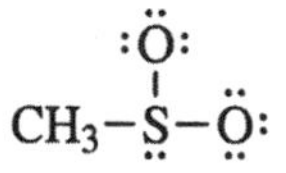 $CH_3-\ddot{S}(-\ddot{O}:)-\ddot{O}:$

(C) 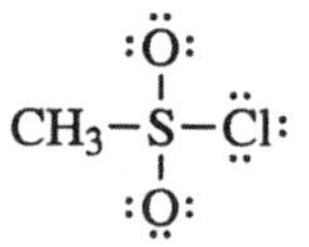 $CH_3-S(-\ddot{O}:)(-\ddot{O}:)-\ddot{Cl}:$ (D) 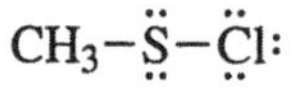 $CH_3-\ddot{S}-\ddot{Cl}:$

5. Which formula is equivalent to the condensed formula $(CH_3)_3C(CH_2)_2CH_3$?

(A)

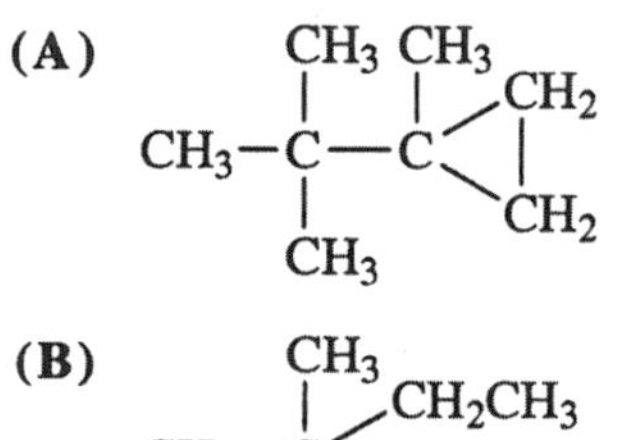

(B) 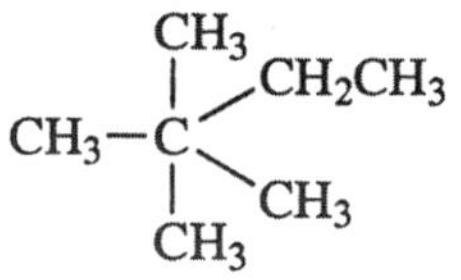 $CH_3-C(CH_3)(CH_3)(CH_2CH_3)$, with CH_3 above, CH_2CH_3 and CH_3 to the right, CH_3 below

(C) $CH_3CH_3CH_3-C(CH_3)$ bonded to a ring of CH_2 and CH_2

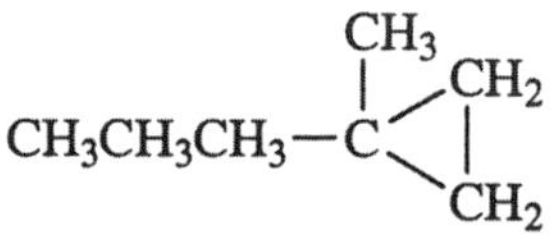

(D) 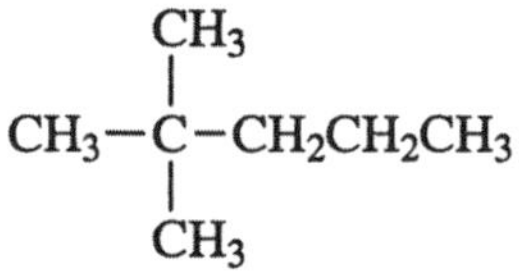 $CH_3-C(CH_3)_2-CH_2CH_2CH_3$

6. What is the hybridization of the orbital containing the unshared pair of electrons on the nitrogen atom in pyridine?

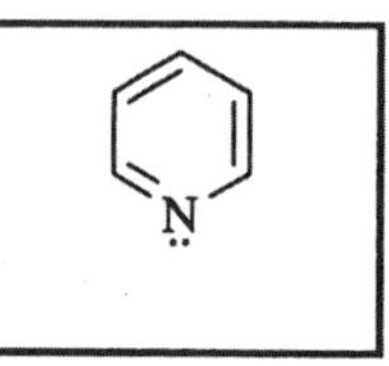

(A) s (B) sp

(C) sp^2 (D) sp^3

7. What are the hybridizations of the atoms marked **I**, **II**, **III**, and **IV** in the molecule?

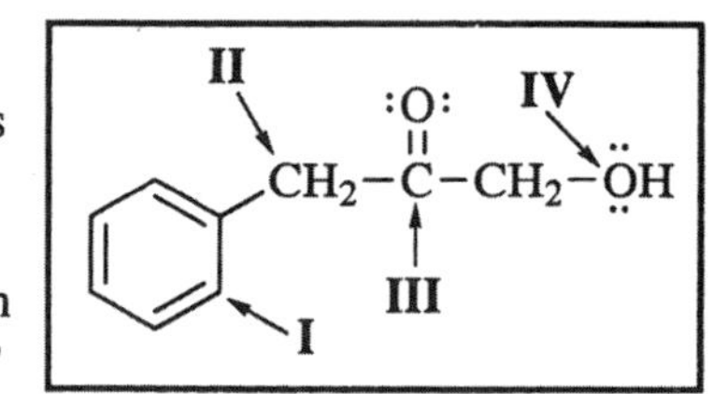

(A) I = sp^2 II = sp^2 III = sp^3 IV = sp^3

(B) I = sp^2 II = sp^3 III = sp^2 IV = sp^3

(C) I = sp^3 II = sp^2 III = sp^3 IV = sp^3

(D) I = sp II = sp^3 III = sp IV = sp^2

8. The heat of combustion (per CH_2) of several cycloalkanes is listed below. Based on the data given, which of these cycloalkanes would be considered *most* stable?

Heat of combustion (kJ/CH_2)	Cycloalkane
–686.5	cyclobutane
–664.0	cyclopentane
–663.6	cyclooctane
–659.0	cyclopentadecane

(A) cyclobutane (B) cyclopentane

(C) cyclooctane (D) cyclopentadecane

9. For which free radical can the most resonance forms be written that show the delocalization of the radical?

(A)  cyclohexyl–$\dot{C}H_2$

(B) phenyl radical (benzene ring with $\cdot$)

(C) benzene ring–$\dot{C}H_2$

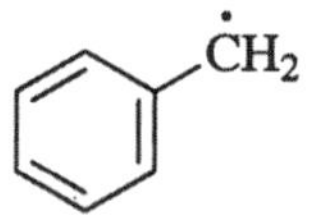

(D) 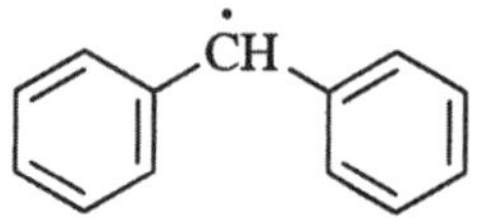 benzene ring–$\dot{C}H$–benzene ring

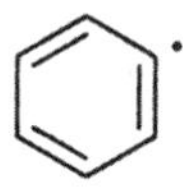

10. Onto which of the numbered atoms in the following structure can the negative charge be moved by drawing resonance structures?

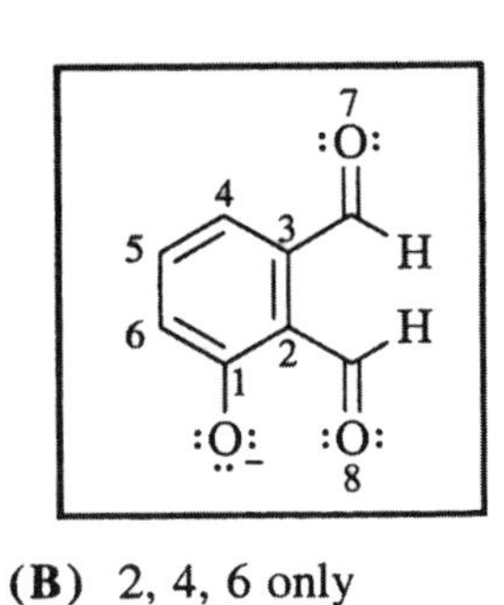

(A) 1, 3, 5 only **(B)** 2, 4, 6 only

(C) 2, 4, 6, 8 **(D)** 1, 3, 5, 7

11. Which pair consists of resonance structures?

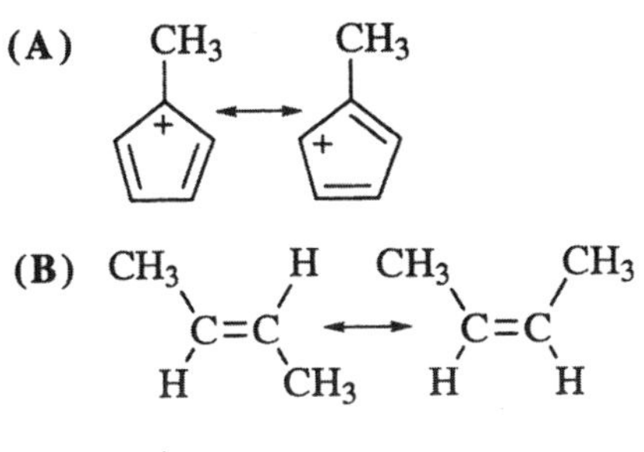

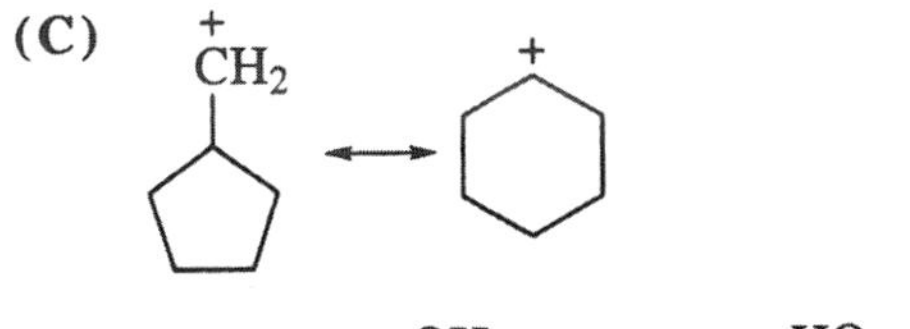

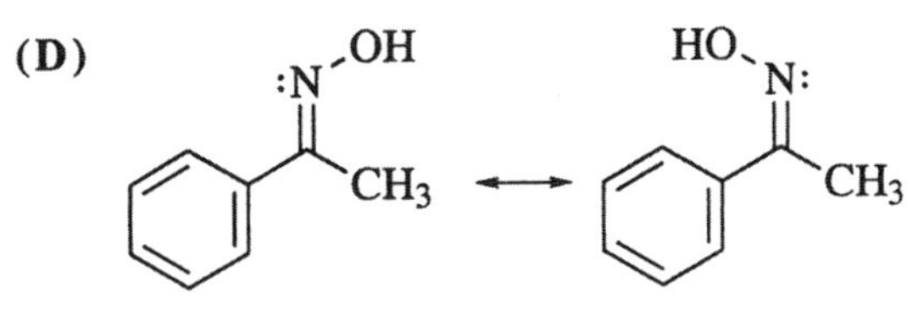

12. Which of these structures have eight π electrons?

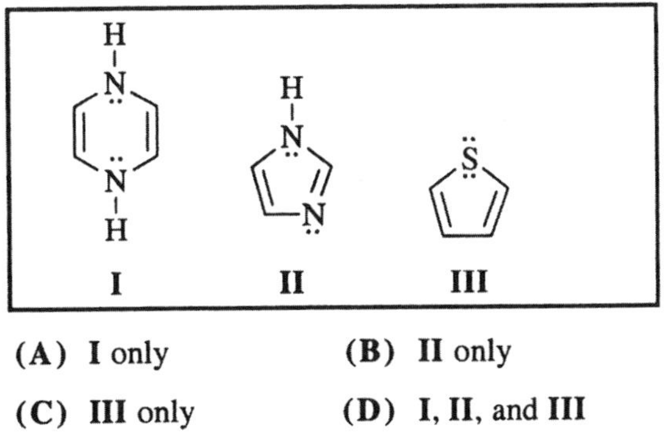

(A) I only **(B)** II only

(C) III only **(D)** I, II, and III

13. Which resonance structure contributes the most to the overall picture of this molecule?

$\ddot{O}=C=\ddot{N}-CH_3$

(A) $^{-}:\ddot{O}-\overset{+}{C}=\ddot{N}-CH_3$ **(B)** $\ddot{O}=\overset{+}{C}-\ddot{N}^{-}-CH_3$

(C) $^{-}:\ddot{O}-C\equiv\overset{+}{N}-CH_3$ **(D)** $:\overset{+}{O}\equiv C-\ddot{N}^{-}-CH_3$

14. Which of these structures can be classified as aromatic?

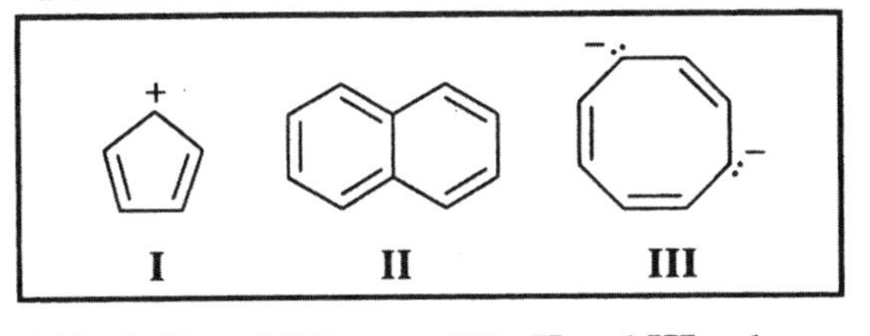

(A) I, II, and III **(B)** II and III only

(C) I and III only **(D)** II only

15. Which structure is *not* aromatic?

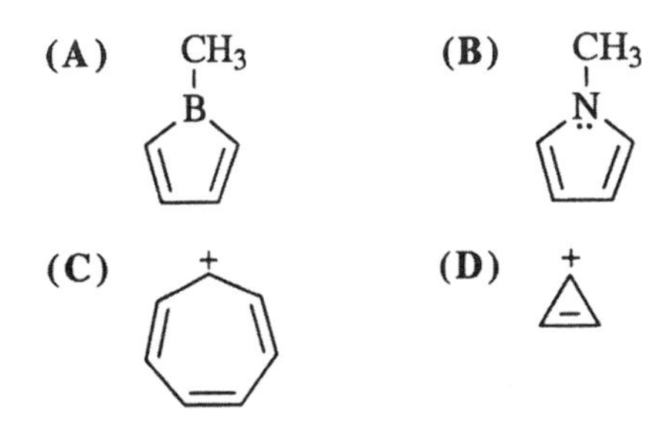

16. Which carbocation is the most stable?

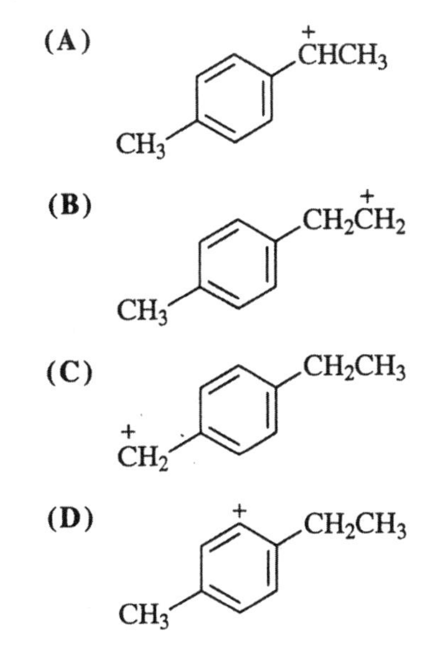

Answers to Study Questions

1. A
2. B
3. D
4. B
5. C
6. B
7. C
8. C
9. C
10. B
11. D
12. A
13. D
14. A

Answers to Practice Questions

1. C
2. C
3. B
4. B
5. D
6. C
7. B
8. D
9. D
10. C
11. A
12. A
13. C
14. B
15. A
16. A

Acids and Bases

One of the fundamental concepts used in understanding organic reactions is the acidic or basic properties of a molecule. It is important to be able to make predictions concerning the relative acidity of hydrogen atoms in a single molecule or the relative acidities of several molecules. Similar arguments must be made concerning the basic nature of molecules. The thermodynamic acidity of a molecule is related to the free energy change ($\Delta G°$) for the reaction of the compound releasing the proton. Factors that stabilize the products relative to the reactants produce a more negative (less positive) value for $\Delta G°$, increasing acidity. Factors that stabilize the reactants relative to the products produce a more positive (less negative) value for $\Delta G°$, decreasing acidity. For two acids **HA** and **HB** the acid reactions (equations *1*, and *2*) are:

$$\mathbf{HA} + H_2O \rightleftharpoons H_3O^+ + \mathbf{A}^- \qquad (1)$$

$$\mathbf{HB} + H_2O \rightleftharpoons H_3O^+ + \mathbf{B}^- \qquad (2)$$

For each reaction, acidity is related to the free energy change for that reaction, as illustrated for **HA** in equation *3*:

$$\Delta G° = -RT\ln\frac{[A^-][H_3O^+]}{[HA][H_2O]} \qquad (3)$$

The two largest factors influencing stability of a species are resonance and induction (electronegativity). When comparing relative acid strength for **HA** and **HB**, always write both reactions showing the compounds (or ions) losing a proton. Then, compare the resonance effect and the inductive effect for each reactant and product. The stronger acid is defined by the reaction in which products are stabilized to the greatest extent relative to the reactants, producing the more negative value for $\Delta G°$.

Study Questions

AB-1. Which species would best act as a Lewis acid?

(A) BF_4^- **(B)** $CH_3C(=O)CH_3$ (:O: double-bonded to C, with two CH_3 groups) **(C)** $CH_3\cdot$ **(D)** $CH_3\ddot{F}:$

Knowledge Required: (1) Lewis acid–base definition. (2) Rules for drawing Lewis structures.

Thinking it Through: The Lewis structure of BF_4^- in choice (**A**) must show 32 electrons around the atoms, with boron being the central atom. The Lewis structure indicates that every atom has a full octet, and therefore there are no low-energy orbitals available to accept an electron pair. Choice (**A**) is therefore eliminated from consideration. Choice (**D**) cannot be correct for the same reason; carbon and fluorine atoms each have full octets of electrons. In choice (**C**), the carbon atom has one unpaired electron. The only possible Lewis structure shows seven valence electrons surrounding the carbon atom. Despite the fact that the species is electron deficient, it cannot accept a *pair* of electrons because the carbon atom cannot accommodate nine valence electrons. Choice (**B**) must be correct, but the reason may not be immediately obvious. The oxygen atom and every carbon atom have an octet of electrons in the Lewis structure as it is normally written. However, if we consider a resonance structure for choice (**B**) where the π electrons are localized on the oxygen atom, then the carbon atom only has 6 valence electrons and a positive charge. This carbon atom can then accept an electron pair, and behave as a Lewis acid.

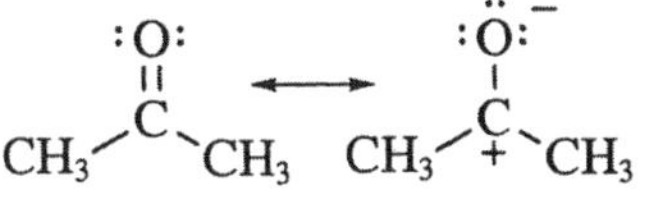

The carbonyl carbon atom is electrophilic (Lewis Acid) and the carbonyl oxygen atom is nucleophilic. Note that carbonyl carbon atoms are made more electrophilic (stronger Lewis acids) if the carbonyl oxygen atom is first protonated.

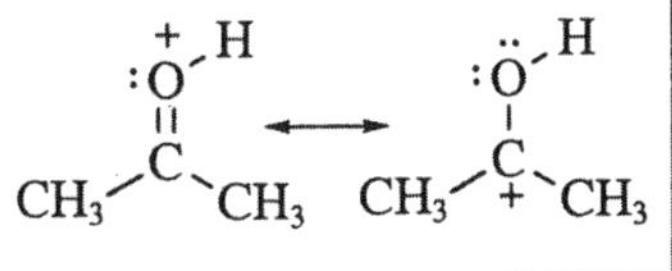

AB-2. Which structure helps explain why anilinium ion is a stronger acid than is ammonium ion?

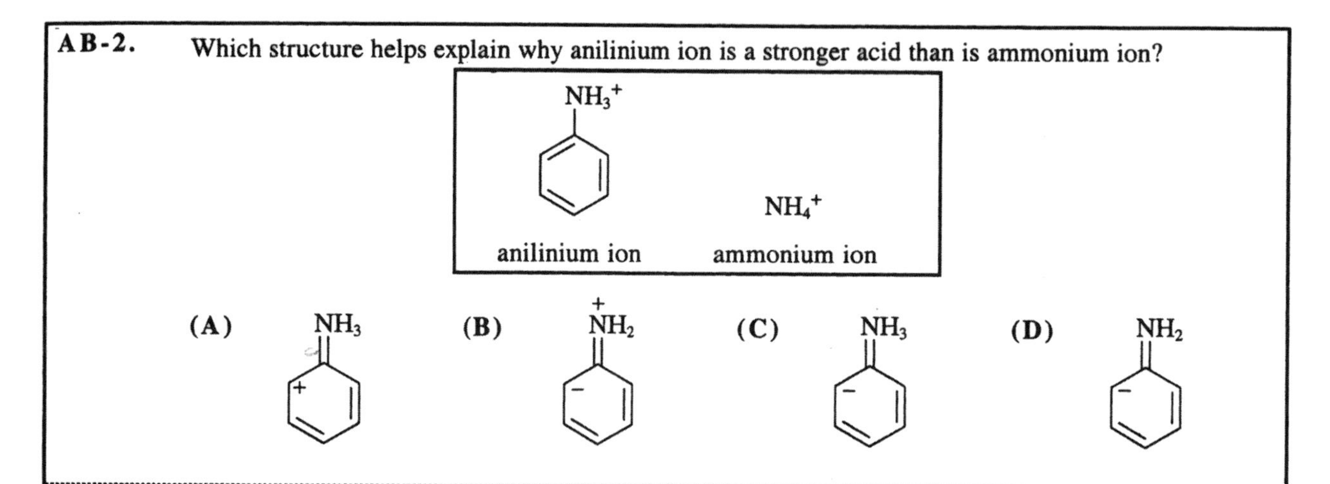

Knowledge Required: (1) How to write the hydrolysis reaction of the conjugate acids of nitrogen bases. (2) Rules for drawing Lewis structures. (3) How resonance stabilizes an acid or its conjugate base.

Thinking it Through: First write the equations for the two species as acids.

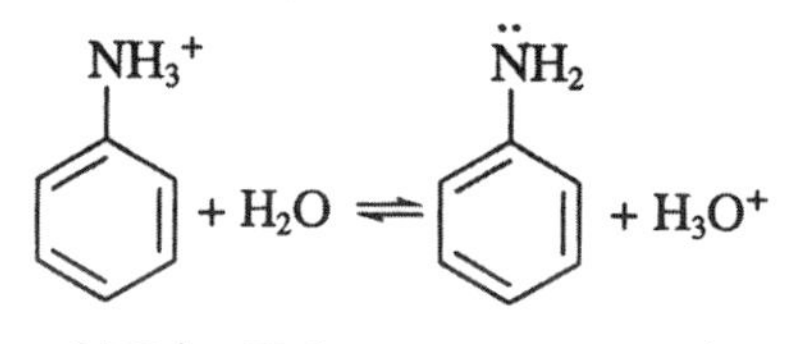

$$NH_4^+ + H_2O \rightleftharpoons :NH_3 + H_3O^+$$

For the ammonium ion reaction, the charge on the NH_4^+ cannot be delocalized. Neither can the electron pair on ammonia. In this equilibrium, the positive charge is best represented as localized on the less electronegative nitrogen atom.

For the anilinium ion, the charge is localized on the nitrogen atom. Reasonable resonance structures cannot delocalize the positive charge into the aromatic ring as implied by structure (**A**) because the nitrogen atom cannot have 5 bonds. This eliminates choice (**A**) from consideration.

Choice (**C**) appears to show delocalization of a negative charge into the aromatic ring, but again, the nitrogen atom is represented with 5 bonds. Choice (**C**) cannot be correct. The resonance structure represented by choice (**D**) indicates that the electron pair on the nitrogen atom in the conjugate base is delocalized into the ring, which would stabilize the products of the reaction. The problem with choice (**D**) is that the nitrogen atom is represented with 4 bonds, and therefore should have a positive charge. Choice (**D**) is thus eliminated from consideration.

Only choice (**B**) remains for consideration. This resonance structure satisfies the octet rule, and shows how aniline, the conjugate base of the anilinium ion, is resonance stabilized. A comparable resonance structure does not exist for anilinium ion, leading to the conclusion that anilinium ion will be a stronger acid than otherwise predicted.

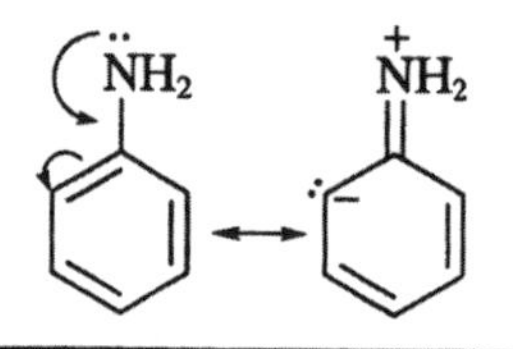

AB-3. If an amino acid is dissolved in water and the pH of the solution is adjusted to 12, which is the predominate species present?

(A) $R-CH(NH_2)-C(=O)-OH$

(B) $R-CH(NH_3^+)-C(=O)-OH$

(C) $R-CH(NH_3^+)-C(=O)-O^-$

(D) $R-CH(NH_2)-C(=O)-O^-$

Knowledge Required: (1) The *approximate* pK_a values of carboxylic acids and of ammonium ion. (2) How relative species concentrations are determined by values for pH and pK_a.

Thinking it Through: Choices (**A**) and (**B**) cannot be correct, because at a pH of 12 the acid would be deprotonated. (The pK_a values for carboxylic acids are about 5.) Choices (**B**) and (**C**) can be eliminated from consideration because hydroxide (HO^-) is a stronger base than NH_3 (pK_a of NH_4^+ is about 9), and hydroxide ion would deprotonate the RNH_3^+ groups. Choice (**D**) must be the correct answer, because both the $-COOH$ and $-NH_3^+$ groups are both deprotonated.

AB-4. What is the order of acid strength from ***strongest*** acid to ***weakest*** acid?

CH_3COOH	CF_3CH_2COOH	CF_3COOH	CCl_3COOH
I	II	III	IV

(A) I > II > IV > III

(B) III > II > IV > I

(C) III > IV > II > I

(D) IV > III > II > I

Knowledge Required: (1) Factors that stabilize the carboxylate ion relative to the carboxylic acid. (2) How the inductive effect of atoms that are more electronegativity than hydrogen atoms affect acid strength. (3) Effect of intervening bonds on the magnitude of the inductive effect.

Thinking it Through: Write the ionization reaction for each acid. Factors that stabilize the carboxylate ion relative to the carboxylic acid will increase the ionization (acid strength).

$$CH_3COOH \rightleftharpoons CH_3CO^- + H^+$$

Chlorine and fluorine atoms are more electronegative than hydrogen atoms, and stabilize the carboxylate ions relative to the carboxylic acid. Both trichloroacetic acid (**IV**) and trifluoroacetic acid (**III**) should be more acidic than acetic acid (**I**). Choice (**A**) is thus eliminated from consideration. Fluorine atoms ($EN \approx 4$) are more electronegative than chlorine atoms ($EN \approx 3.2$), so trifluoroacetic acid (**III**) should be a stronger acid than trichloroacetic acid (**IV**), eliminating choice (**D**). The inductive effect decreases quickly with distance, leading to the conclusion that 3,3,3-trifluoropropanoic acid (**II**) should be less acidic than trichloroacetic acid (**IV**), which eliminates choice (**B**). Choice (**C**) has the four acids arranged from strongest to weakest, and is the correct answer.

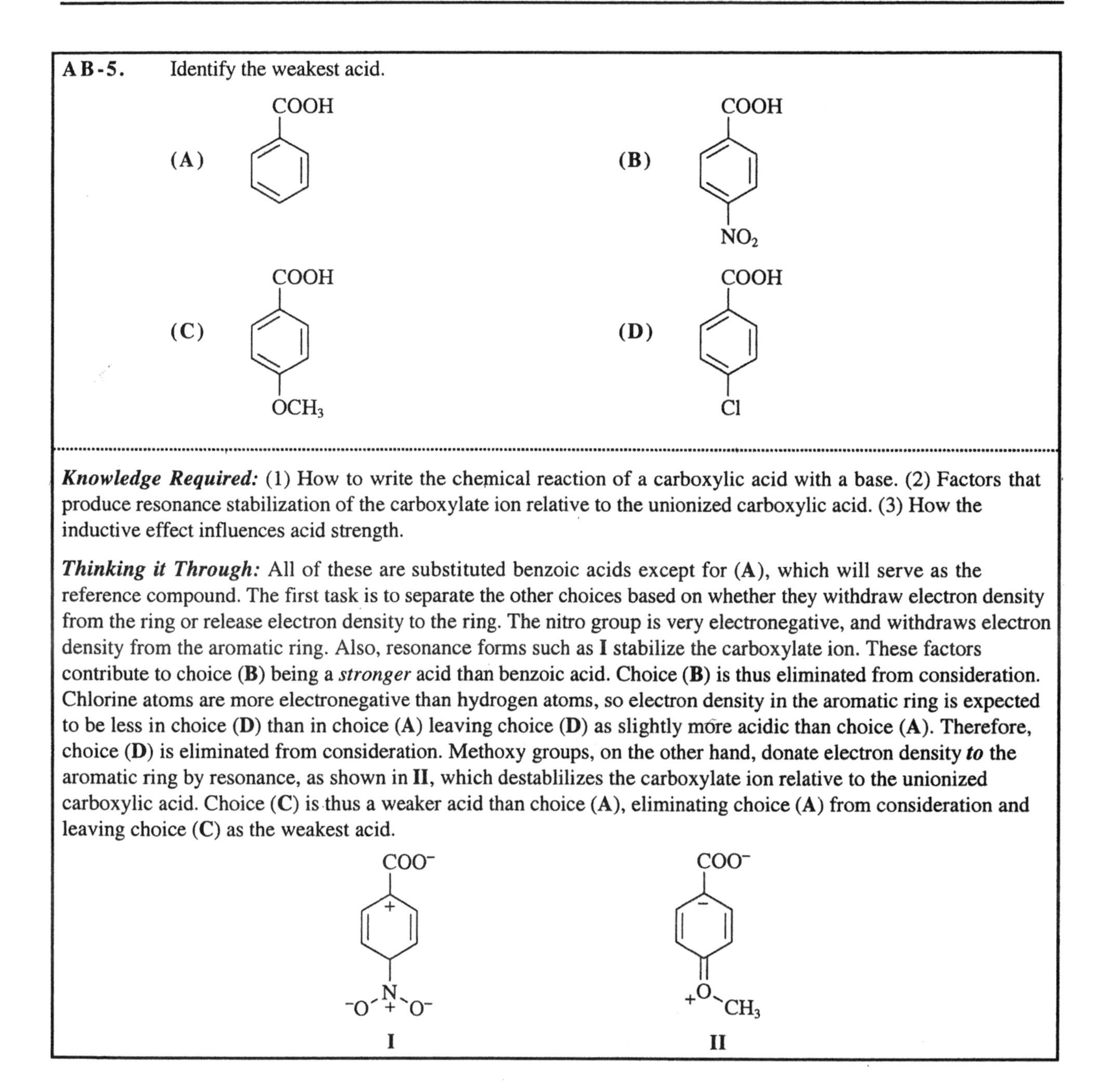

AB-5. Identify the weakest acid.

Knowledge Required: (1) How to write the chemical reaction of a carboxylic acid with a base. (2) Factors that produce resonance stabilization of the carboxylate ion relative to the unionized carboxylic acid. (3) How the inductive effect influences acid strength.

Thinking it Through: All of these are substituted benzoic acids except for **(A)**, which will serve as the reference compound. The first task is to separate the other choices based on whether they withdraw electron density from the ring or release electron density to the ring. The nitro group is very electronegative, and withdraws electron density from the aromatic ring. Also, resonance forms such as **I** stabilize the carboxylate ion. These factors contribute to choice **(B)** being a *stronger* acid than benzoic acid. Choice **(B)** is thus eliminated from consideration. Chlorine atoms are more electronegative than hydrogen atoms, so electron density in the aromatic ring is expected to be less in choice **(D)** than in choice **(A)** leaving choice **(D)** as slightly more acidic than choice **(A)**. Therefore, choice **(D)** is eliminated from consideration. Methoxy groups, on the other hand, donate electron density ***to*** the aromatic ring by resonance, as shown in **II**, which destablilizes the carboxylate ion relative to the unionized carboxylic acid. Choice **(C)** is thus a weaker acid than choice **(A)**, eliminating choice **(A)** from consideration and leaving choice **(C)** as the weakest acid.

AB-6. Which is the strongest base?

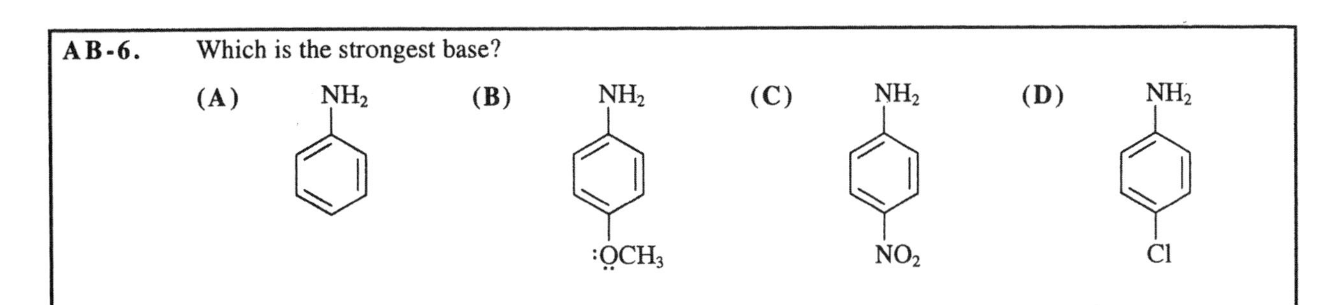

Knowledge Required: (1) How to write the chemical reaction of an acid with an organic base. (2) Factors that produce resonance stabilization of the anilinium ions relative to their conjugate aniline bases. (3) How the inductive effect influences base strength. (4) Relative tendencies of second and third period elements to form multiple bonds.

Thinking it Through: Write the reaction for a substituted aniline (**I**) acting as a base.

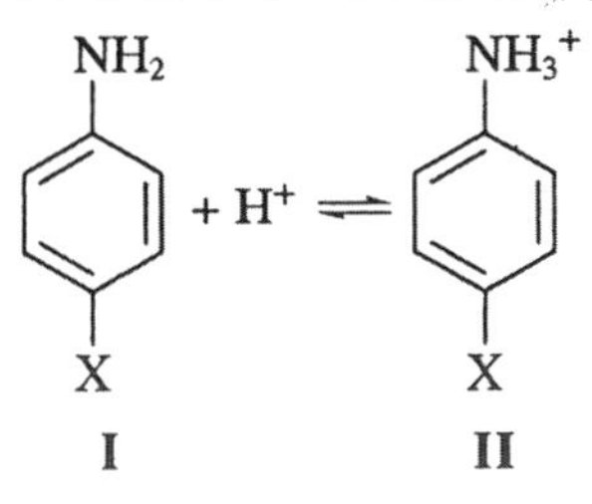

In aniline (X = H), the electron pair on the nitrogen atom forms **II** when it is protonated. In **I**, the electron pair on a nitrogen atom can be delocalized into the ring by resonance structures such as **III**. This resonance delocalization stabilizes **I**. Once the electron pair is protonated, as shown in **II**, there is no resonance stabilization. Consequently, aniline is a weak base. Our concern now is what effect the substituents, X, will have on the stability of **I** and **II**. Electron withdrawing groups such as NO_2 would stabilize **I** more than **II** both through induction and through resonance stabilization, as shown in **IV**. Consequently, nitroaniline, choice (**C**), is much less basic than aniline, choice (**A**). The chloro and methoxy anilines pose a very interesting situation. Both chlorine and oxygen atoms are more electronegative than hydrogen atoms. Both would stabilize **I** more than **II** if induction were the only factor. However, both chlorine and oxygen atoms can donate electron density to the ring through resonance forms such as **V** and **VI**, which would make the substituted aniline more basic. Based on these two resonance forms, methoxyaniline and chloroaniline should be stronger bases than aniline. We have a dilemma. Considering induction, choices (**B**) and (**D**) should both be weaker bases than aniline, and considering resonance, they both should be stronger bases than aniline. However, there is an important difference between structures **V** and **VI**. Oxygen and carbon are both in the second period, and their atoms are about the same size. Chlorine is in the third period, and its atoms are much larger. Third period elements are much less prone to form double bonds than second period elements. **V** is a very important resonance structure that releases electron density to the ring. Induction is a minor contributor to withdrawing electron density. Structure **VI**, on the other hand, is a minor resonance contributor, because chlorine is forced to form a double bond. Induction, by default, is the major factor. Methoxyaniline is a stronger base than aniline, and chloroaniline is slightly weaker. Choice (**B**) is the correct answer.

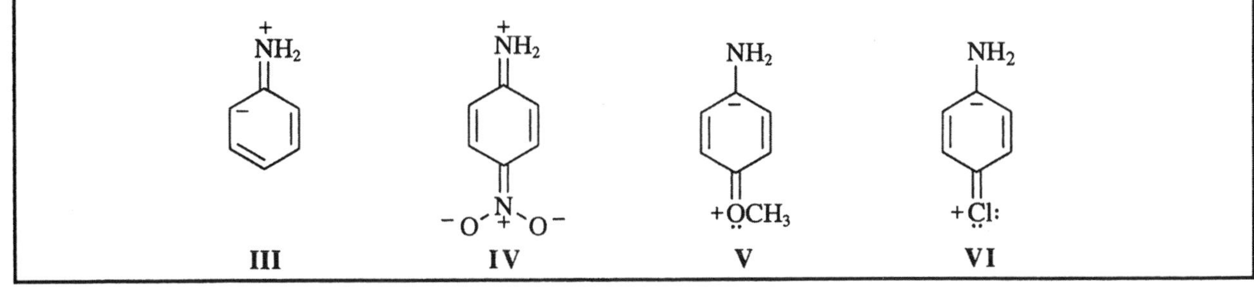

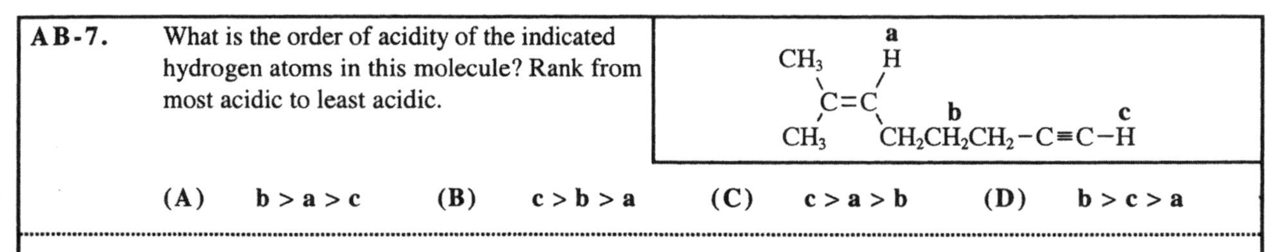

AB-7. What is the order of acidity of the indicated hydrogen atoms in this molecule? Rank from most acidic to least acidic.

(A) b > a > c (B) c > b > a (C) c > a > b (D) b > c > a

Knowledge Required: (1) Percent *s* and *p* character in hybrid orbitals. (2) How hybridization affects acidity of C–H bonds. (3) How percentage of *s* character in an orbital affects stability of an anion.

Thinking it Through: When a C–H bond acts as an acid, the resulting negative charge resides on the carbon atom. The strength of the acid depends on the stability of the resultant carbanion.

$$\text{>C–H} \rightleftharpoons \text{>C}^- + H^+$$

Since *s* orbitals are lower in energy than *p* orbitals, the stability of a lone pair increases with increasing *s* character of its orbital. Consequently, the acid strength of a C–H bond increases as the *s* character of the bond hybrid orbital increases.

For hydrogen atom **a**, the bond is to a sp^2 hybrid orbital on a carbon atom. Upon ionization, the resulting electron pair would be in the sp^2 hybrid orbital, which is 1/3 *s* and 2/3 *p*. For hydrogen atom **b**, the resulting electron pair would be in an sp^3 hybrid orbital, which is 1/4 *s* and 3/4 *p*. For hydrogen atom **c**, the resulting electron pair would be in an *sp* hybrid orbital, which is 1/2 *s* and 1/2 *p*. The percentage of *s* character for the orbitals containing electron pairs is **c** > **a** > **b**, and this is the order of acidity of the hydrogen atoms. Choice **(C)** is correct.

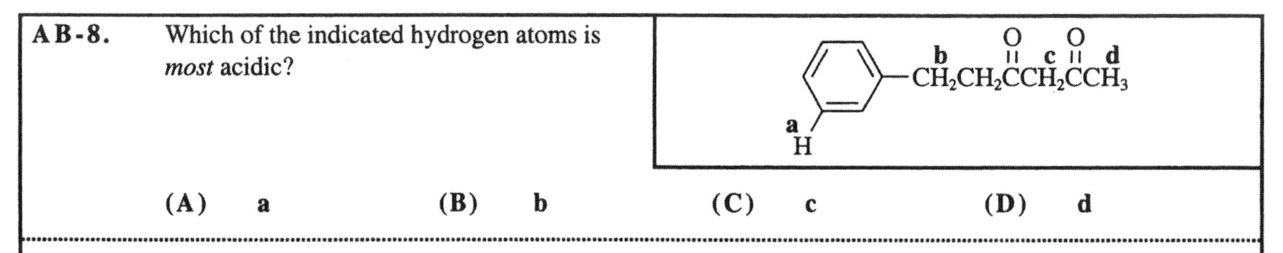

AB-8. Which of the indicated hydrogen atoms is *most* acidic?

(A) a (B) b (C) c (D) d

Knowledge Required: (1) Percent *s* and *p* character in hybrid orbitals. (2) How hybridization affects acidity of C–H bonds. (3) How resonance structures affect stability of an anion.

Thinking it Through: The acidity of the C–H bond will depend on the stability of the resulting anions following loss of the proton. The four ions **I** through **IV** must be evaluated to determine their relative stability.

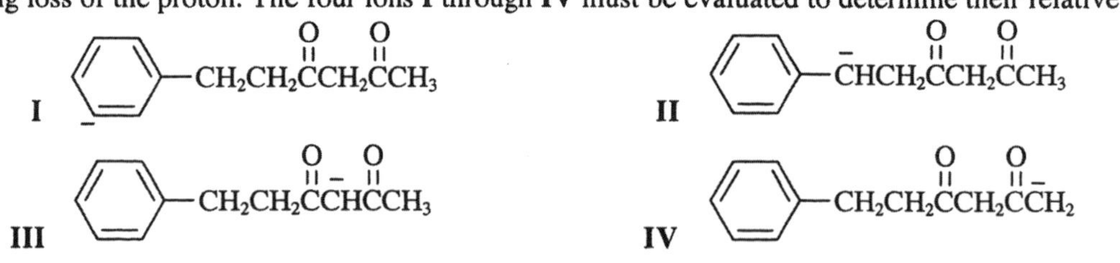

In anion **I**, the negative charge is in an sp^2 hybridized orbital, which is 1/3 *s* and 2/3 *p*. For anion **II**, the negative charge is in an sp^3 hybridized orbital, which is 1/4 *s* and 3/4 *p*. Based on percentage *s* in the hybrid, **I** is expected to be more stable than **II**. However, we must also consider resonance. In **I**, the negative charge is localized on the ring carbon atom, and cannot be delocalized. In **II**, the charge *can* be delocalized into the aromatic ring by resonance. Consequently, hydrogen atom **b** should be more acidic than hydrogen atom **a**, eliminating choice **(A)**. The anions **III** and **IV** also have negative charges in sp^3 hybrid orbitals that are similar to **II**. However, both of these ions have resonance forms that place the negative charge on an oxygen atom. Resonance forms with the negative charge on an oxygen atom stabilize ions more than delocalization of a negative charge onto an aromatic ring. Structure **V** is the resonance structure associated with **IV**. **VI** and **VII** are corresponding structures associated with **III**. Choice **(B)** is eliminated because hydrogen atom **b**, having no resonance structure with a negative oxygen atom, must be less acidic than either hydrogen atom **c** or **d**.

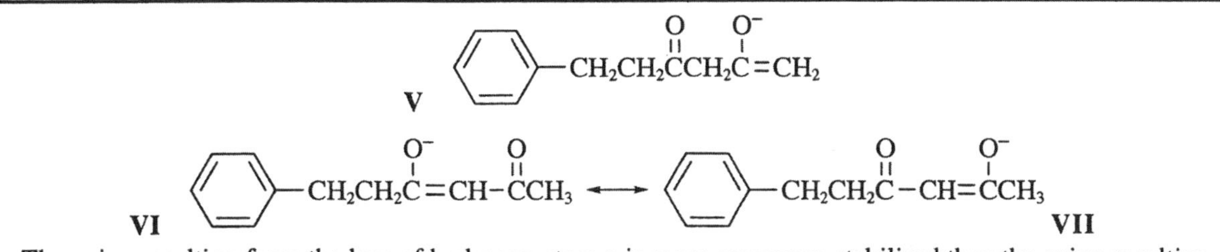

The anion resulting from the loss of hydrogen atom **c** is more resonance stabilized than the anion resulting from the loss of hydrogen atom **d**, because **III** has two resonance forms (**VI** and **VII**) that place the negative charge on different oxygen atoms. Consequently, choice **(D)** is eliminated, leaving choice (**C**) as the correct answer.

Note: $pK_a = 43$ for hydrogen **a**; $pK_a = 41$ for hydrogen **b**; $pK_a = 10$ for hydrogen **c**; $pK_a = 20$ for hydrogen **d**.

AB-9. Arrange these compounds in order from strongest to weakest base.

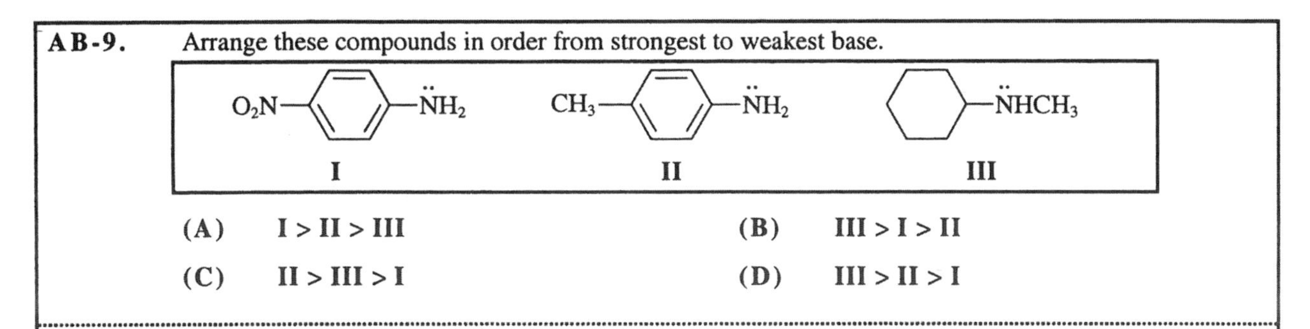

(A) I > II > III

(B) III > I > II

(C) II > III > I

(D) III > II > I

Knowledge Required: (1) The ability to write the chemical reaction between a base and an acid. (1) How resonance stabilizes a base or its conjugate acid.

Thinking it Through: To evaluate the base strength of compounds **I–III**, one needs to write the reactions for protonation and then evaluate the relative stability of the reactants and products. We will initially consider the aliphatic amine, **III**. The electron pair on the nitrogen atom is localized, and cannot be delocalized by resonance. In the protonated form, **IIIa**, the charge is on the nitrogen atom and it also cannot be delocalized. Aliphatic amines have base strengths comparable to ammonia, and we will use it as the reference compound.

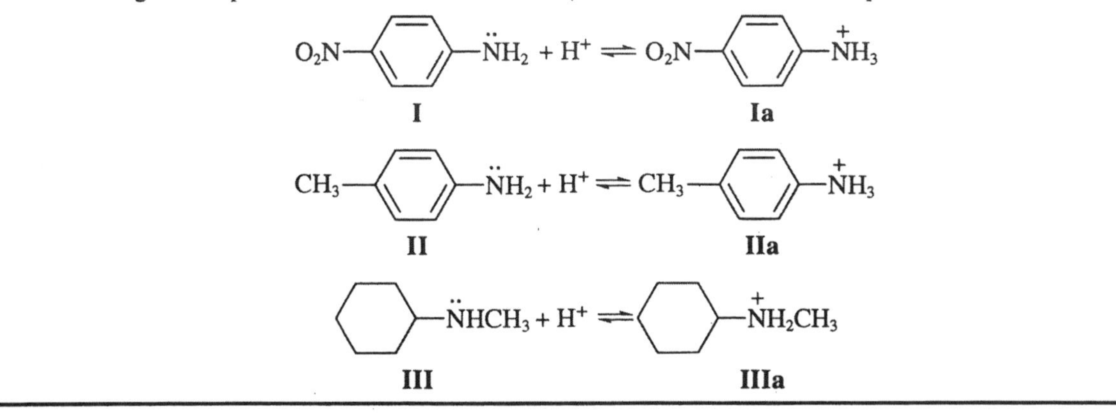

For compounds **I** and **II**, the electron pair on the nitrogen atom is protonated to form **Ia** and **IIa**, much as **IIIa** is formed from **III**. The protonated forms of all three compounds are similar in that the charge is localized on the nitrogen atom and cannot be delocalized. The difference comes in a comparison of **I** and **II** with the aliphatic amine, **III**. In **III** the electron pair cannot be delocalized, whereas in **I** and **II**, delocalization of the electron pair into the aromatic ring stabilizes the base form. Resonance stabilization suggests that **III** should be more basic than either **I** or **II**, eliminating choices (**A**) and (**C**). Now considering the relative stabilities of **I** and **II**, note that **I** has two resonance structures (**IV** and **V**), while **I** has only one resonance structure (**VI**). Consequently, *p*-nitroaniline, **I**, should be a weaker base than *p*-methylaniline, **II**, and the correct answer must be choice (**D**).

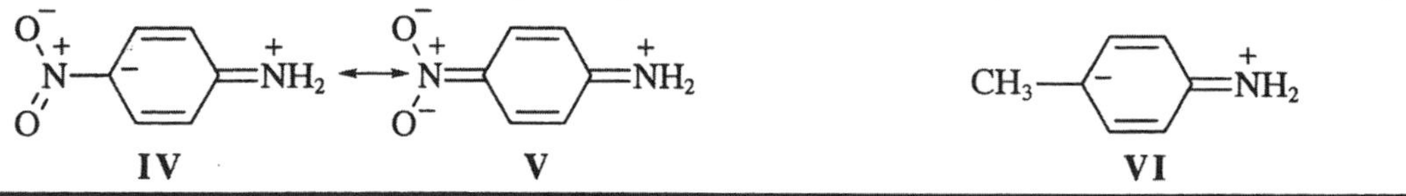

AB-10. Which is the correct order of **decreasing** acid strength for these species? (most acidic → least acidic)

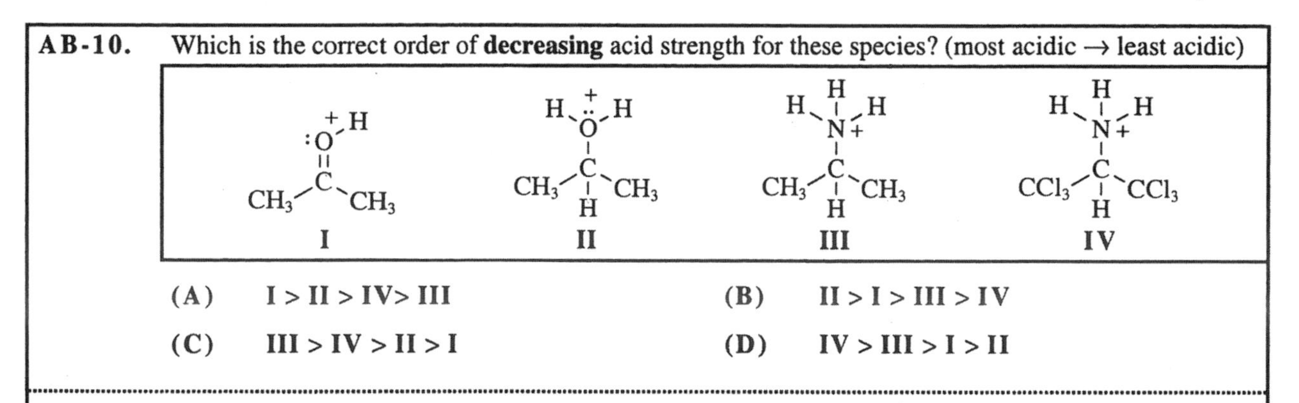

(A) I > II > IV> III

(B) II > I > III > IV

(C) III > IV > II > I

(D) IV > III > I > II

Knowledge Required: (1) How to write the ionization reaction of an acid. (2) How inductive effects in an acid and its conjugate base affects acid strength. (3) How hybridization affects the base strength of an electron pair.

Thinking it Through: Write the equation for each compound behaving as an acid.

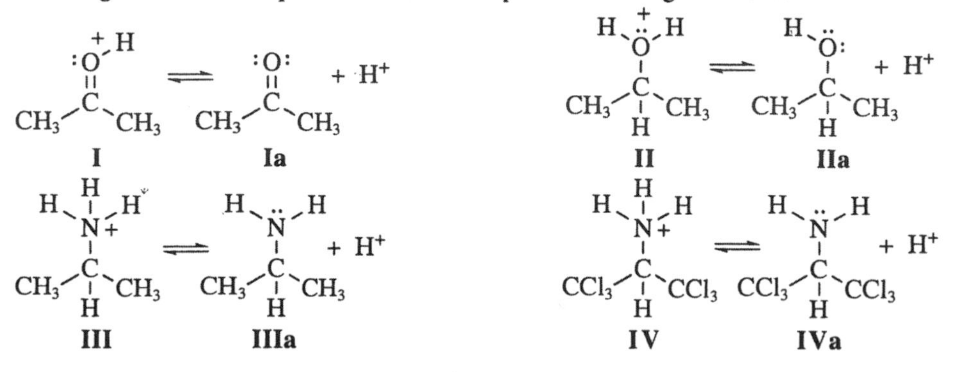

Comparison of **I** and **II** with **III** and **IV** reveals that the first two have the positive charge on the oxygen atoms and the second two have the positive charges on the nitrogen atoms. Oxygen atoms are more electronegative than nitrogen atoms, so **III** and **IV** would be more stable (less acidic) than either **I** or **II**. Choices (**C**) and (**D**) are eliminated. Structures **III** and **IV** are similar except that two trichloromethyl groups in **IV** replace methyl groups in **III**. Chlorine atoms are electron withdrawing relative to hydrogen atoms, so **IVa** should be more stable than **IIIa**, making **IV** a stronger acid than **III**. Choice (A) is the only possible correct answer. Note that the answer could also have been reached by showing that **I** must be a stronger acid than **II**. The electron pair on the carbonyl oxygen atom in **Ia** is in an sp^2 hybrid orbital that is 1/3 *s* character. In **IIa** the electron pair is an sp^3 hybrid orbital that is 1/4 *s* character. The more *s* character the more stable the electron pair, and thus the less basic.

(pK_a values: protonated ketones ~ –7, protonated alcohols ~ –2, protonated amines ~ 10. The protonated hexachloroisopropylamine would be less than 10 but greater than –2)

AB-11. What is the order of acidity from weakest to strongest acid for these compounds?

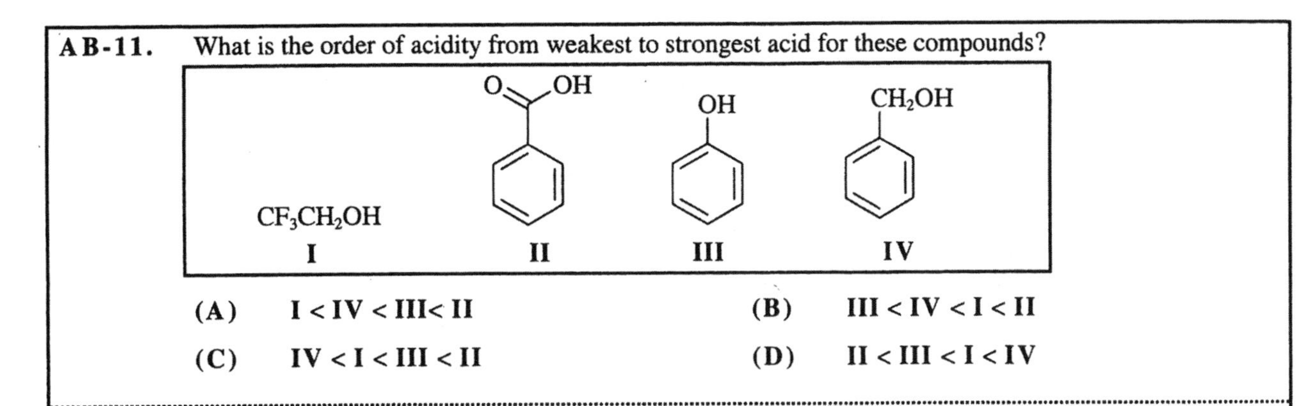

(A) I < IV < III< II (B) III < IV < I < II

(C) IV < I < III < II (D) II < III < I < IV

Knowledge Required: (1) How to write the ionization reaction of an acid. (2) How acid strength depends on resonance stabilization and inductive effects for both the acid and its conjugate base.

Thinking it Through: Write the ionization reaction for each compound behaving as an acid.

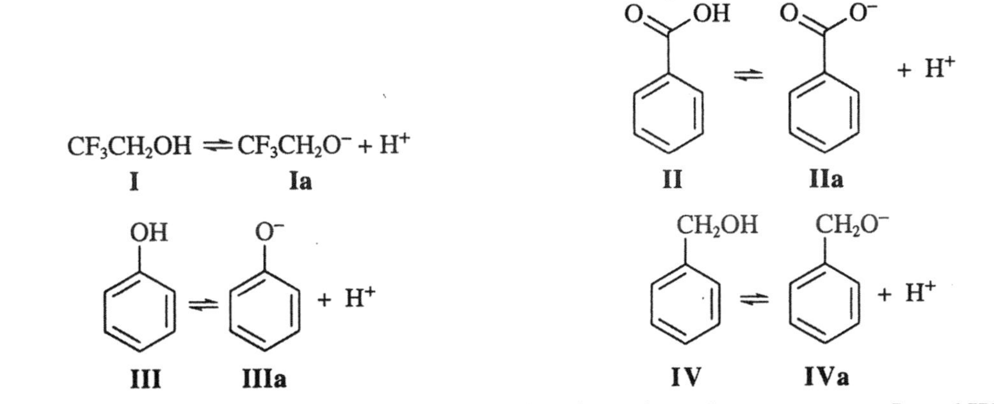

In all four anions (**Ia**, **IIa**, **IIIa**, and **IVa**) the negative charge is on the oxygen atom. **Ia** and **IVa** each have a $-CH_2-$ group attached to the negative oxygen atom, so no resonance stabilization of these ions is possible. In **IIa**, the negative charge can be delocalized by resonance to the second oxygen atom as shown in **IIb**. **II** must be the more acidic than **I** or **IV**, which eliminates choice (**D**).

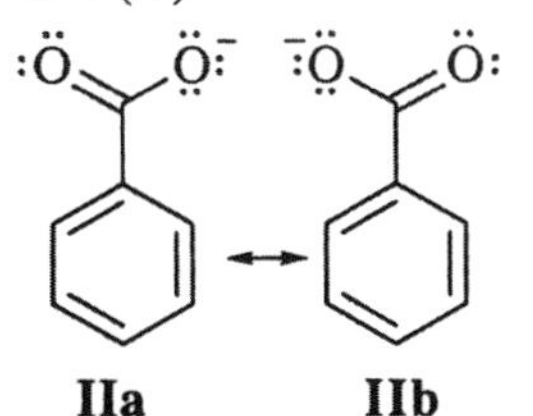

The anion **IIIa** can also delocalize electron density, but not as effectively as **IIa**, because aromaticity of the ring is lost in the resonance structures. **IIIb**, and **IIIc** show the delocalization onto the ring. Structure **III** should be more acidic than either **I** or **IV**, but less acidic than **II**. Choice (**B**) is eliminated.

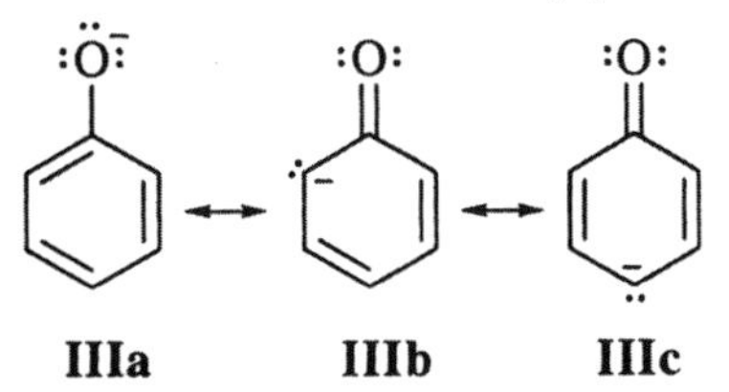

The difference in acid strength between **I** and **IV** is explained by the inductive effect of fluorine atoms on the stability of the anion. Fluorine atoms are more electronegative than either hydrogen atoms or phenyl groups, so replacing the two hydrogen atoms and the phenyl group of **IVa** with three fluorine atoms (**Ia**) produces a more stable anion. Consequently, **I** must be more acidic than **IV**, making choice (**C**) the correct answer.

(pK_a values: carboxylic acids ~ 5; phenols ~10; trifluoroethanol ~12; other alcohols ~18)

Practice Questions

1. Which structure corresponds to the predominant form of this molecule near pH 7?

$CH_3-CH(NH_2)-C(=O)-OH$

(A) $CH_3-CH(NH_3^+)-C(=O)-OH$

(B) $CH_3-CH(NH_2)-C(=O)-O^-$

(C) $CH_3-CH(NH_3^+)-C(=O)-O^-$

(D) $CH_3-CH(NH_2)-C(=O)-OH$

2. Which is the ***strongest*** base?

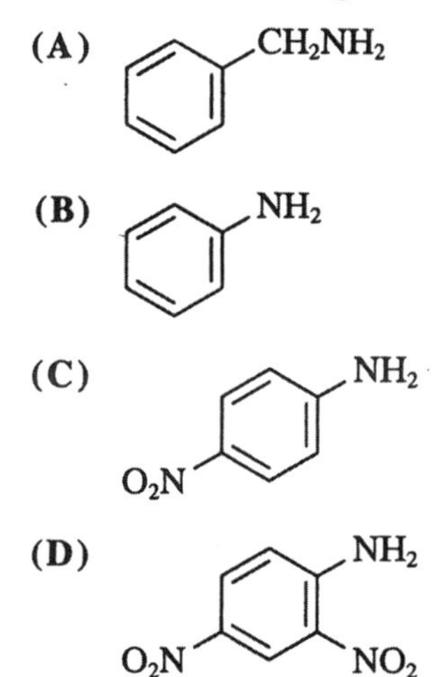

3. Which carboxylate is the ***strongest*** base?

(A) $CH_3CH_2CH_2C(=O)O^-$

(B) $CH_3CH_2CH(Cl)C(=O)O^-$

(C) $CH_3CH(Cl)CH_2C(=O)O^-$

(D) $CH_3CH_2C(Cl)_2C(=O)O^-$

4. Which is the order from the ***strongest*** acid to the ***weakest*** acid for these species?

CH_3OH	$CH_3OH_2^+$	CH_3NH_2	$CH_3NH_3^+$
I	II	III	IV

(A) II > IV > I > III

(B) III > I > IV > II

(C) III > IV > I > II

(D) II > I > IV > III

5. Which is the order from ***most*** acidic to ***least*** acidic?

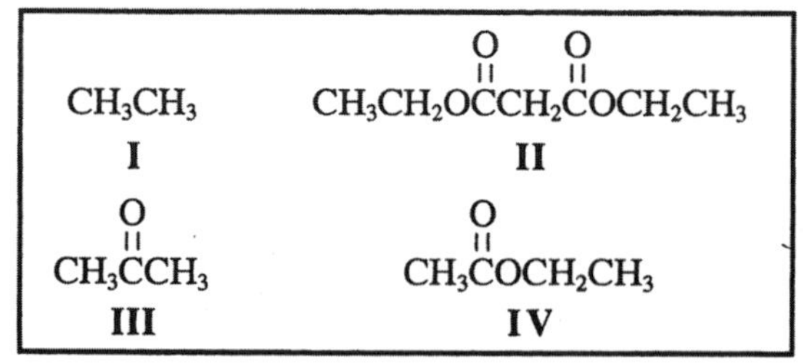

(A) III > IV > II > I

(B) II > IV > I > III

(C) II > III > IV > I

(D) IV > II > III > I

6. Which resonance form best explains the greater acidity of *p*-nitrophenol relative to phenol?

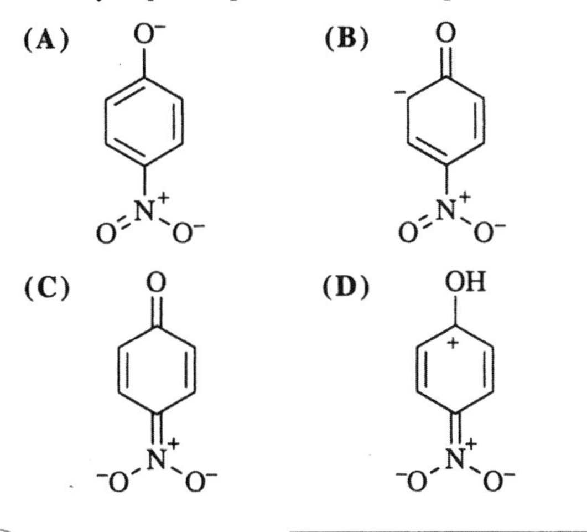

7. Which proton in this compound would be most acidic?

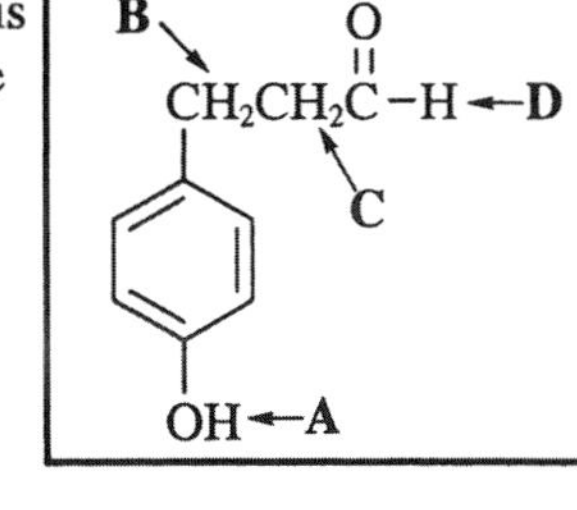

(A) A (B) B (C) C (D) D

8. Which of the indicated protons in this compound would have the smallest pK_a value?

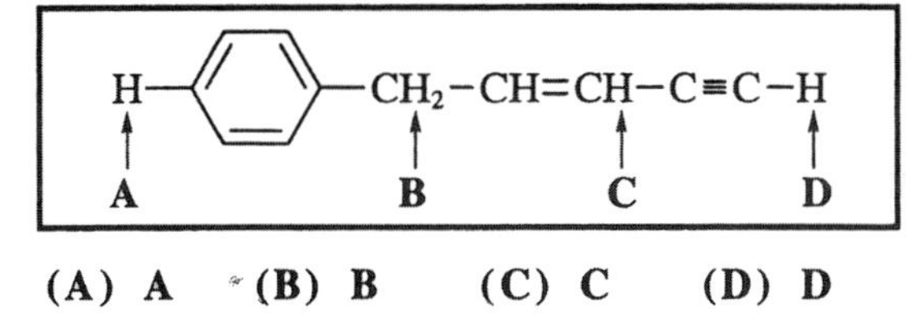

(A) A (B) B (C) C (D) D

9. The imidazole ring system, which contains two nitrogen atoms, is found in the amino acid histidine. Which statement is true regarding the basicity of the two ring nitrogen atoms?

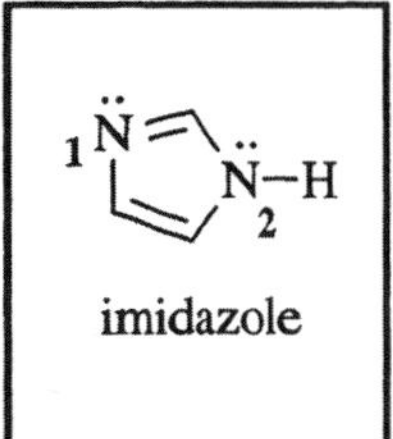

(A) Both are strongly basic since the ring is aromatic.

(B) The nitrogen atom labeled (**1**) is more basic than (**2**).

(C) The nitrogen atom labeled (**2**) is more basic than (**1**).

(D) Neither of the nitrogen atoms would be considered basic.

10. Which choice ranks these compounds from ***strongest*** to ***weakest*** acid?

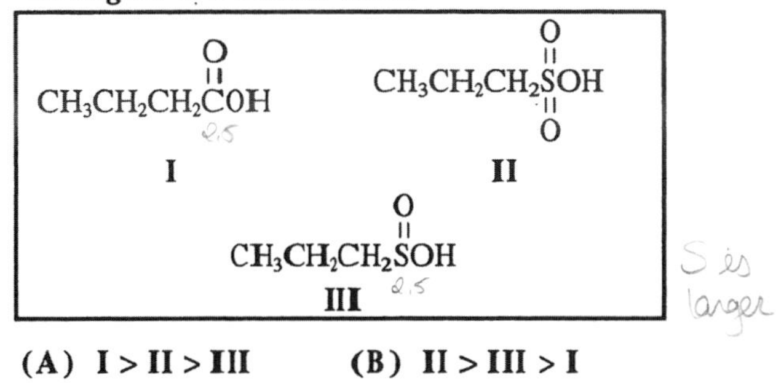

(A) I > II > III (B) II > III > I

(C) III > I > II (D) III > II > I

Answers to Study Questions

1. B
2. B
3. D
4. C
5. C
6. B
7. C
8. C
9. D
10. A
11. C

Answers to Practice Questions

1. C
2. A
3. A
4. A
5. C
6. C
7. A
8. D
9. B
10. B

Stereoisomerism

Appreciating three-dimensional structure is fundamental to understanding organic chemistry. A concept map showing the classification of stereoisomer descriptions is shown in the figure.

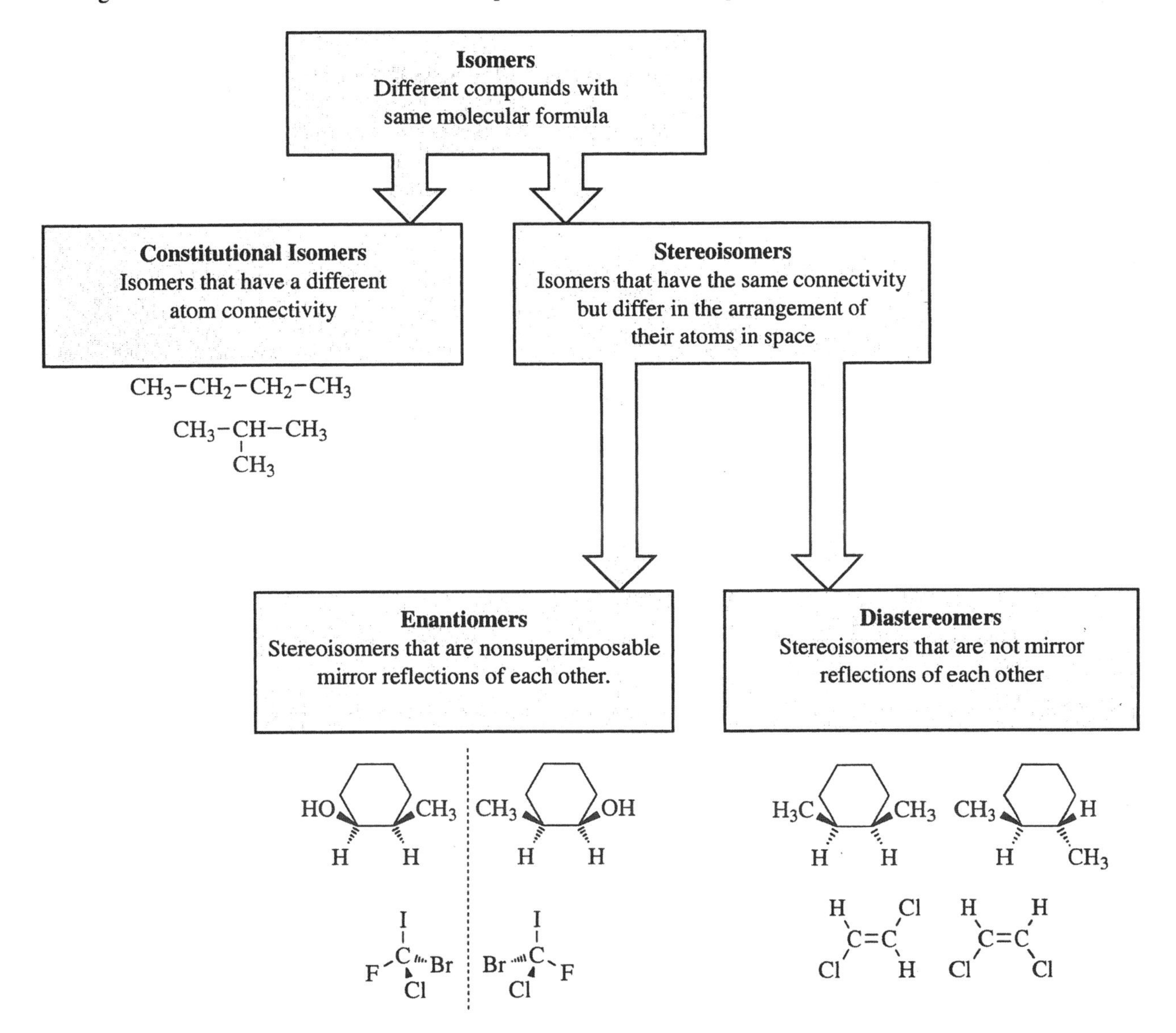

A single enantiomer of a mirror-image pair is said to be optically active because it can rotate plane-polarized light. A mixture of two enantiomers in equal concentrations is called a racemic mixture (or a racemate), and is optically inactive. The absolute configuration of stereoisomers must be specified in a complete nomenclature.

The Cahn–Ingold–Prelog rules are used for alkenes and for carbon atoms that are tetrahedral stereocenters (chiral centers or stereogenic centers). Groups are assigned priorities according to the atomic number of the atoms closest to the stereocenter. For example:

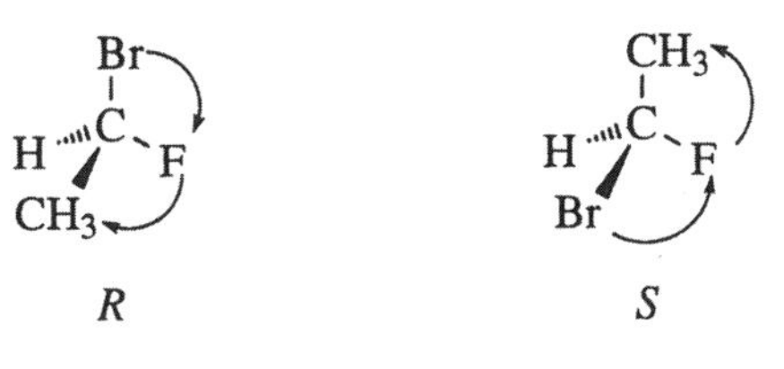

The same principle of assigning priorities is used to designate configuration about a double bond.

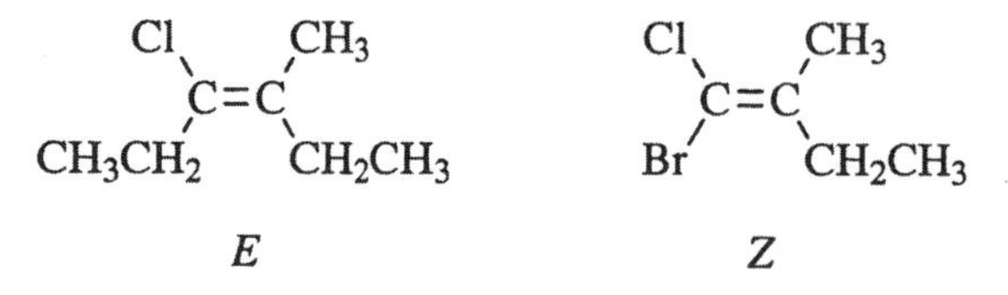

Also intriguing is the concept of a *meso* compound, which is a compound with two (or any even number of) stereocenters that is constitutionally symmetrical and contains an internal mirror plane or point of symmetry.

An important distinction is that between conformations and configurations. Generally, conformations are different stereochemical representations of the same molecule that may be interconverted by rotations about single bonds, whereas configurations are different stereoisomers that are interconverted by breaking and reforming bonds.

Study Questions

SI-1. Which is the most stable conformation for *cis*-1-bromo-3-methylcyclohexane?

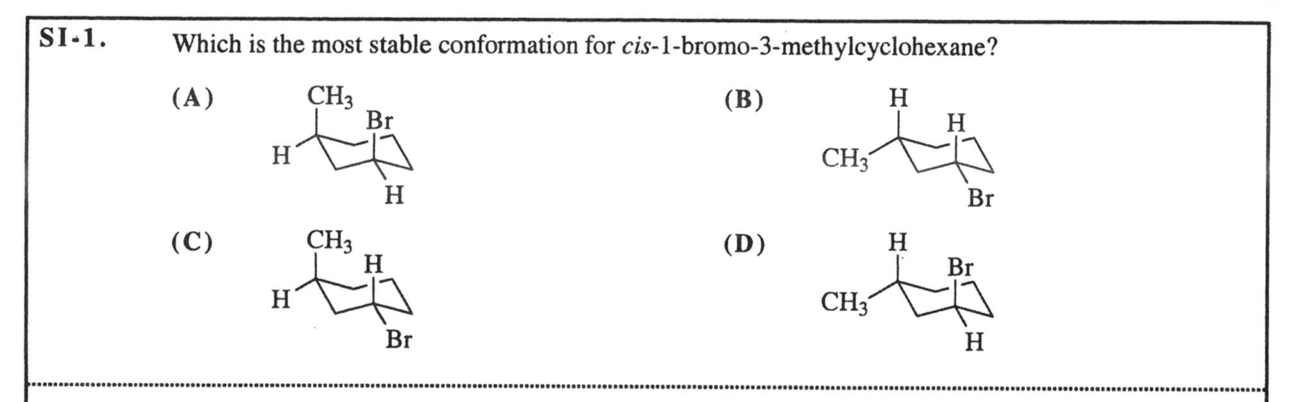

Knowledge Required: (1) Nomenclature of organic compounds. (2) Cyclohexane chair conformations. (3) Axial versus equatorial positions.

Thinking it Through: The two substituents in all four structures are in 1,3-positions relative to each other so only the three-dimensional geometry needs to be analyzed. The *cis-* designation specifies that the substituents are both on the same side of the ring. Inspection of the choices reveals that only in choices **(A)** and **(B)** are the methyl and bromine *cis* to each other. Therefore choices **(C)** and **(D)** are eliminated. Substituents in equatorial positions are more stable than in axial positions because of the torsional strain of 1,3-diaxial interactions. The illustration style of cyclohexane substituents depicts the axial position as vertical lines pointing above or below the ring while equatorial substituents are inclined slightly from horizontal. In choice **(A)** both substituents are axial, whereas in choice **(B)** both substituents are equatorial. Consequently, choice **(B)** is the correct answer.

SI-2. Which Newman projection represents the most stable conformation of 3-methylpentane when viewed down the 2–3 carbon–carbon bond?

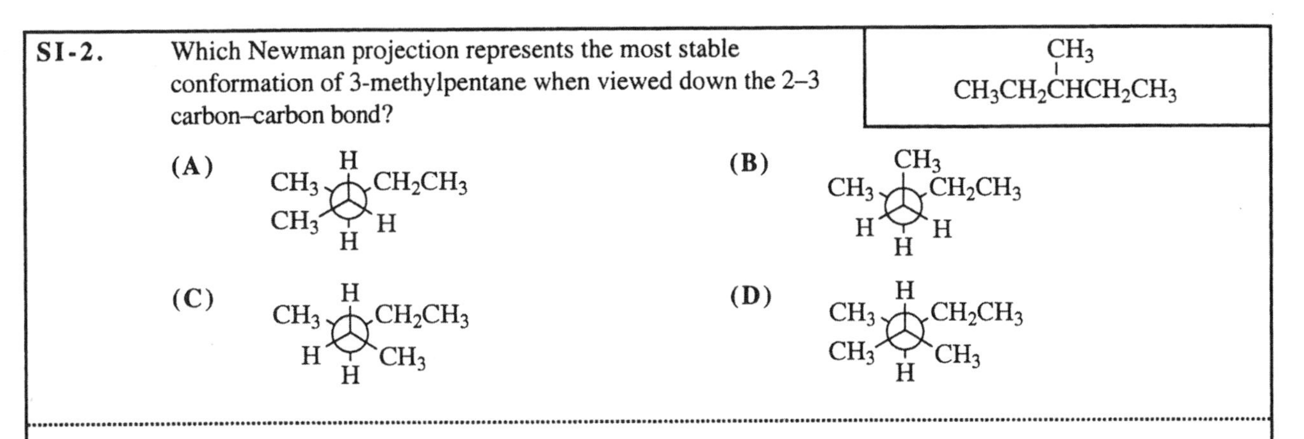

Knowledge Required: (1) The meaning and significance of rotational isomers (rotamers) and gauche interactions. (2) Use of Newman projections. (3) Practice with molecular models and understanding of essential connections. (4) Permitted and forbidden motions.

Thinking it Through: A Newman projection is a view down a bond joining two carbon atoms and showing the spatial relationship of the three substituents on the front carbon atom that radiate from the center of the circle to the three substituents on the rear carbon atom, which appear starting at the edge of that circle. Choice **(A)** shows a position of rotation about the 2–3 carbon–carbon bond in which two methyl groups are adjacent to each other. This adjacency is referred to as a gauche interaction and it increases the energy of the rotational conformation because the groups interfere with each other. Choice **(B)** has two gauche interactions and is even less stable than choice **(A)**. Choice **(C)** has a single gauche interaction but it is between a methyl group and an ethyl group, which constitutes greater interference than the one in choice **(A)**. Choice **(D)** represents a different compound, 2,3-dimethylpentane, so it is eliminated from consideration. Choice **(A)**, having the smallest gauche interaction, is the correct answer.

SI-3. Which property of the two enantiomers of 1-phenylethanol would you expect to be different in principle, even though it may be difficult to measure in practice?

(A) boiling point

(B) melting point

(C) density

(D) strength of interaction with membrane proteins

Knowledge Required: (1) The concept of chirality. (2) Physical properties of enantiomers and diastereomers. (3) The dominance of a single enantiomer of biologically important molecules.

Thinking it Through: Enantiomers are nonsuperimposable mirror images of the same isomer. Enantiomers exhibit identical responses to achiral tests (tests that do not distinguish between mirror image forms). For choices **(A)**, **(B)**, and **(C)**, you can imagine doing the experiment on either side of a mirror on the mirror images of the compound and reasoning that the determination will yield the same measurement for each enantiomer. The molecules important in plants and animals (proteins, carbohydrates, hormones and so on) exist in only one enantiomeric form. Choice **(D)** is correct because the interaction of both enantiomers with a single enantiomer of a protein yields a pair of diastereomers, which do have different physical properties, such as nonbonding interactions. Forming a diastereomer is analogous to putting a right-hand glove on a left hand and another right-hand glove on a right hand. The physical difference is immediately noticeable.

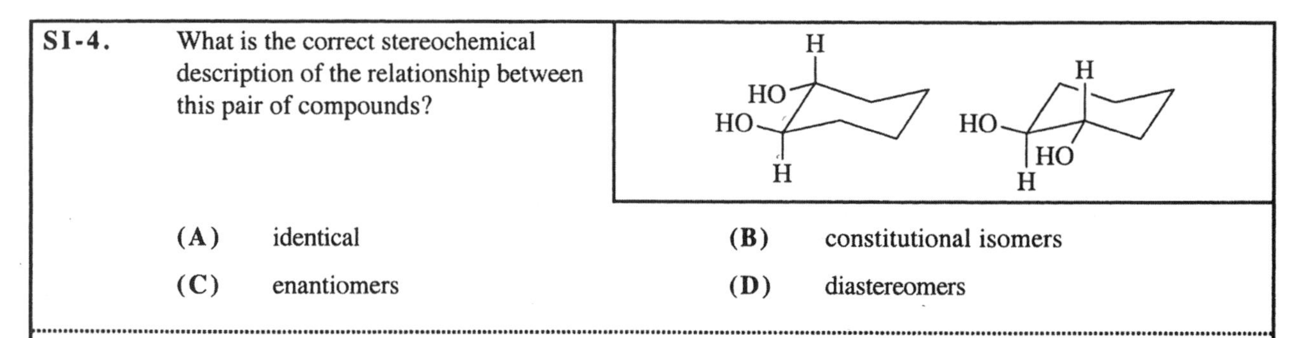

SI-4. What is the correct stereochemical description of the relationship between this pair of compounds?

(A) identical **(B)** constitutional isomers

(C) enantiomers **(D)** diastereomers

Knowledge Required: (1) Stereochemical terminology. (2) Three-dimensional geometrical significance of enantiomers and diastereomers. (3) Chair conformations of cyclohexane.

Thinking it Through: Questions of this kind require mentally rotating the molecules in space or around single bonds. Cyclohexane structures also may have to be flipped from one chair conformation to another to compare their essential 3-dimensional identity for classification in the stereochemical categories **(A)** through **(D)**. In both structures the OH substituents are on adjacent carbon atoms so they are not constitutional isomers. Choice **(B)** is eliminated.

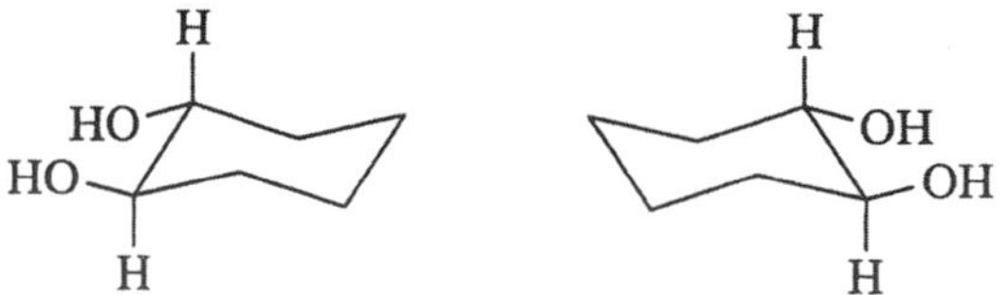

Rotation of the right molecule 180° about a vertical axis reveals that the molecules are nonsuperimposable mirror images of each other. Since the structures are neither identical, choice **(A)**, or diastereomers, choice **(D)**, those choices are eliminated. Nonsuperimposable mirror images are called enantiomers, which is choice **(C)**, the correct answer.

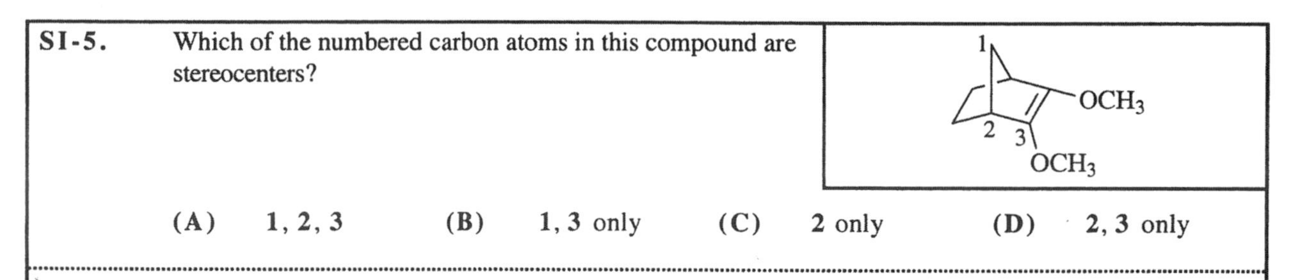

SI-5. Which of the numbered carbon atoms in this compound are stereocenters?

(A) 1, 2, 3 **(B)** 1, 3 only **(C)** 2 only **(D)** 2, 3 only

Knowledge Required: (1) Concept of chirality. (2) Meaning of stereocenter.

Thinking it Through: A tetrahedral stereocenter is a stereogenic center comprising an sp^3 carbon atom bonded to four different substituents. C–**1** is bonded to two hydrogen atoms and C–**3** is sp^2 hybridized. Consequently, neither qualifies as a tetrahedral stereocenter. C–**2** is a tetrahedral stereocenter bonded to four different substituents: a hydrogen atom, an sp^2 carbon atom, a one-carbon-atom bridge, and a two-carbon-atom bridge. Thus choice **(C)** indicates the only tetrahedral stereocenter among the three numbered carbon atoms.

SI-6. What would be the stereochemical classification of the major product of this reaction?

$CH_3CH{=}C(CH_3)_2 \xrightarrow{HBr}$

(A) *R*-enantiomer **(B)** *S*-enantiomer **(C)** achiral **(D)** racemic

Knowledge Required: (1) The meaning of the stereochemical descriptors, *R*, *S*, achiral, and racemic. (2) The regiochemistry of electrophilic addition of hydrogen halides to alkenes.

Thinking it Through: HBr will add to the double bond of the alkene according to Markovnikov's rule to give the product below. The product has no tetrahedral stereocenter and is not otherwise chiral.

$$CH_3CH_2{-}C(CH_3)_2{-}Br$$

Since the product is achiral, choice **(C)** is the correct answer. Choices **(A)** and **(B)** apply to a single enantiomer of a chiral compound. Choice **(D)** refers to the presence of equimolar quantities of both enantiomers of a chiral compound.

SI-7. Which statement about these Fischer projections is correct?

	I	II	III	IV
top	CH_3	CH_3	CH_3	CH_3
C2	HO—H	HO—H	H—OH	H—OH
C3	HO—H	H—OH	H—OH	HO—H
bottom	CH_3	CH_3	CH_3	CH_3

(A) **I** and **III** are enantiomers. **(B)** **II** and **IV** are identical.

(C) **II** is a *meso* isomer. **(D)** **I** and **II** are diastereomers.

Knowledge Required: (1) The stereochemical terms: enantiomer, diastereomer, and *meso*. (2) The stereochemical interpretation of Fischer projections. (3) Allowed manipulations of Fischer projections.

Thinking it Through: These Fischer projections must be compared to see if they are superimposable (identical), nonsuperimposable mirror images (enantiomers), or non-mirror image stereoisomers (diastereomers), or if they have an internal mirror plane (*meso* forms). Fischer projections represent stereoisomers in which the vertical bonds point *away* from the observer and the horizontal bonds point *towards* the observer. Consequently, they may be compared by rotating the structures180° or 360° in the plane of the projection. If projection **I** is rotated 180°, it is seen to be identical to projection **III**. Consequently, choice **(A)** is eliminated. Projections **II** and **IV** are nonsuperimposable mirror images, and are enantiomers. Thus, choice **(B)** is also incorrect. Neither projection **II** nor any of its conformers has an internal mirror plane, and is therefore not a *meso* form. This eliminates choice **(C)**. (Projections **I** and **III** *do* have an internal mirror plane and are the same *meso* form.) This leaves choice **(D)** as the only possible correct answer. Projections **I** and **II** are indeed non-mirror-image stereoisomers (diastereomers).

SI-8. What are the absolute configurations of these two molecules?

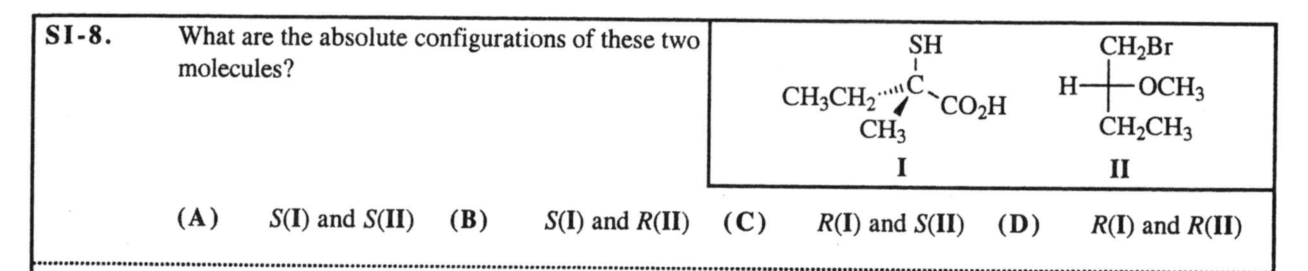

(A) S(I) and S(II) (B) S(I) and R(II) (C) R(I) and S(II) (D) R(I) and R(II)

Knowledge Required: (1) How to assign absolute (*R* or *S*) configurations using the Cahn–Ingold–Prelog sequence rules. (2) The stereochemical meaning of Fischer projections. (3) Manipulations of stereochemical formulas.

Thinking it Through: Start with molecule **I** and assign priorities to the four groups attached to the stereocenter using the Cahn–Ingold–Prelog sequence rules. The order in decreasing priority is: –SH, $-CO_2H$, $-CH_2CH_3$, $-CH_3$. If the molecule does not have the lowest priority group (the CH_3 in this example) pointing away from the observer, the molecule is turned so that it is. This maneuver is difficult for some persons to do mentally so another method is often used. In this method, the lowest priority group is exchanged with the group that is pointing away. Then, exchange the other pair (see example below). Exchanging the first pair converts the molecule into its mirror image (opposite configuration), but exchanging the second pair reverts to the original configuration.

SH / CH_3CH_2–C–CO_2H / CH_3 ⟶ SH / CH_3–C–CO_2H / CH_3CH_2 ⟶ CO_2H / CH_3–C–SH / CH_3CH_2

The next step is to see whether a clockwise or counterclockwise motion is required to go from the highest priority to the next highest priority group. If clockwise the configuration is *R*; if counterclockwise the configuration is *S*. In the right-most structure above, going from the SH to the CO_2H is counterclockwise and the configuration of molecule **I** is *S*. Thus choices **(C)** and **(D)**, which have *R*–**I**, may be eliminated. To choose between choices **(A)** and **(B)**, the configuration of molecule **II** must be determined. Molecule **II** is shown as a Fischer projection. The configuration could be assigned by converting the Fischer projection into a dimensional formula and applying the procedure used for molecule **I**. However it is easier to work with the Fischer projection directly. Remember that the vertical bonds point away from the observer and the horizontal bonds point towards the observer. Thus one wants the lowest priority group to be on one of the vertical bonds. If not already on a vertical bond, exchange it with one that is, and exchange any other two groups to preserve the original configuration. On going from $-OCH_3$ (highest priority) to $-CH_2Br$ (next highest priority), a clockwise motion occurs. The resulting configuration is *R*, and choice **(B)** is the correct answer.

Practice Questions

1. Which molecule has the *R* configuration?

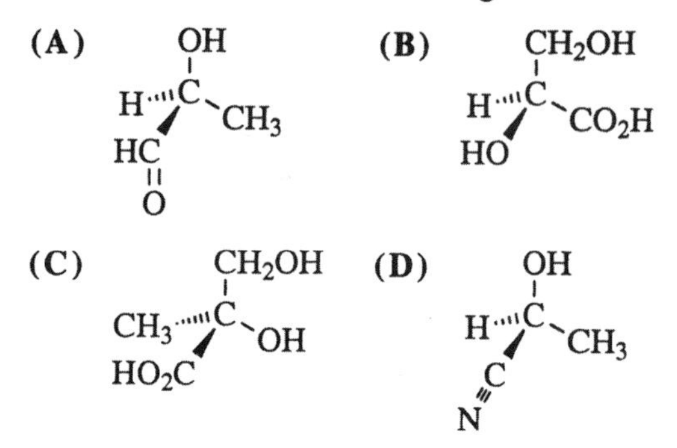

2. The absolute configuration of the two carbon atoms in this compound are

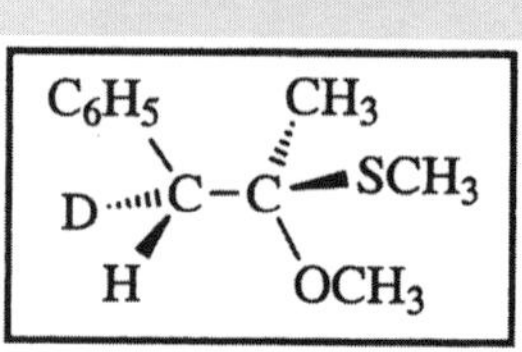

(A) *R*–left carbon atom; *R*–right carbon atom.

(B) *S*–left carbon atom; *S*–right carbon atom.

(C) *S*–left carbon atom; *R*–right carbon atom.

(D) *R*–left carbon atom; *S*–right carbon atom.

3. Which is the enantiomer of this compound?

(Newman projection: front carbon OH, $HOCH_2$, CH_2Br; back carbon H, CH_3, H)

(A) HO, $HOCH_2$, CH_2Br / CH_3, H, H

(B) H, H, CH_3 / OH, CH_2Br, CH_2OH

(C) CH_3, H, H / CH_2OH, OH, CH_2Br

(D) CH_3, H, H / OH, CH_2OH, CH_2Br

4. Which compound will have three peaks of equal height and three valleys of equal depth in a diagram of potential energy *vs.* angle of rotation for one complete rotation around the C–C bond?

(A) $ClCH_2-CH_2Cl$ (H–C(Cl)(H)–C(Cl)(H)–H)

(B) Cl–C(H)(H)–C(Br)(Br)–Br

(C) H–C(Cl)(Cl)–C(Cl)(Cl)–H

(D) H–C(H)(Cl)–C(Cl)(Br)–H

5. Which of these molecules are *meso* isomers?

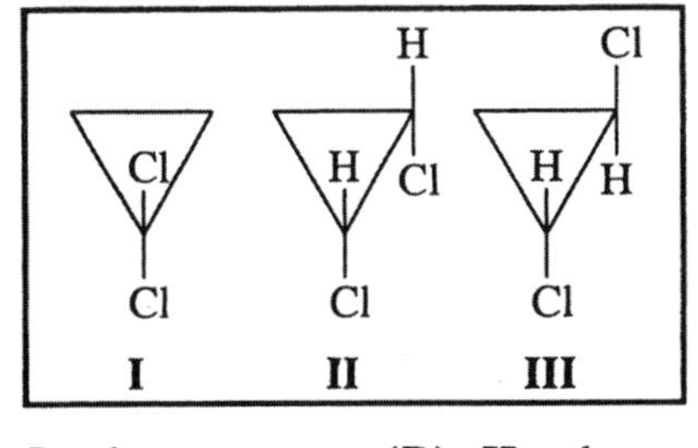

(A) I only
(B) II only
(C) III only
(D) I and II only

6. Which Newman projection represents the most stable conformation of $(CH_3)_2CHCH(CH_3)_2$?

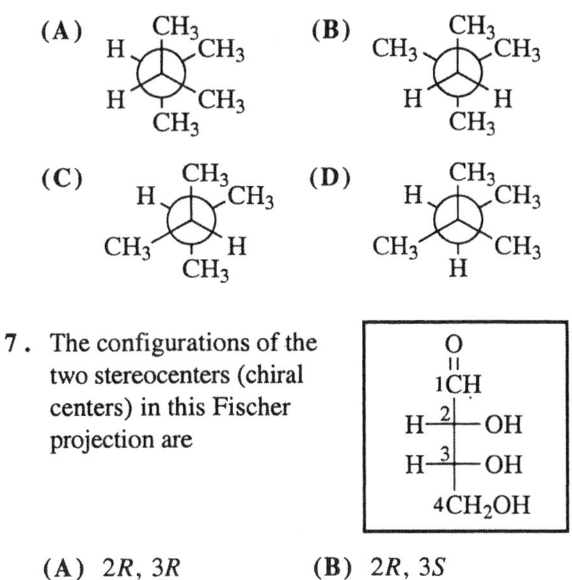

7. The configurations of the two stereocenters (chiral centers) in this Fischer projection are

(Fischer projection: C1 = CHO; C2: H–C–OH; C3: H–C–OH; C4 = CH_2OH)

(A) 2*R*, 3*R*
(B) 2*R*, 3*S*
(C) 2*S*, 3*R*
(D) 2*S*, 3*S*

8. What is the correct stereochemical description of the relationship between this pair of molecules?

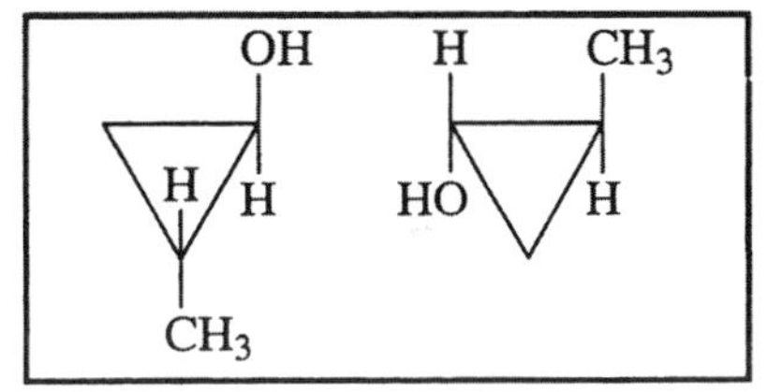

(A) identical
(B) constitutional isomers
(C) enantiomers
(D) diastereomers

9. What would be the stereochemical classification of the product of this reaction?

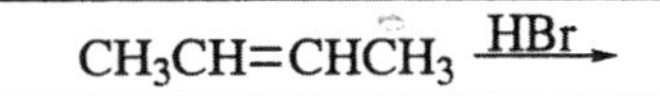

(A) *R*-enantiomer
(B) *S*-enantiomer
(C) *meso* compound
(D) racemate

10. Which diastereoisomer is most stable?

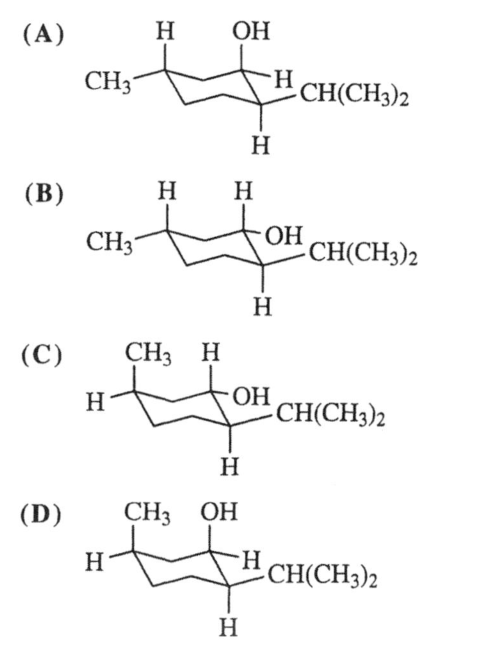

11. The product of this reaction will be

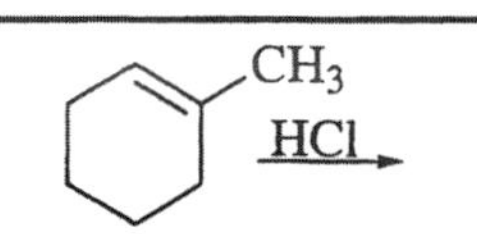

(A) achiral but not *meso*.

(B) a mixture of diastereomers.

(C) a *meso* compound.

(D) a racemate

12. What is the correct stereochemical description for these two Fischer projections?

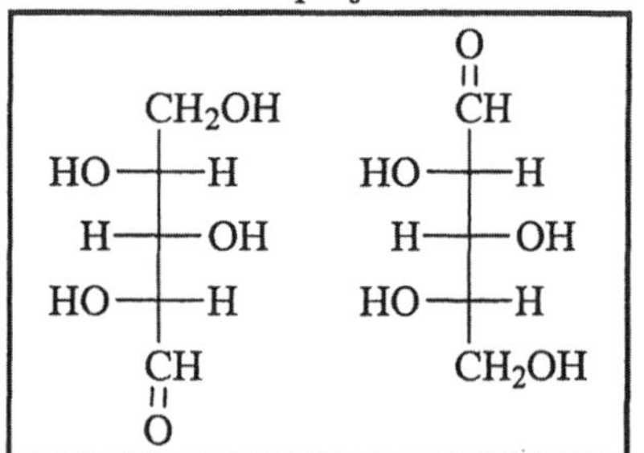

(A) identical

(B) *meso* isomers

(C) diastereomers

(D) enantiomers

13. What is the order from most stable to least stable for these conformations of propylene glycol?

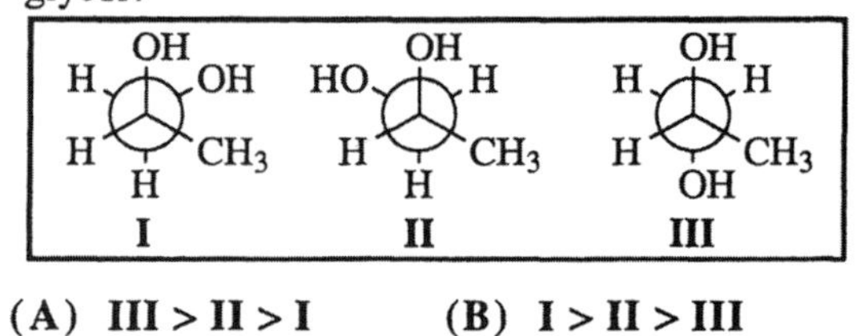

(A) III > II > I (B) I > II > III

(C) I > III > II (D) II > III > I

14. How many total possible stereoisomers are there for 1,2-dimethylcyclopropane? Count pairs of enantiomers (if any) as two different stereoisomers.

(A) 3 (B) 4 (C) 5 (D) 6

15. What is the stereochemical classification of (1*S*,2*S*)-1,2-cyclohexanediol and (1*R*,2*S*)-1,2-cyclohexanediol?

(A) enantiomers

(B) diastereomers

(C) *meso* compounds

(D) racemates

16. Which statement is true?

(A) Compounds with *R* stereocenters rotate plane-polarized light clockwise.

(B) For equal concentrations and equal path lengths, solutions of (+) and (–) enantiomers rotate plane-polarized light equally, but in opposite directions.

(C) Racemic mixtures can rotate plane-polarized light either clockwise or counterclockwise.

(D) *Meso* compounds can rotate plane-polarized light either clockwise or counterclockwise.

17. How many stereoisomers will be formed from the addition of phenyllithium to this molecule?

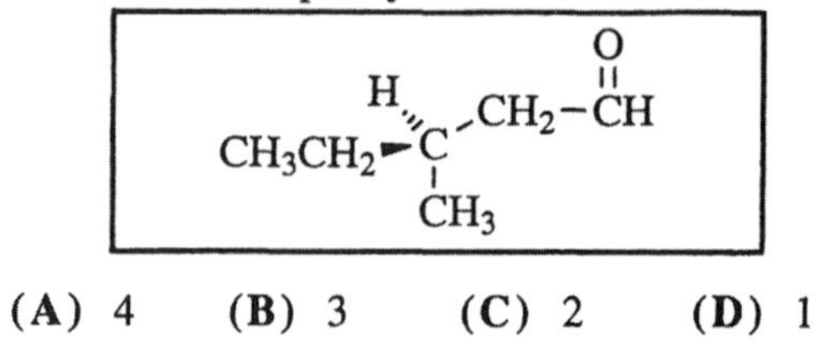

(A) 4 (B) 3 (C) 2 (D) 1

18. Which of these tartaric acid isomers will have the same melting point?

(A) (+)-tartaric acid and (–)-tartaric acid only

(B) (+)-tartaric acid and (–)-tartaric acid and racemic (±)-tartaric acid only

(C) (+)-tartaric acid and (–)-tartaric acid and *meso*-tartaric acid only

(D) *meso*-tartaric acid and racemic (±)-tartaric acid only

19. What is the stereochemical relationship between the salts formed by (+)-tartaric acid with racemic 1-phenylethanamine?

(A) enantiomers

(B) diastereomers

(C) *meso* compounds

(D) racemates

20. Cyclohexene reacts with bromine to yield 1,2-dibromocyclohexane. Molecules of the product would be

(A) a racemic form and, in their most stable conformation, they would have both bromine atoms equatorial.

(B) a racemic form and, in their most stable conformation, they would have one bromine atom equatorial and one axial.

(C) a *meso* compound and, in its most stable conformation, it would have both bromine atoms equatorial.

(D) a *meso* compound and, in its most stable conformation, it would have one bromine atom equatorial and one axial.

21. What is the relationship between these two structures?

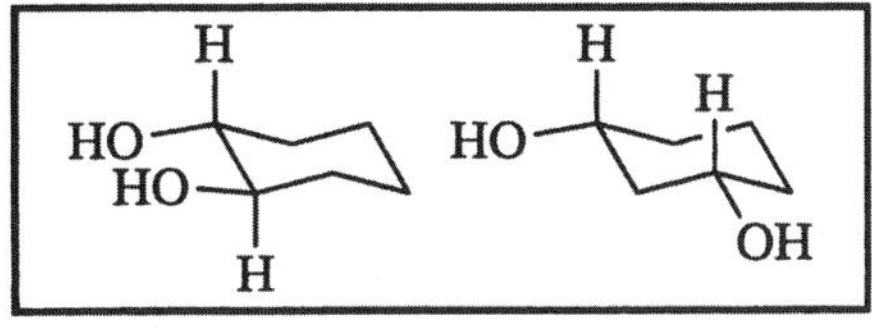

(A) identical molecules

(B) conformational isomers

(C) constitutional isomers

(D) stereoisomers

22. Which is capable of exhibiting *cis-trans* isomerism?

(A) $CH_3C{\equiv}CCH_3$

(B)

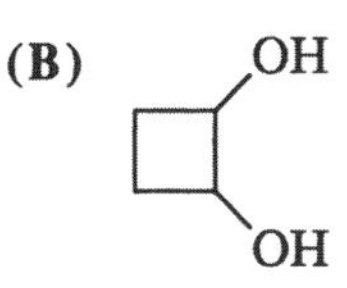

(C) $(CH_3)_2C{=}CHCl$

(D) $CH_3CH{=}C{=}CHCH_3$

23. Which molecules are optically active (chiral)?

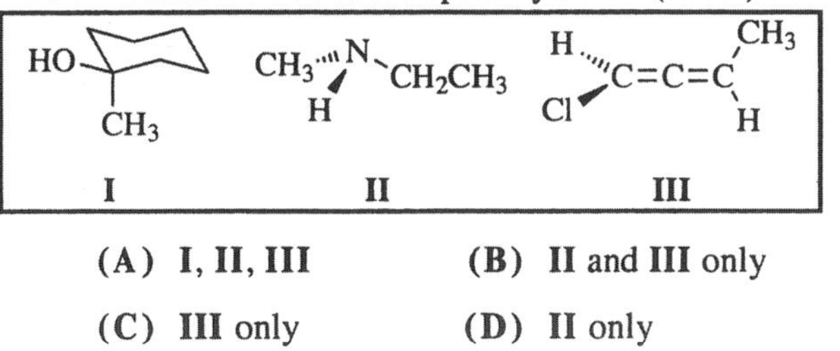

(A) I, II, III **(B)** II and III only

(C) III only **(D)** II only

24. Which would form a racemic mixture upon hydrogenation of the double bond with H_2/Pt?

(A) $C_6H_5{-}CH{=}C(CH_3){-}C_6H_5$

(B) $C_6H_5{-}CH{=}CH{-}C_6H_5$

(C) $C_6H_5{-}CH{=}CH{-}CH_3$

(D) 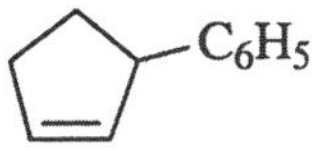 C_6H_5

25. Which reaction would yield a *meso* product?

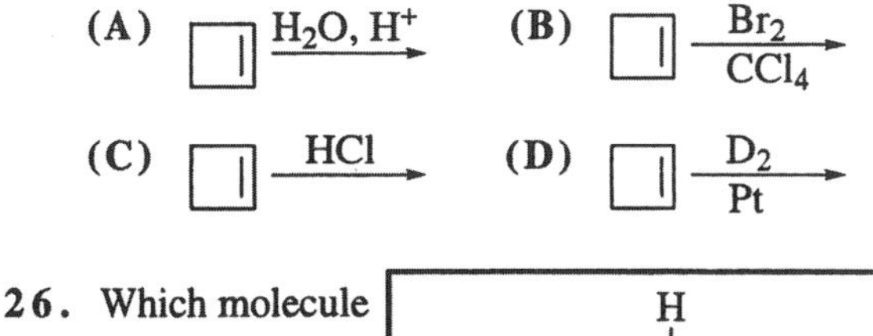

26. Which molecule is the enantiomer of this molecule?

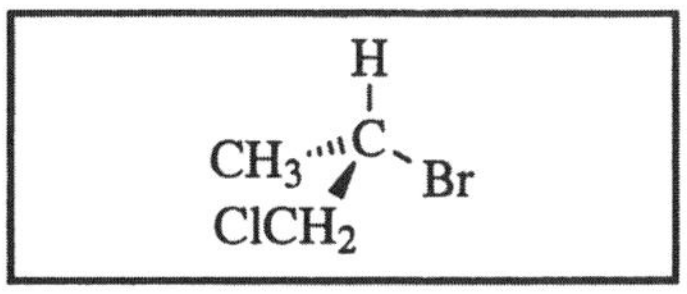

H, CH_3, C, Br, $ClCH_2$

(A) 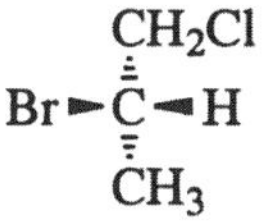 CH_2Cl, Br, C, H, CH_3

(B) 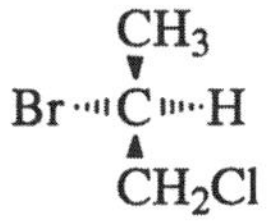 CH_3, Br, C, H, CH_2Cl

(C) CH_3, $ClCH_2$, C, H, Br

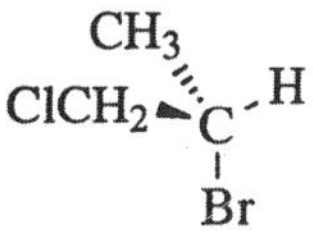

(D) 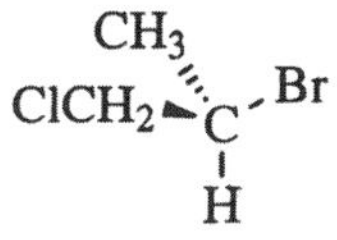 CH_3, $ClCH_2$, C, Br, H

27. Which of these molecules could have an enantiomer?

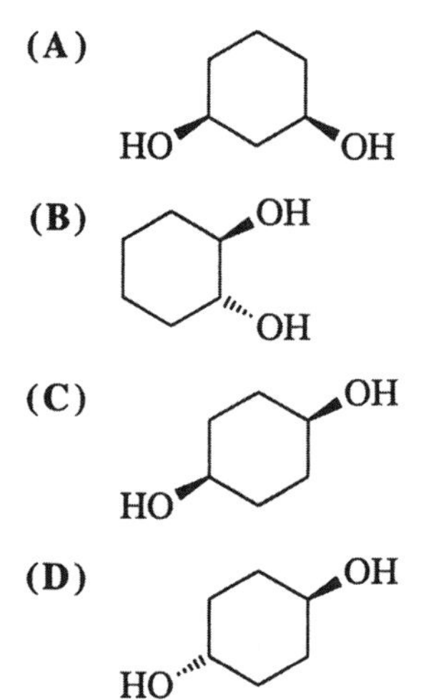

28. Which molecule is the same stereoisomer as the one in this Fischer projection?

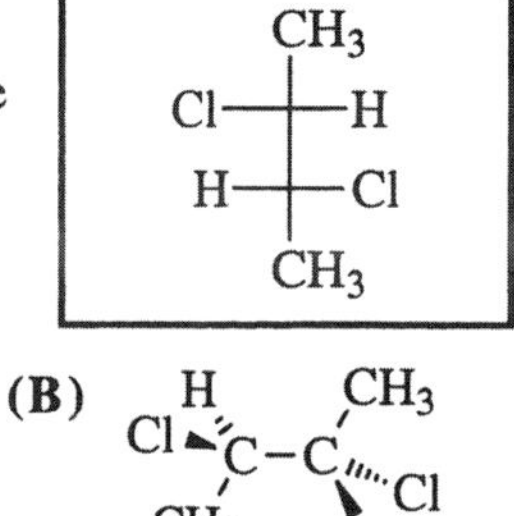

(A) Fischer projection: CH_3 top; H–C–Cl; Cl–C–H; CH_3 bottom

(B) H, Cl, CH_3 on C; C–C; CH_3, Cl, H on C

(C) Newman projection: front H, CH_3, Cl; back CH_3, Cl, H

(D) Newman projection: front Cl, CH_3, Cl; back CH_3, H, H

29. Which Newman projection corresponds to point A on the graph of potential energy *vs.* rotation about the C_2–C_3 bond?

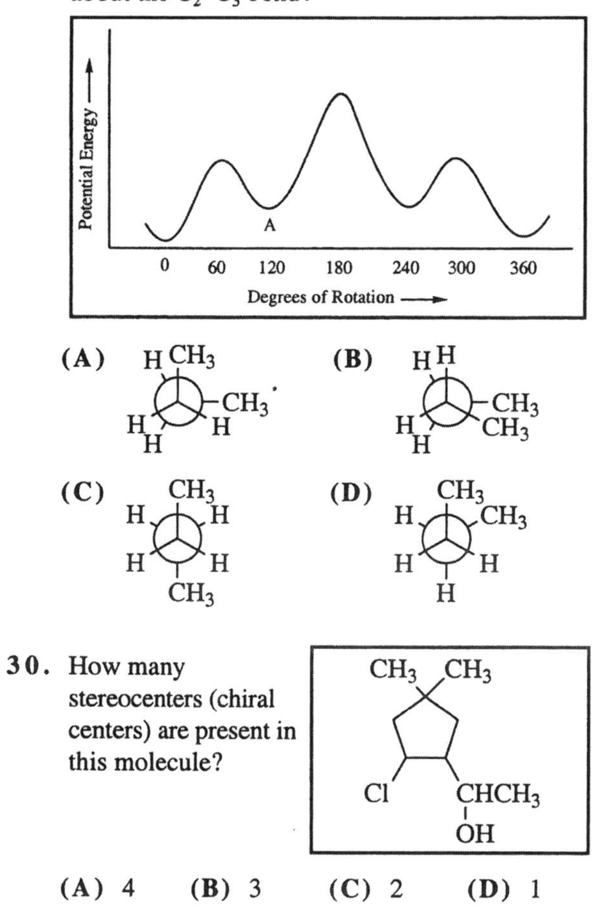

30. How many stereocenters (chiral centers) are present in this molecule?

(A) 4 (B) 3 (C) 2 (D) 1

Answers to Study Questions

1. B
2. A
3. D
4. C
5. C
6. C
7. D
8. B

Answers to Practice Questions

1. C
2. A
3. D
4. B
5. B
6. C
7. A
8. C
9. D
10. B
11. A
12. D
13. D
14. A
15. B
16. B
17. C
18. A
19. B
20. A
21. C
22. B
23. C
24. A
25. D
26. D
27. B
28. C
29. D
30. B

Nucleophilic Substitutions and Eliminations

Substitution Reactions

In these reactions, an electron pair on the nucleophile (which is usually negatively charged) attacks the carbon atom with partial-positive charge that is bound to an electronegative atom. The electronegative atom is displaced, and is referred to as a 'leaving group.' Equation *1* is a symbolic representation of the reaction.

$$:Nu^- + \overset{\delta^+}{R}-\overset{\delta^-}{X} \longrightarrow R-Nu + :X^- \qquad (1)$$

A specific example is the displacement of bromine from ethyl bromide by hydroxide ion (equation *2*). Hydroxide ion is the nucleophile, and bromide ion is the leaving group.

$$^-OH + CH_3CH_2Br \longrightarrow CH_3CH_2OH + Br^- \qquad (2)$$

Nucleophilic substitution can occur as the result of either a one-step mechanism or a two-step mechanism. The one-step mechanism is bimolecular, and the rate equation is first order in both concentration of nucleophile and concentration of alkyl substrate. Substitution reactions that occur via this mechanism (equation *3*) are called S_N2 reactions (substitution, nucleophilic, bimolecular).

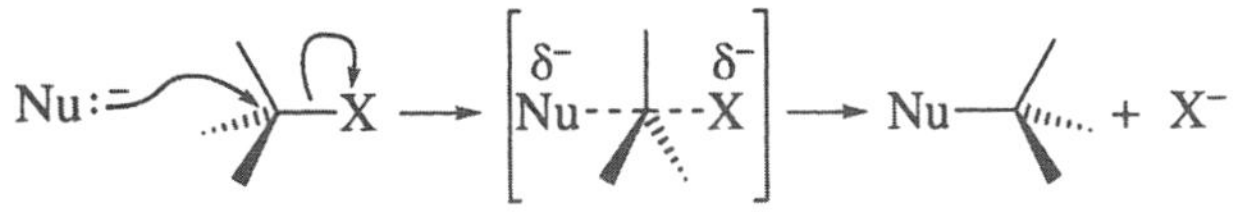

$$Nu:^- + \text{R}_3C-X \longrightarrow \left[\overset{\delta^-}{Nu}\cdots C \cdots \overset{\delta^-}{X}\right] \longrightarrow Nu-C\text{R}_3 + X^- \qquad (3)$$

The two-step mechanism involves initial ionization of the alkyl substrate to form a high-energy carbocation (equation *4*), which then undergoes rapid reaction with the nucleophile (equation *5*). The ionization reaction is the rate-determining step, and the rate equation is first order in the concentration of the alkyl substrate only. Reactions occurring via this mechanism are abbreviated S_N1 (substitution, nucleophilic, unimolecular).

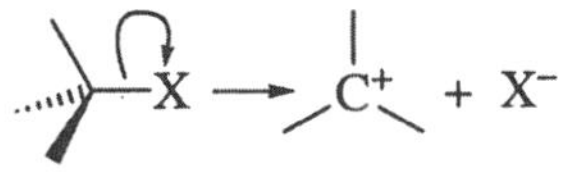

$$\text{R}_3C-X \longrightarrow \text{R}_3C^+ + X^- \qquad (4)$$

$$\text{R}_3C^+ + :Nu^- \longrightarrow Nu-C\text{R}_3 + \text{R}_3C-Nu \qquad (5)$$

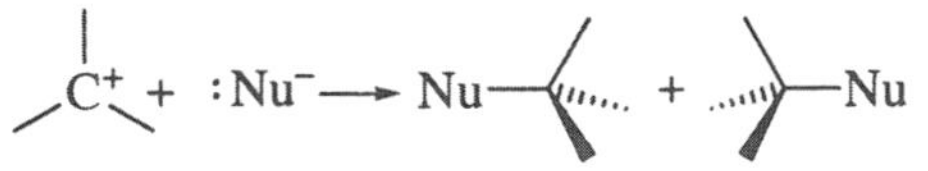

The stereochemical outcomes of the unimolecular and bimolecular reactions are distinctly different. Unimolecular reactions produce racemized tetrahedral stereocenters bearing the nucleophile because the intermediate carbocation is planar sp^2-hybridized, and the nucleophile can attack from either side. Bimolecular reactions, on the other hand, proceed by inversion of configuration at the carbon center.

Elimination Reactions

Because of their unshared pair of electrons, nucleophiles can behave as bases, reacting with acidic protons of the alkyl substrate. The result is that elimination reactions compete with nucleophilic substitution reactions. In these reactions, the proton is removed from the alkyl substrate as it becomes bound to the base, and the departure of both the proton and the electronegative 'leaving group' yields an alkene. As with substitution reactions, the mechanism is one-step (E2) or two-step (E1). The E2 mechanism is shown in equation *6*, and the E1 mechanism in equations *7* and *8*.

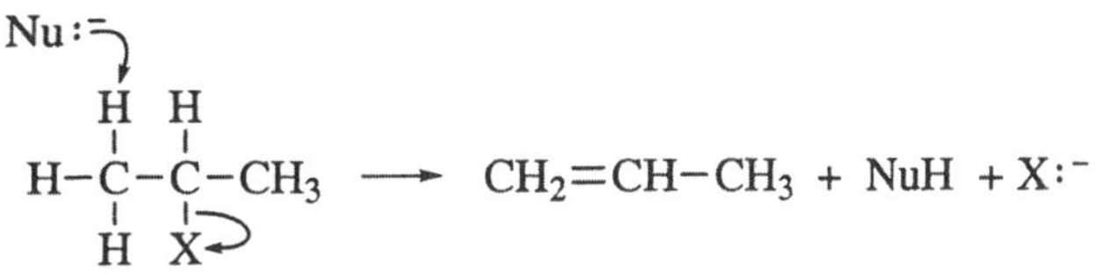

$$Nu:^- + H-CH_2-CH(X)-CH_3 \longrightarrow CH_2=CH-CH_3 + NuH + X:^- \qquad (6)$$

$$CH_3-C(CH_3)_2-X \xrightarrow{slow} CH_3-\overset{+}{C}(CH_3)_2 + X:^- \qquad (7)$$

$$Nu:^- + H-CH_2-\overset{+}{C}(CH_3)_2 \longrightarrow CH_2=C(CH_3)_2 + NuH \qquad (8)$$

Since every nucleophile has basic character, and every base has nucleophilic character, substitution and elimination reactions are both possible (Table 1). Two factors determine mainly whether substitution or elimination will be the major product.

- *The primary, secondary, or tertiary character of the alkyl group bearing the leaving group.* The major product is substitution if the leaving group is on a primary (1°) carbon atom. The major product is elimination if the leaving group is on a tertiary (3°) carbon atom. Another important trend associated with the structural character of the alkyl group is that bimolecular reactions are faster on primary carbon atoms and unimolecular reactions are faster on tertiary carbon atoms.
- *The nature of the base/nucleophile.* Strong bases, such as alkoxides, favor the elimination product from 2° and 3° carbon atoms. Weaker bases such as alcohol and water produce more substitution product, mainly by the unimolecular mechanism. Hindered bases favor the elimination product, even from 1° carbon atoms.

Table 1

	Good Nucleophile		**Poor Nucleophile**
Alkyl halide	**Strong base** (such as OH^- or *t*-butoxide)	**Weak base** (such as Cl^-)	(such as H_2O or HOR)
1°	S_N2 favored; E2 with strong, non-nucleophilic bases	S_N2	no reaction
2°	Mostly E2	Mostly S_N2	S_N1/E1 (slow) in polar, protic solvents
3°	E2	S_N1/E1 in polar, protic solvents	S_N1/E1 in polar, protic solvents

The rates of both substitution and elimination are favored by "good" leaving groups—those that are weaker bases. Protonation of a poor leaving group turns it into a weaker base, making it a better leaving group. ***Protonation is the most common mechanistic step in all of organic chemistry.***

A potential energy diagram for bimolecular mechanisms is the familiar one-hump variety with a single transition state. Potential energy diagrams for unimolecular mechanisms (Figure 1) show the two steps with two activation energies and an intermediate in the energy "valley," illustrated here for the reaction described by equations *7* and *8*. The first step, heterolytic dissociation to the carbocation, is rate determining. The rate-determining first step is fastest (lowest activation energy) for the formation of tertiary carbocations and slowest (highest activation energy) for the formation of primary carbocations.

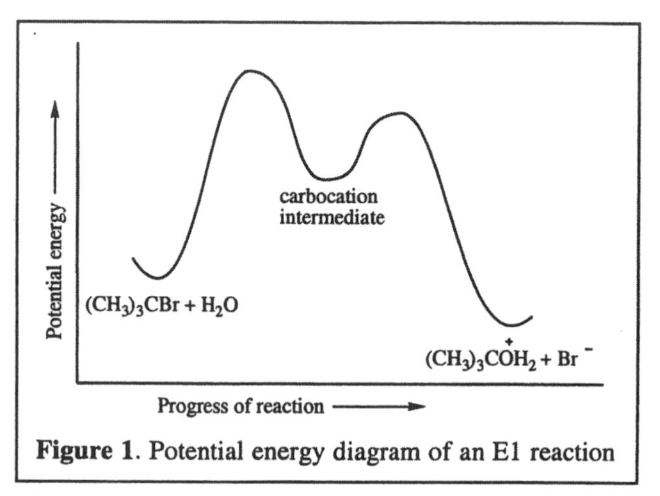

Figure 1. Potential energy diagram of an E1 reaction

Study Questions

NS-1. The reaction of benzyl bromide ($C_6H_5CH_2Br$) with azide ion (N_3^-) is as shown in the box. When the benzyl bromide concentration is constant and azide concentration doubles, the reaction rate is observed to increase by a factor of two. When the azide concentration is held constant and the benzyl bromide concentration doubled, the rate of the reaction doubles. What is the correct rate law for this reaction?

$$C_6H_5CH_2Br + N_3^- \rightarrow C_6H_5CH_2N_3 + Br^-$$

(A) Rate = $k\,[C_6H_5CH_2Br]^2\,[N_3^-]^2$ **(B)** Rate = $k\,[C_6H_5CH_2Br]^4\,[N_3^-]^2$

(C) Rate = $k\,[C_6H_5CH_2Br]^2\,[N_3^-]$ **(D)** Rate = $k\,[C_6H_5CH_2Br]\,[N_3^-]$

Knowledge Required: The kinetic rate law.

Thinking it Through: Reaction rates depend in some way on the concentration of some or all of the reactants. Rate = $k\,[C_6H_5CH_2Br]^a\,[N_3^-]^b$. This makes sense because reactions depend on how much of a reactant is present to react or how frequently the reactants contact each other. Consider the exponent **a**. The rate equation simplifies to Rate = $k'\,[C_6H_5CH_2Br]^a$, where k' includes the concentration of azide if it remains constant. You can reason that **a** must be one for the rate to increase by a factor of two when $[C_6H_5CH_2Br]$ is doubled. The same reasoning applies to exponent **b** of the azide concentration. All choices except for **(D)** are removed from consideration because only choice **(D)** shows the correct exponents of one. Reaction rates also depend on temperature and the presence or absence of catalysts; however, these variables are constant in the data supplied in this question.

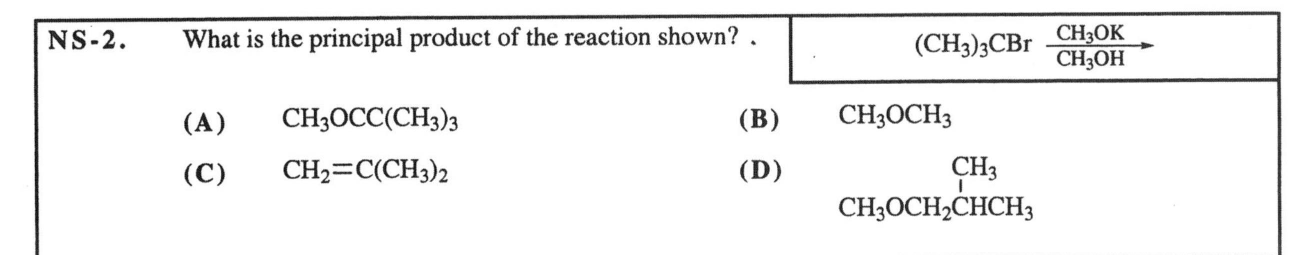

NS-2. What is the principal product of the reaction shown?

$$(CH_3)_3CBr \xrightarrow[CH_3OH]{CH_3OK}$$

(A) $CH_3OCC(CH_3)_3$ **(B)** CH_3OCH_3

(C) $CH_2{=}C(CH_3)_2$ **(D)** $CH_3OCH_2CH(CH_3)CH_3$

Knowledge Required: (1) The meaning and nature of nucleophiles, bases, and leaving groups. (2) The structure of primary, secondary, and tertiary alkyl groups. (3) Reactants, products, and mechanisms of substitution and elimination reactions. (4) Factors that control substitution versus elimination.

Thinking it Through: Observe that this is a reaction of a tertiary alkyl halide with the strongly basic nucleophile, methoxide ion, in methanol solution. This nucleophile/solvent system is one that favors an S_N2 displacement or E2 elimination. Choices **(B)** and **(D)** are eliminated because they cannot be derived as the result of any feasible displacement reaction with the alkyl halide. Tertiary alkyl halides are too sterically hindered to undergo an S_N2 displacement (choice **(A)**) and readily undergo E2 eliminations (choice **(C)**) with a basic nucleophile. Thus, choice **(C)** is left as the only possible correct answer.

NS-3. What is the structure of the reactant that would yield the two products shown?

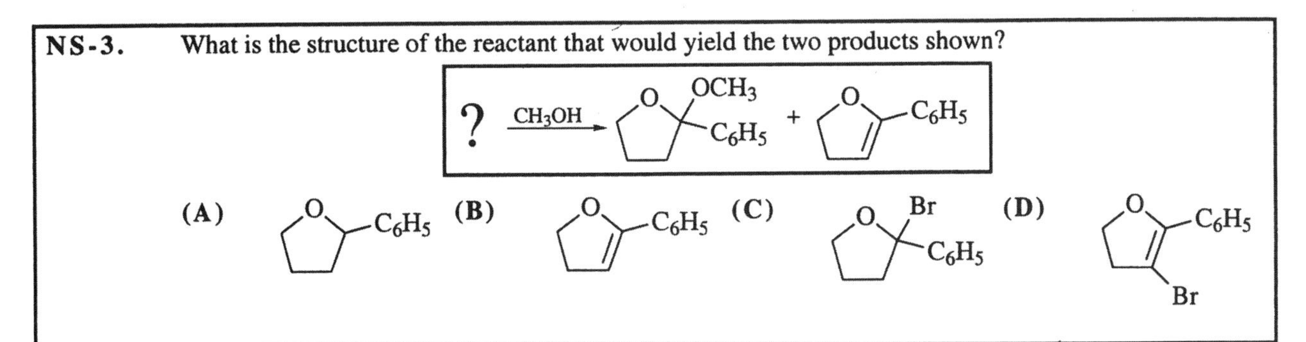

Knowledge Required: (1) The meaning and nature of nucleophiles, bases, and leaving groups. (2) Reactants, products, and mechanisms of substitution and elimination reactions.

Thinking it Through: It is important to recognize that the CH_3OH solvent written above the reaction arrow appears in the product (without its easily abstracted proton). The CH_3O– group has become attached to the carbon atom that is bonded to both the phenyl group and the ring oxygen atom. This means that, unless rearrangements have taken place, a leaving group must have originally also been bound to that carbon atom. Choices **(A), (B),** and **(D)** do not contain a leaving group, and are eliminated. Only choice **(C)** contains a leaving group, bromine. Choices **(B)** and **(D)** each have a double bond, which is reactive with electrophiles but not with just a solvent. A bromine attached to a double-bond carbon atom is not reactive as a leaving group.

NS-4. What is the expected major reaction pathway for the reaction shown?

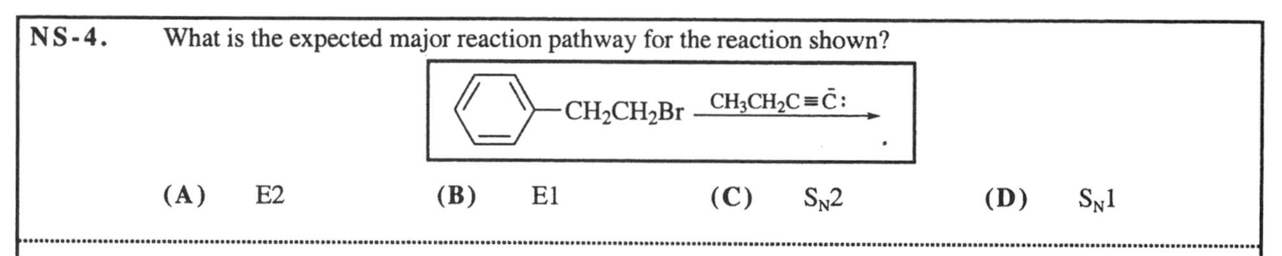

(A) E2 **(B)** E1 **(C)** S_N2 **(D)** S_N1

Knowledge Required: E1, E2, S_N1, and S_N2 reaction mechanisms.

Thinking it Through: The Br^- leaving group is attached to a primary carbon atom. Leaving groups on a primary carbon atom are susceptible to displacement by a nucleophile in a one-step, or bimolecular, collision. Such a direct collision resulting in a substitution is labeled as S_N2, which is choice **(C)**. Choices **(B)** and **(D)**, unimolecular mechanisms, are possible only for unique primary carbon atoms, such as allylic and benzylic carbon atoms. Those unique carbon atoms permit resonance stabilization of the primary carbocation that forms when the Br^- leaves. Choice **(A)** would be a possibility if the base were bulky, such as *tert*-butoxide.

NS-5. What is the configuration of the product in the base-catalyzed hydrolysis of (*R*)-1-chloro-1-deuteriobutane ($CH_3CH_2CH_2CHDCl$)?

(A) (*S*)-1-deuterio-1-butanol **(B)** (*R*)-1-deuterio-1-butanol

(C) *meso*-1-deuterio-1-butanol **(D)** a racemic mixture of **(A)** and **(B)**

Knowledge Required: (1) Assignment of *R* and *S* absolute configurations. (2) Stereochemical consequences of S_N2 and S_N1 mechanisms. (3) Reactants and products of hydrolysis reactions.

Thinking it Through: In a basic hydrolysis reaction, a leaving group is replaced by hydroxide forming a neutral alcohol. The chloride leaving group in the substrate of this question is on a primary carbon atom. Displacement of chloride by hydroxide then would proceed by an S_N2 mechanism. As the hydroxide oxygen atom is forming a new bond 180° opposite to the leaving chloride, the bonds to the other three substituents, H, D, and $CH_3CH_2CH_2$, change to the opposite configuration as shown in the diagram.

$$CH_3CH_2CH_2-C(H)(D)(Cl) \xrightarrow{^-OH} CH_3CH_2CH_2-C(OH)(H)(D)$$

The *R* configuration changes to *S* (since the hydroxide and chloride both have highest priority on the carbon stereocenter that is reacting). Consequently, choice **(A)** must be correct. None of the *R* is formed, eliminating choice **(B)**. Since a racemic mixture is equimolar quantities of both enantiomers, the elimination of choice **(B)** also eliminates choice **(D)**. A *meso* compound, choice **(C)**, requires at least two stereocenters and is overall achiral. The substituted butane in this problem has only one stereocenter.

NS-6. What is the order of rates from fastest to slowest for the reactions of the three nucleophiles with propyl bromide?

$$CH_3CH_2CH_2Br \xrightarrow{Nu} CH_3CH_2CH_2Nu$$

$Nu = CH_3OH, CH_3O^-, CH_3NH_2$

(A) $CH_3OH > CH_3NH_2 > CH_3O^-$

(B) $CH_3NH_2 > CH_3O^- > CH_3OH$

(C) $CH_3O^- > CH_3NH_2 > CH_3OH$

(D) $CH_3O^- > CH_3OH > CH_3NH_2$

Knowledge Required: The relative strengths of nucleophiles (nucleophilicity).

Thinking it Through: Three generalizations concerning relative strengths of nucleophiles are important in determining the reaction-rate sequence: (1) Negatively charged nucleophiles react faster in substitution reactions than neutral nucleophiles. (2) Nucleophilic strength is in the same order as base strength when the nucleophilic atoms are in the same period of the periodic table. (3) Nucleophilic strength increases with increasing atomic size when the nucleophilic atoms are in the same column of the periodic table. Criterion (1) comes into play here, and the negatively charged methoxide ion should be the strongest nucleophile and react the fastest. This eliminates choices **(A)** and **(B)**. Of the remaining nucleophiles, methylamine is more basic, and more nucleophilic, than methanol because the nitrogen atom is less electronegative than the oxygen atom (criterion (2)). This eliminates choice **(D)**, leaving choice **(C)** with the correct order of reaction rates.

NS-7. Which alkyl halide would you expect to undergo S_N1 hydrolysis most rapidly?

(A) $(CH_3)_3CI$ (B) $(CH_3)_3CBr$ (C) $(CH_3)_3CCl$ (D) $(CH_3)_3CF$

Knowledge Required: (1) The course of an S_N1 hydrolysis mechanism. (2) Relative leaving group tendencies.

Thinking it Through: For all choices, the first rate-determining step is the formation of the same *tertiary* carbocation and a halide leaving group, as shown in this reaction. X stands for any of the halogens in the question.

$$(CH_3)_3C{-}X \longrightarrow (CH_3)_3C^+ + X^-$$

Leaving group ability, that is the rate of the mechanistic step shown, increases with decreasing base strength of the leaving group (or increasing strength of the leaving group's conjugate acid). Since iodide ion is the weakest base in the halide series (HI is the strongest acid in the HI, HBr, HCl, and HF series), iodide ion is the best leaving group, and choice **(A)** is expected to react most rapidly.

NS-8. Which of these anions would be the most nucleophilic towards methyl iodide in an ethanol solution?

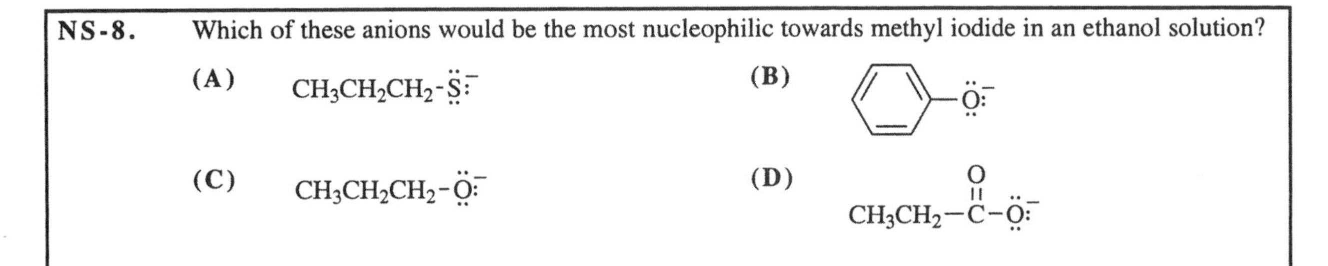

Knowledge Required: (1) How structure affects base strength. (2) Relationships of basicity and atom size to nucleophilicity.

Thinking it Through: Notice that the nucleophilic atoms of the anions are either oxygen or sulfur atoms. Of the three oxygen anions, the anion in choice **(C)** is the strongest base (and most nucleophilic) because it is not stabilized by resonance. The anions in choices **(B)** and **(D)** are resonance stabilized. The question now comes down to whether the anion in choice **(C)** or **(A)** is more nucleophilic. Oxygen and sulfur lie in the same column of the periodic table with sulfur below oxygen. Thus, a sulfur atom is larger than an oxygen atom, which results in sulfur being more nucleophilic. Choice **(A)** is the correct answer.

NS-9. From the data given in the right-hand box, what is the specific rotation of the alcohol produced in the bimolecular nucleophilic displacement shown below?

$$\mathrm{Br{-}C(H)(CH_3)(C_6H_{13})} \xrightarrow{^{-}OH} \mathrm{C_6H_{13}CH(OH)CH_3}$$

$[\alpha]_D = -34.6$ $\qquad$ $[\alpha]_D = ?$

Data

$\mathrm{HO{-}C(H)(CH_3)(C_6H_{13})}$

$[\alpha]_D = -9.9$

(A) +34.6 **(B)** +9.9 **(C)** 0.0 **(D)** –9.9

Knowledge Required: (1) Stereochemical outcome of bimolecular nucleophilic displacement reactions. (2) Relationship of the optical rotations of enantiomers.

Thinking it Through: Since this is a bimolecular nucleophilic (S_N2) displacement, the reaction occurs with inversion of configuration at the chiral center, with retention of optical activity in the product. Choice **(C)** is eliminated because that would be the result of racemization *via* an S_N1 reaction. Choice **(A)** may also be eliminated, because that value is the specific rotation for the enantiomer of the starting bromide. The alcohol for which the specific rotation is –9.9 has the same configuration as the starting bromide, and is the mirror image (enantiomer) of the alcohol produced in the reaction, so choice **(D)** is eliminated. The specific rotation of the product would be +9.9, which is choice **(B)**.

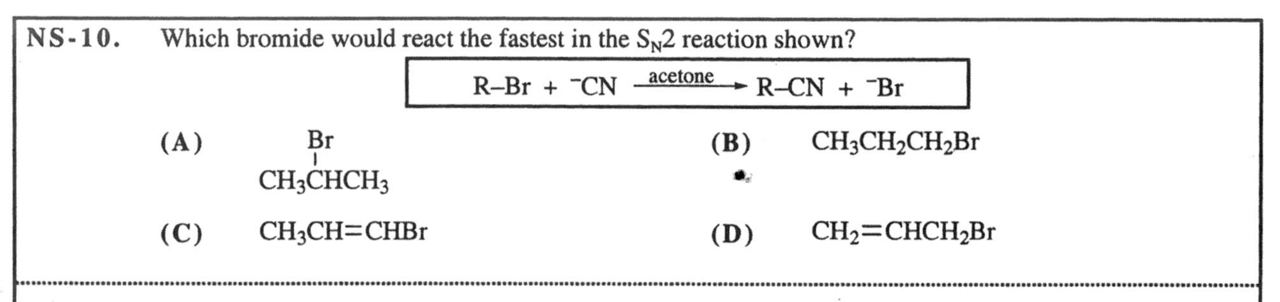

NS-10. Which bromide would react the fastest in the S_N2 reaction shown?

$$\mathrm{R{-}Br} + {}^{-}\mathrm{CN} \xrightarrow{\text{acetone}} \mathrm{R{-}CN} + {}^{-}\mathrm{Br}$$

(A) $CH_3CH(Br)CH_3$ **(B)** $CH_3CH_2CH_2Br$

(C) $CH_3CH{=}CHBr$ **(D)** $CH_2{=}CHCH_2Br$

Knowledge Required: (1) Differentiation of classes of alkyl halides. (2) Effects of steric and electronic factors in the substrate on S_N2 reactions.

Thinking it Through: Recognition that choice **(C)** is a vinylic bromide immediately eliminates choice **(C)** because vinylic halides do not undergo S_N2 reactions. Choice **(A)** is a secondary alkyl bromide and has more steric hindrance to attack by the nucleophile than do the primary alkyl halides represented in choices **(B)** and **(D)**. To choose between choices **(B)** and **(D)**, the critical factor is that, in addition to being a primary alkyl bromide, choice **(D)** is also an allylic bromide. Allylic and benzylic halides undergo S_N2 reactions faster than the corresponding simple alkyl halides. Consequently, choice **(D)** is the correct answer.

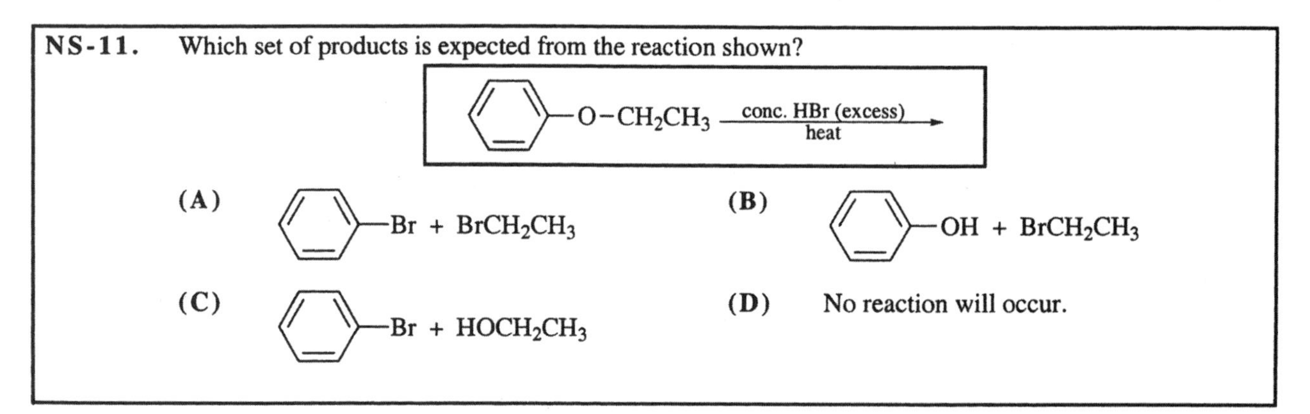

NS-11. Which set of products is expected from the reaction shown?

$$\mathrm{C_6H_5{-}O{-}CH_2CH_3} \xrightarrow[\text{heat}]{\text{conc. HBr (excess)}}$$

(A) $C_6H_5{-}Br + BrCH_2CH_3$ **(B)** $C_6H_5{-}OH + BrCH_2CH_3$

(C) $C_6H_5{-}Br + HOCH_2CH_3$ **(D)** No reaction will occur.

Knowledge Required: (1) Mechanisms and reactivities of ethers and alcohols with hydrobromic and hydroiodic acids. (2) Aryl halides and phenols do not undergo either S_N2 or S_N1 reactions.

Thinking it Through: Cleavage of an ether usually requires a strong acid that has a conjugate base which is a strong nucleophile, such as HBr and HI. Depending on the type of alkyl group, the acid reaction with the halide ion produces an alkyl halide and an alcohol. Under the usual reaction conditions, an excess of the halogen acid is used, which converts the initial alcohol into a secondary aryl halide. The present ether is an alkyl phenyl ether, and since phenols and phenyl halides do not undergo either S_N2 or S_N1 reactions, the cleavage will occur at the alkyl–oxygen bond. An S_N2 attack on the primary alkyl group will produce ethyl bromide and phenol. No further reaction will occur, even in excess HBr, because the phenol is inert to S_N2 or S_N1 reactions. Thus, choices **(A)** and **(C)** are incorrect and the correct answer is choice **(B)**. Since diaryl ethers are the only type of ethers that are not cleaved by HBr or HI, choice **(D)** is not valid.

NS-12. What is the product of the reaction shown?

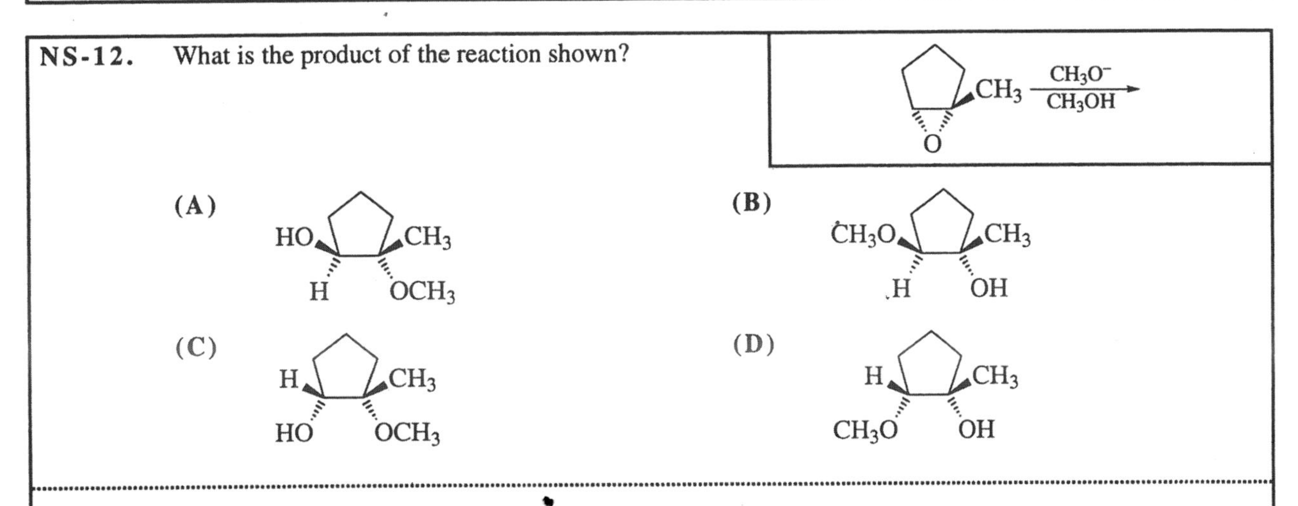

Knowledge Required: (1) The importance of ring strain in epoxides, which allows them to undergo nucleophilic carbon–oxygen bond cleavage reactions that normal ethers will not undergo. (2) Mechanism and regiochemistry of ring opening of epoxides by basic nucleophiles (as opposed to neutral nucleophiles with acid catalysis).

Thinking it Through: Ring opening reactions of epoxides occur by an S_N2 mechanism whether or not the ring oxygen atom is protonated. However, the regiochemistry of the nucleophilic attack *does* depend upon whether or not the ring oxygen atom is protonated prior to the attack by the nucleophile. When the oxygen atom is not protonated, the nucleophile behaves as in a normal S_N2 reaction, and attacks the less substituted carbon atom. When the oxygen atom is protonated, the nucleophile attacks the more highly substituted carbon atom if it is a tertiary carbon atom. Otherwise, attack occurs at both carbon atoms to give mixtures of products. Since the ring openings are S_N2 reactions, the nucleophile must attack from the backside of the carbon–oxygen bond. This inverts the configuration at the reaction site and retains the configuration at the site that becomes the alcohol. Fused-ring epoxides will then give a product in which the nucleophile and the OH group are *trans* to each other. Choices **(C)** and **(D)** are immediately eliminated because the $-OCH_3$ and –OH groups are *cis*. The $-OCH_3$ and –OH groups are *trans* in both choices **(A)** and **(B)**, but the regiochemistry is different. Choice **(B)** has the correct regiochemistry, with the $-OCH_3$ group on the less substituted carbon atom, and is the correct answer.

NS-13. Which bromide will undergo hydrolysis by an S_N1 mechanism the fastest?

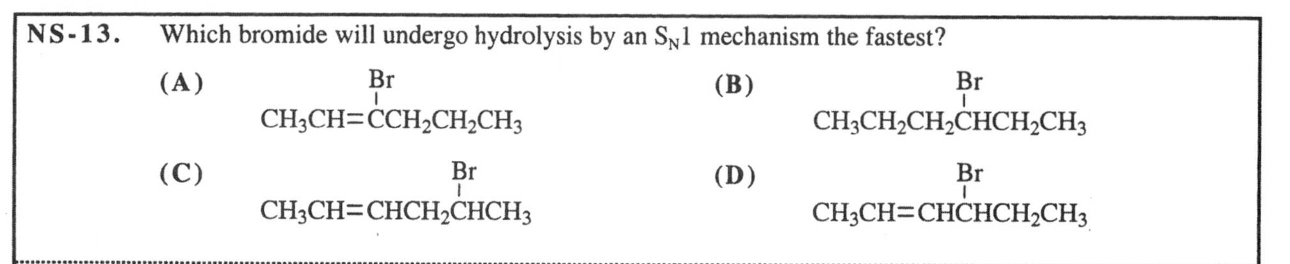

Knowledge Required: (1) Formation of a carbocation in the rate-determining step of S_N1 reactions. (2) Differentiation of allylic and vinylic halides.

Thinking it Through: The rate of an S_N1 reaction is a function of which compound will form the most stable (or least stable) carbocation. Choice **(A)** differs from the other three choices in that it is a vinylic bromide. Vinylic cations are much less stable that alkyl cations, and vinylic halides do not normally undergo S_N1 reactions. Choice **(A)** is eliminated. Choices **(B)**, **(C)** and **(D)** are all secondary alkyl bromides, but choice **(D)** is also allylic. The carbocation from choice **(D)** is stabilized by resonance, whereas the carbocations from choices **(B)** and **(C)** are not. Consequently, choice **(D)** forms the most stable carbocation, and is the correct answer.

NS-14. What is the principal substitution product of the reaction shown?

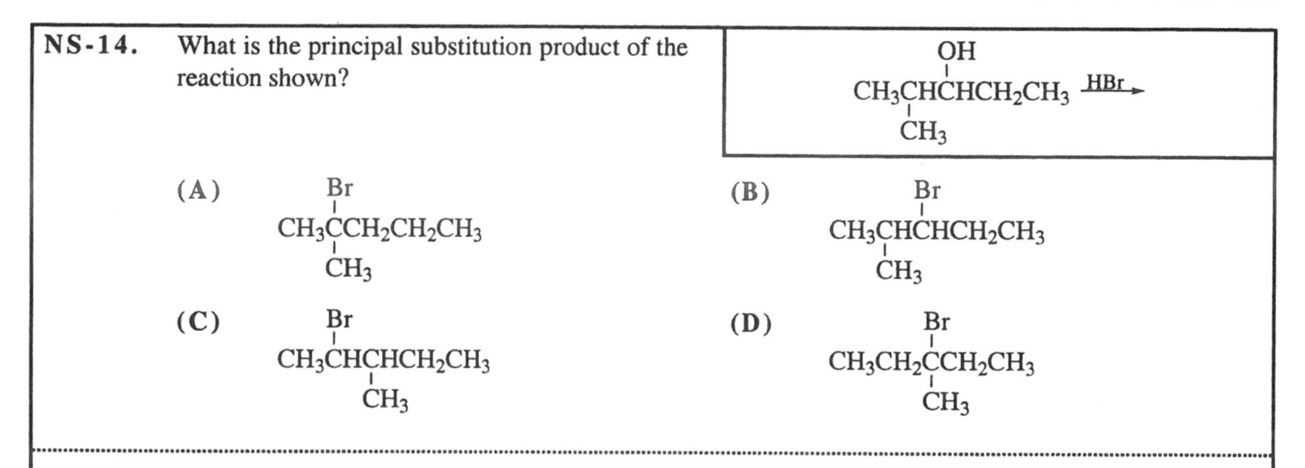

Knowledge Required: (1) Mechanisms for the reactions of alcohols with the hydrogen halides (HX). (2) Carbocation rearrangements.

Thinking it Through: Secondary and tertiary alcohols react with the hydrogen halides *via* an S_N1 mechanism. 1,2-alkyl or 1,2-hydride rearrangements often occur when a more stable carbocation can be formed from the initial carbocation. When rearrangement occurs, the alkyl halide that is produced is derived from the rearranged carbocation. Here, a more stable tertiary carbocation forms from the initial secondary carbocation as the result of a 1,2-hydride shift. This rearranged carbocation then reacts with bromide ion to give the product. Choice **(A)** has the structure that would result from bromide ion attacking this rearranged carbocation, and is the correct answer. Choice **(B)** corresponds to the product of the non-rearranged carbocation. Choice **(C)** could come from a rearranged carbocation (*via* a 1,2-methyl shift), but this rearrangement will not occur because the second carbocation is not more stable than the first one. Choice **(D)** corresponds to the product of a 1,2-methyl shift without rearrangement of the carbocation.

$$CH_3{-}\underset{CH_3}{\overset{H}{C}}{-}\overset{+}{C}HCH_2CH_3 \longrightarrow CH_3{-}\underset{CH_3}{\overset{+}{C}}{-}CH_2CH_2CH_3$$

NS-15. What is the major elimination product from the reaction shown?

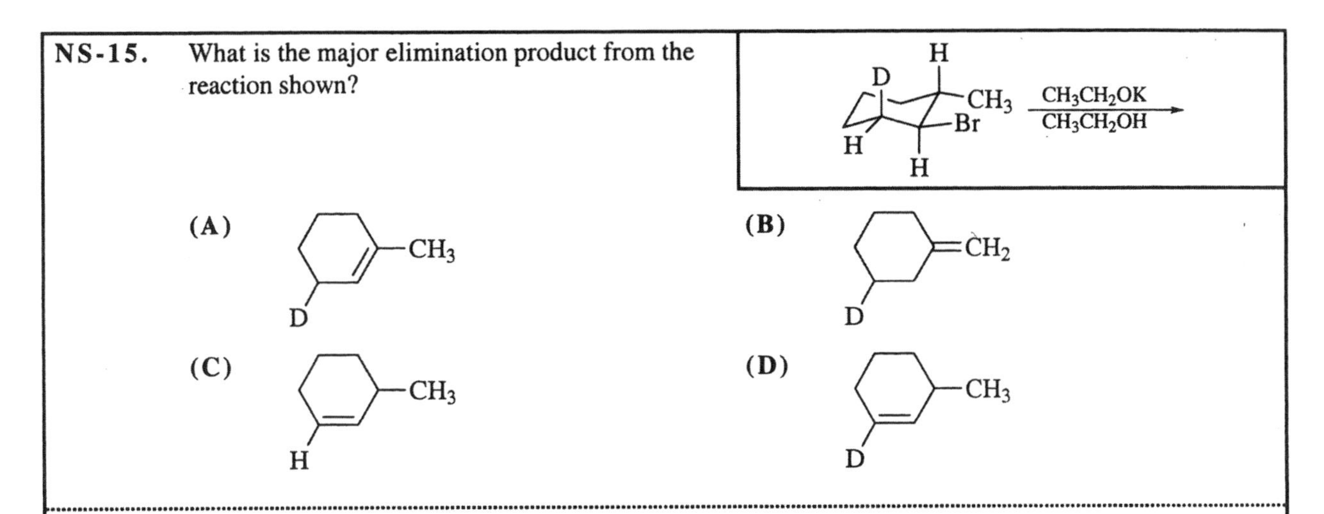

Knowledge Required: (1) The stereochemical requirements for β-eliminations that proceed *via* the E2 mechanism. (2) Conformational changes for cyclohexanes.

Thinking it Through: The majority of E2 eliminations require a transition state conformation in which the β-hydrogen atom and the leaving group are *anti*. When the elimination occurs in a cyclohexane ring this requirement means that the hydrogen atom and leaving group are *trans* and diaxial. In the chair conformation shown, the bromine is equatorial, so the ring must flip to the other chair conformation before an elimination can occur.

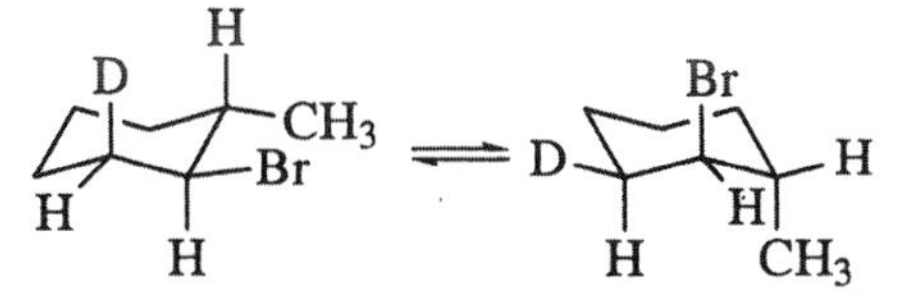

Inspection of this conformation reveals that there is only one β-hydrogen atom that is axial and *trans* to the bromine. Choice **(D)** is the only alkene that can be produced, and is the correct answer. Choices **(A)** and **(C)** would have to form *via* a *syn*-elimination, requiring a boat conformation in the transition state. The higher energy of the boat versus a chair conformation makes this unlikely. Choice **(B)** is not possible, because the hydrogen and bromine atoms are not on adjacent carbon atoms.

NS-16. What is the major product of the reaction shown?

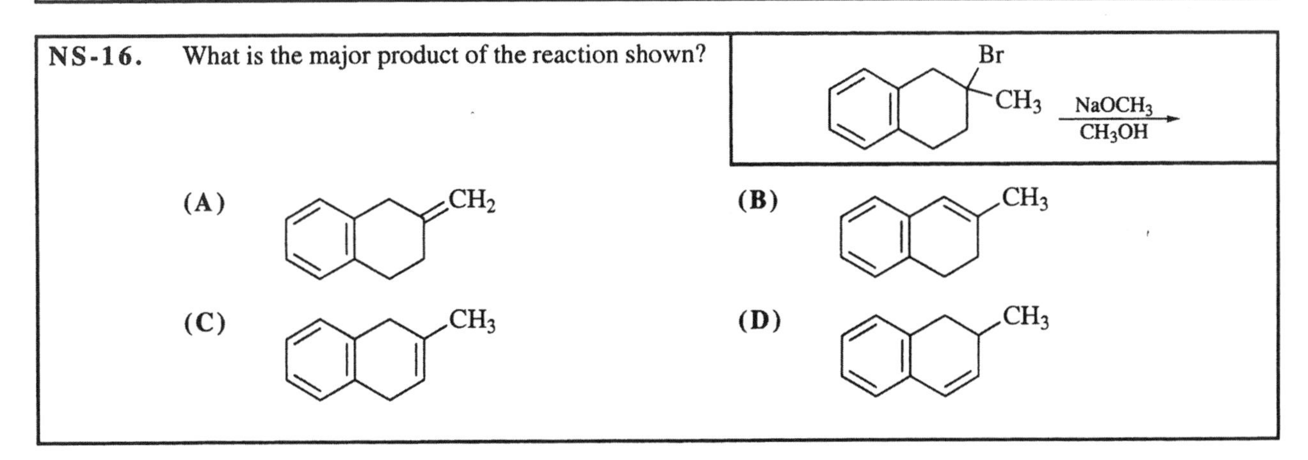

Knowledge Required: Factors that control the regiochemistry of β-eliminations.

Thinking it Through: When more than one type of β-hydrogen atom is present in the alkyl halide, electronic and steric factors control which β-hydrogen atom is preferentially abstracted by the base. When the alkyl halide and base are relatively unhindered, electronic factors dominate, and control which β-hydrogen atom is abstracted. The transition state for an E2 reaction has some double-bond character; therefore, alkene stability affects the transition state energy. The base preferentially abstracts the β-hydrogen atom that leads to the more stable alkene. When bulky substituents in the base (and/or the alkyl halide) produce steric hindrance in the transition state, steric factors override the electronic factor. Then, the base preferentially abstracts the least sterically hindered β-hydrogen atom. Here, the reactant contains three types of β-hydrogen atoms, which can give rise to the alkenes in choices **(A)**, **(B)** and **(C)**. Choice **(D)** cannot be produced by a β-elimination and is eliminated. There are no overriding steric effects from the alkyl bromide or the base, so the most stable alkene is formed. Choices **(B)** and **(C)** are both tri-substituted alkenes, whereas choice **(A)** is a di-substituted alkene. Choice **(A)** is the least stable alkene, and is eliminated. The new double bond in choice **(B)** is conjugated with the aromatic ring, but the new double bond in choice **(C)** is not. Hence, choice **(B)** is the more stable alkene, and is the correct answer.

Practice Questions

1. What reagents could be used to accomplish the synthesis of this compound?

$CH_3CH_2CH_2C{\equiv}CCH_3$

(A) $CH_3CH_2CH_2Br$ + Na^+ $^-C{\equiv}CCH_3$

(B) CH_3CH_2Br + Na^+ $^-C{\equiv}CCH_3$

(C) $CH_3CH_2CH_2CH_2Br$ + Na^+ $^-C{\equiv}CCH_3$

(D) $CH_3CH_2CH_3$ + $HC{\equiv}CCH_3$

2. When 2-bromo-2-methylbutane is treated with a base, a mixture of 2-methyl-2-butene and 2-methyl-1-butene is produced.

$$CH_3CH_2-\underset{CH_3}{\overset{CH_3}{C}}-Br \xrightarrow{\text{base}} CH_3CH{=}\overset{CH_3}{C}CH_3 + CH_3CH_2\overset{CH_3}{C}{=}CH_2$$

When potassium hydroxide is the base, 2-methyl-1-butene accounts for 45% of the mixture. But when potassium tert-butoxide is the base, 2-methyl-1-butene accounts for 70% of the mixture. What would you predict for the percent of 2-methyl-1-butene in the mixture if potassium propoxide were the base?

(A) less than 45%

(B) 45%

(C) between 45% and 70%

(D) more than 70%

3. Which reaction would proceed the fastest?

(A) $CH_3CH_2OCH_3$ + ^-OH → CH_3CH_2OH + $^-OCH_3$

(B) $CH_3CH_2SCH_3$ + ^-OH → CH_3CH_2OH + $^-SCH_3$

(C) $CH_3CH_2N(CH_3)_2$ + ^-OH → CH_3CH_2OH + $^-N(CH_3)_2$

(D) $CH_3CH_2O\overset{O}{\underset{O}{S}}C_6H_4CH_3$ + ^-OH → CH_3CH_2OH + $^-O\overset{O}{\underset{O}{S}}C_6H_4CH_3$

4. Which is a possible product of the reaction shown?

(Reactant: H, D, CH_3 on one carbon; Br, H, CH_3 on the other) $\xrightarrow{^-OH}$

(A) H and CH_3 on one carbon; Br and CH_3 on the other: $(H)(CH_3)C{=}C(Br)(CH_3)$

(B) $(H)(CH_3)C{=}C(H)(CH_3)$

(C) $(D)(CH_3)C{=}C(CH_3)(H)$

(D) $(H)(CH_3)C{=}C(CH_3)(H)$

5. What would be the major product of the dehydrohalogenation of 2-chloropentane by KOH?

(A) $CH_2{=}CHCH_2CH_2CH_3$

(B) $CH_3CH{=}CHCH_2CH_3$

(C) $CH_2{=}C(CH_3)CH_2CH_3$

(D) $CH_3C(CH_3){=}CHCH_3$

6. What would be the major product of the dehydrohalogenation of 3-chloropentane by KOH?

(A) $CH_2{=}C(CH_3)CH_2CH_3$

(B) $CH_3C(CH_3){=}CHCH_3$

(C) $CH_2{=}CHCH_2CH_2CH_3$

(D) $CH_3CH{=}CHCH_2CH_3$

7. Identify possible product(s) of dehydrohalogenation of *cis*-1- bromo-2-methylcyclohexane.

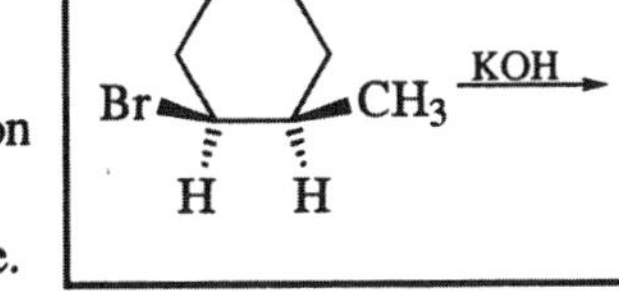

(A) 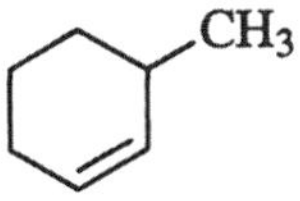 CH_3 only

(B)

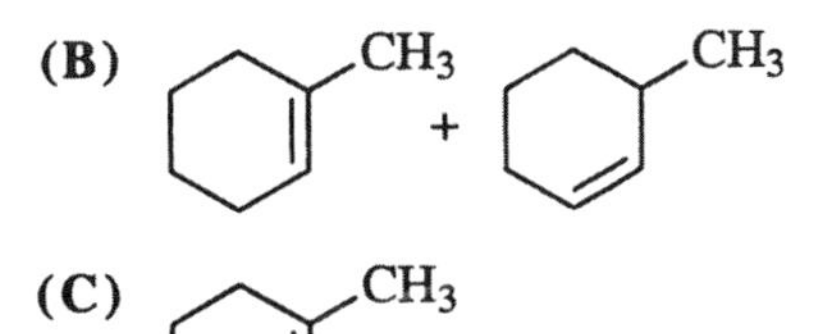

(C) CH_3 only

(D)

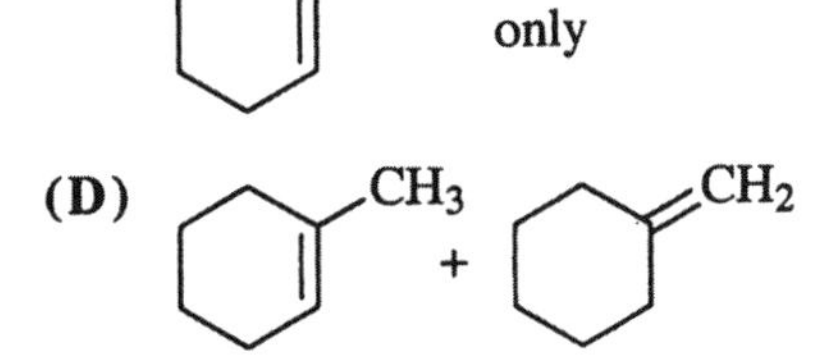

8. What would be the major product of dehydrohalogenation of 2-chloropentane by potassium hydroxide?

(A) $CH_2{=}CHCH_2CH_2CH_3$

(B) $CH_3CH{=}CHCH_2CH_3$

(C) $CH_2{=}C(CH_3)CH_2CH_3$

(D) $CH_3C(CH_3){=}CHCH_3$

9. Identify possible product(s) of dehydrohalogenation of *trans*-1- bromo-2-methylcyclohexane.

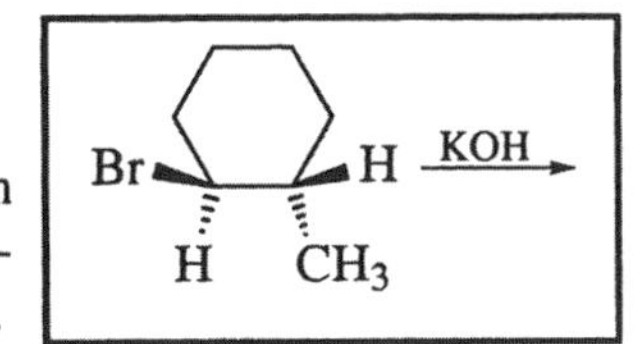

(A)

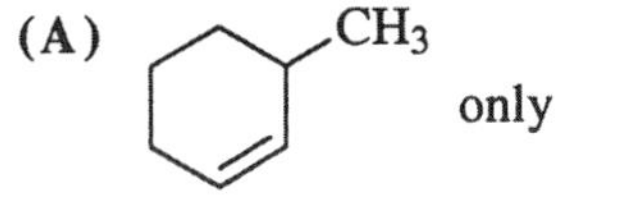

(B) 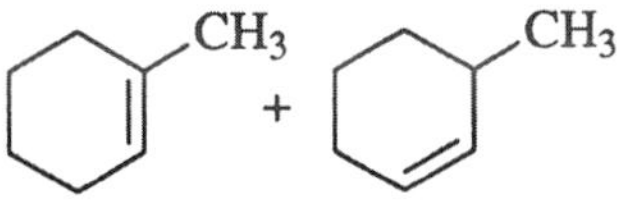 CH_3 + CH_3

(C) 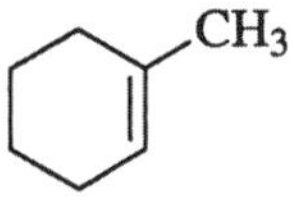 CH_3 only

(D) 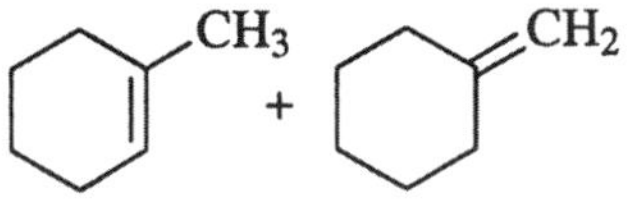 CH_3 + CH_2

10. In the dehydrohalogenation of 2-bromo-2-methylbutane with potassium hydroxide, which hydrogen atom is preferentially abstracted?

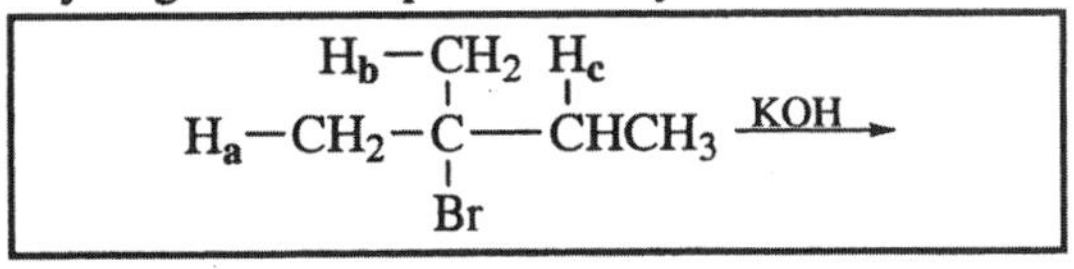

$$H_b{-}CH_2\ H_c$$
$$H_a{-}CH_2{-}C(Br){-}CHCH_3 \xrightarrow{KOH}$$

(A) H_a **(B)** H_b

(C) H_c **(D)** H_a or H_b

11. Why would concentrated hydrobromic acid be an inappropriate catalyst for the dehydration of alcohols?

(A) HBr is too weakly acidic to protonate the alcohol.

(B) The conjugate base, Br^-, is a good nucleophile and it would attack the carbocation to form an alkyl bromide.

(C) HBr is strongly acidic, so the water molecule would not be a good leaving group after protonation of the alcohol.

(D) HBr would be more likely to promote rearrangement of the carbocation intermediate.

12. Which carbocation would *not* be likely to undergo rearrangement?

(A) $CH_3\overset{+}{C}H\overset{CH_3}{\overset{|}{C}}HCH_3$

(B) $CH_3\overset{+}{C}H\underset{CH_3}{\overset{CH_3}{C}}CH_3$

(C) $CH_3\underset{+}{\overset{CH_3}{C}}CH_2CH_3$

(D) $CH_3\overset{CH_3}{C}H\underset{+}{C}H_2$

13. Which alcohol is dehydrated fastest in concentrated H_2SO_4?

(A) $CH_3CH_2\underset{CH_3}{\overset{OH}{C}}CH_3$

(B) $CH_3CH_2\underset{CH_3}{C}HCH_2OH$

(C) $CH_3\underset{CH_3}{C}H\overset{OH}{C}HCH_3$

(D) $CH_3\underset{CH_3}{C}HCH_2CH_2OH$

14. What reagent could accomplish this reaction?

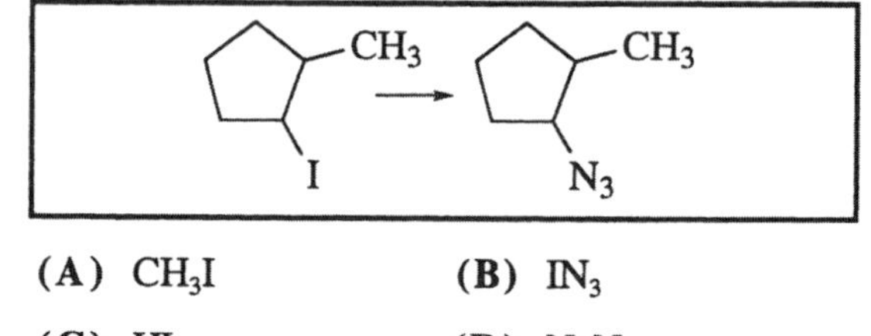

(A) CH_3I **(B)** IN_3

(C) HI **(D)** NaN_3

15. Consider this prototypical nucleophilic substitution shown in the box. The effect of doubling the volume of solvent would be to multiply the reaction rate by a factor of

$$CH_3Br + {}^-OH \rightarrow CH_3OH + {}^-Br$$

(A) 1/4 **(B)** 1/2 **(C)** 2 **(D)** 4

16. If this alcohol is dehydrated, which alkene is likely to be formed in the largest quantity?

$$CH_3-\underset{CH_3}{\overset{CH_3}{C}}-\underset{OH}{C}H-CH_2CH_3 \xrightarrow{H_2SO_4}$$

(A) $CH_3-\underset{CH_3}{\overset{CH_3}{C}}-CH=CHCH_3$

(B) $CH_3-\underset{CH_3}{C}=\overset{CH_3}{C}-CH_2CH_3$

(C) $CH_3-\underset{CH_2}{\overset{}{C}}{\!\!}-\overset{CH_3}{C}H-CH_2CH_3$ (with $C=CH_2$)

(D) $CH_3-\underset{CH_3}{C}H-\overset{CH_3}{C}=CHCH_3$

17. What would be the first step in the dehydration of cyclohexanol in sulfuric acid?

(A) loss of ^-OH

(B) loss of H^+ by the alcohol

(C) formation of a sulfite ester

(D) protonation of the alcohol

18. A bimolecular nucleophilic substitution (S_N2) is

(A) a two-step process in which a bond is broken then a new bond is formed and there is inversion of configuration.

(B) a two-step process in which a bond is broken then a new bond is formed and there is retention of configuration.

(C) a one-step process with inversion of configuration.

(D) a one-step process with retention of configuration.

19. Which is the order from fastest to slowest for the rates of the S_N2 reactions of these alkyl bromides with CH_3S^-/DMSO?

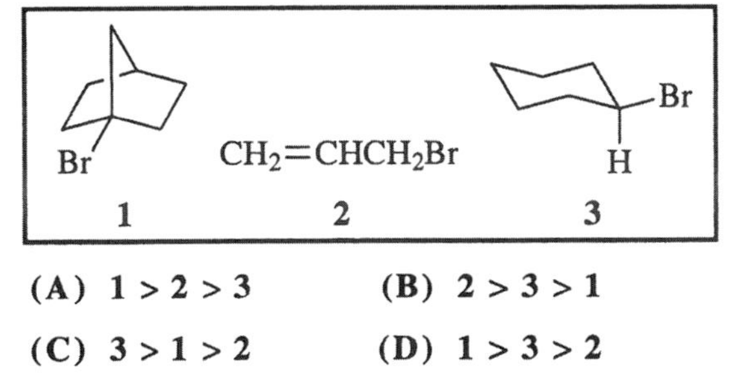

(A) 1 > 2 > 3 **(B)** 2 > 3 > 1

(C) 3 > 1 > 2 **(D)** 1 > 3 > 2

20. Which of the mechanistic steps shown is ***not*** a reasonable one in the mechanism to describe the formation of 2-methyl-1-butene by dehydration of 3-methyl-2-butanol?

$$CH_3-\underset{H}{\overset{CH_3}{C}}-\overset{OH}{C}HCH_3 \longrightarrow CH_2{=}\overset{CH_3}{C}-CH_2CH_3$$

(A) $CH_3-\underset{H}{\overset{CH_3}{C}}-\overset{:\overset{+}{O}H_2}{C}HCH_3 \rightarrow CH_3-\underset{H}{\overset{CH_3}{C}}-\overset{+}{C}HCH_3$

(B) $CH_3-\underset{H}{\overset{CH_3}{C}}-\overset{+}{C}HCH_3 \rightarrow CH_3-\overset{CH_3}{\overset{+}{C}}-CH_2CH_3$

(C) $CH_3-\underset{H}{\overset{CH_3}{C}}-\overset{+}{C}H-\underset{H}{C}H_2 \rightarrow CH_3-\underset{H}{\overset{CH_3}{C}}-CH_2-\overset{+}{C}H_2$

(D) $H_2\ddot{O}:$ + $\underset{H}{CH_2}-\overset{CH_3}{\overset{+}{C}}-CH_2CH_3 \rightarrow CH_2{=}\overset{CH_3}{C}-CH_2CH_3$

21. Which is the *weakest* nucleophile?

(A) $CH_3\overset{O}{\overset{\|}{C}}ONa$ **(B)** CH_3CH_2ONa

(C) $(CH_3CH_2)_2NLi$ **(D)** CF_3CH_2ONa

22. What is the stereochemistry of the nitrile produced in the reaction shown?

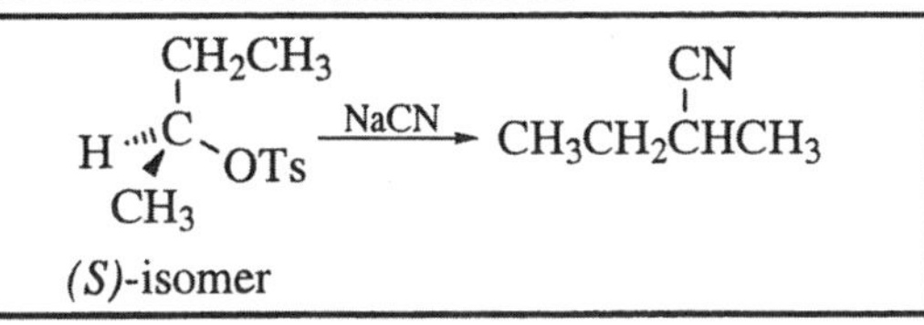

(A) 1:1 mixture of (*R*)- and (*S*)-isomers

(B) (*R*)-isomer only

(C) unequal mixture of (*R*)- and (*S*)-isomers

(D) (*S*)-isomer only

23. Which of these structures best depicts the transition state for the reaction of CH_3I with CH_3OK in CH_3OH?

(A) $\left[CH_3\text{--}\overset{H\ \ H}{\underset{H}{C}}\text{--}I\right]^-$

(B) $\left[KO\text{--}\overset{H\ \ H}{\underset{H}{C}}\text{--}I\right]^-$

(C) $\left[CH_3O\text{--}\overset{H\ \ H}{\underset{H}{C}}\text{--}I\right]^-$

(D) $\left[OH_3C\text{--}H\text{--}\overset{H}{\underset{H}{C}}\text{--}I\right]^-$

24. Which reaction is *best* suited for the preparation of the ether shown?

$(CH_3)_3COCH_2CH_3$

(A) $(CH_3)_3CBr \xrightarrow{CH_3CH_2OK}$

(B) $CH_3CH_2MgBr \xrightarrow{(CH_3)_3COH}$

(C) $CH_3CH_2Br \xrightarrow{(CH_3)_3COK}$

(D) $(CH_3)_3CMgBr \xrightarrow{CH_3CH_2OH}$

25. Which reaction would produce phenyl propyl ether?

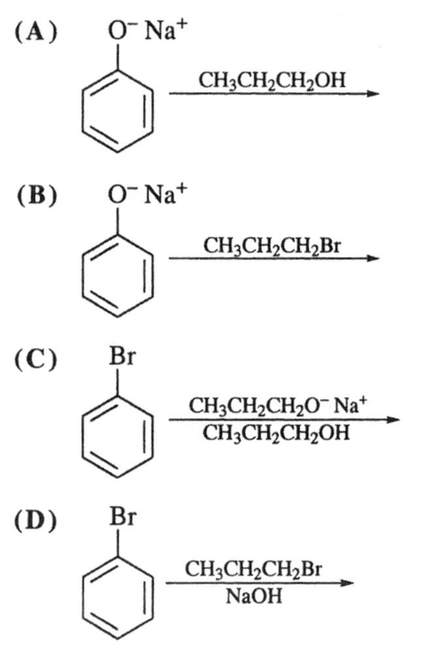

26. What is the expected product of this reaction?

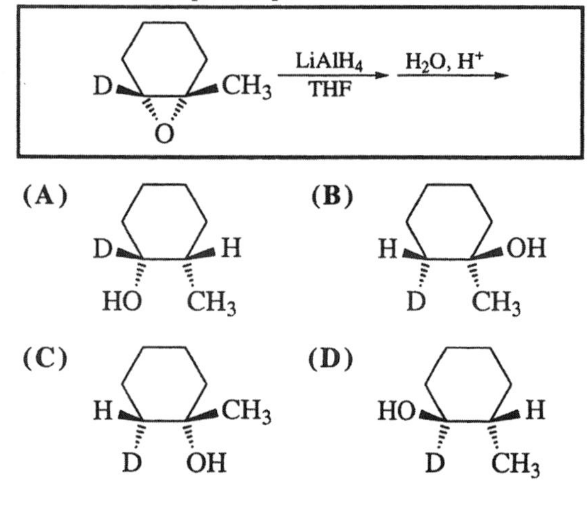

27. Which bromide will most rapidly undergo solvolysis in aqueous ethanol?

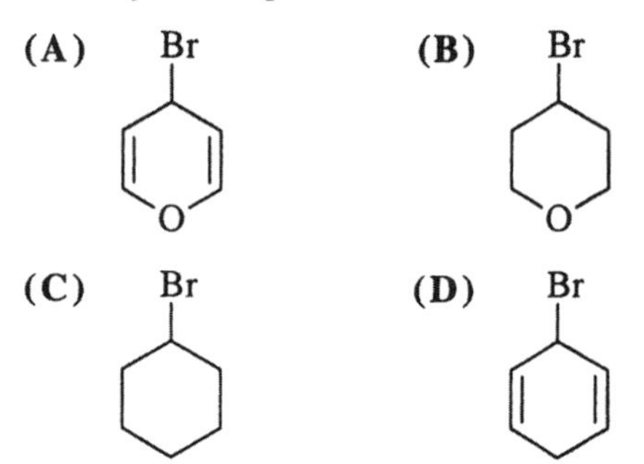

28. What is the principal product of this reaction?

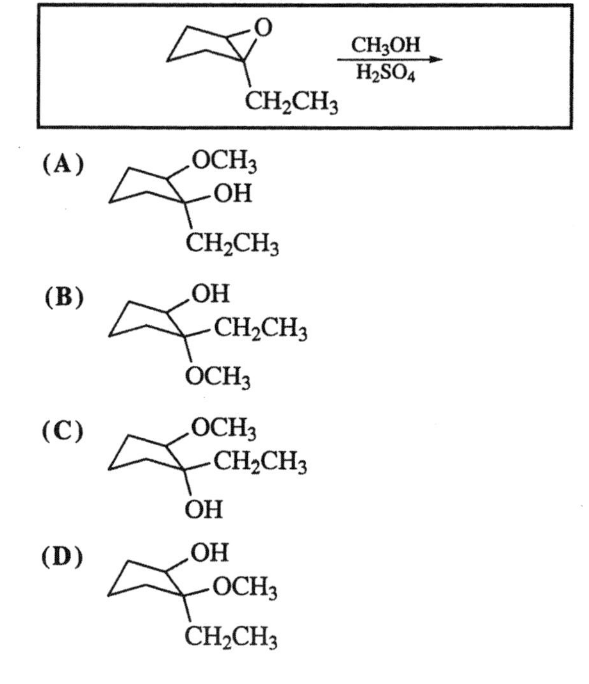

29. Which two compounds ionize with loss of bromide ion to form the same carbocation?

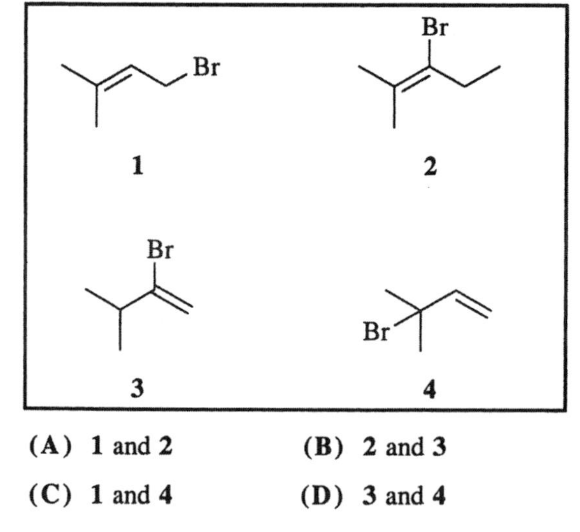

(A) 1 and 2 **(B)** 2 and 3

(C) 1 and 4 **(D)** 3 and 4

30. Predict the major product of the reaction shown.

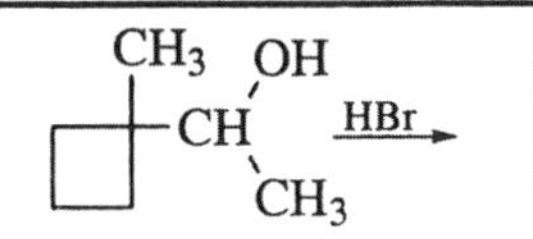

(A) 1-methyl-1-(CH_2CH_2Br)cyclobutane: CH_3 and $-CH_2CH_2Br$ on the same ring carbon

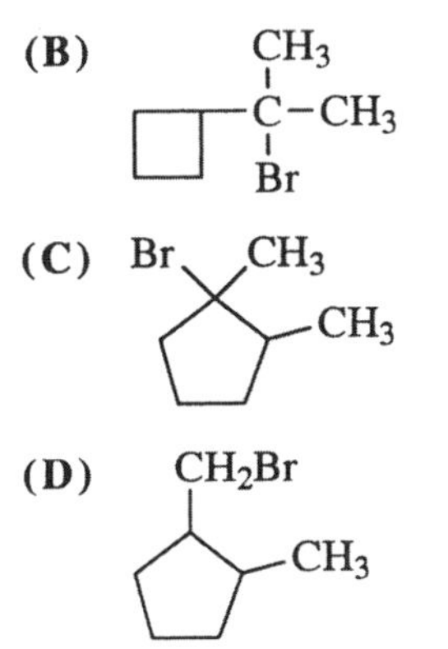

31. Which is the proper representation of the "flow" of electrons in this E2 elimination?

$$CH_3CH_2CH(OTs)CH_3 \xrightarrow{C_2H_5ONa} CH_3CH{=}CHCH_3 + NaOTs + C_2H_5OH$$

(A) $C_2H_5O^-$ → H; H–CH bond → C–C; $CH_3-CH(H)-CH(OTs)-CH_3$; arrow from OTs returning to C

(B) $C_2H_5O^-$ → H; $CH_3-CH(H)-CH(OTs)-CH_3$; arrow from C–C bond to OTs

(C) $C_2H_5O^-$ → C; $CH_3-CH(H)-CH(OTs)-CH_3$; arrow from C–H bond to C–C; arrow from C–OTs bond to OTs

(D) $CH_3-CH_2-CH(OTs)-CH_3$ → (arrow to OTs) $CH_3-CH-\overset{+}{C}H-CH_3 + {}^-OTs$; $C_2H_5O^-$ → H; arrow from C–H bond to C–C

32. What is the major product of this reaction?

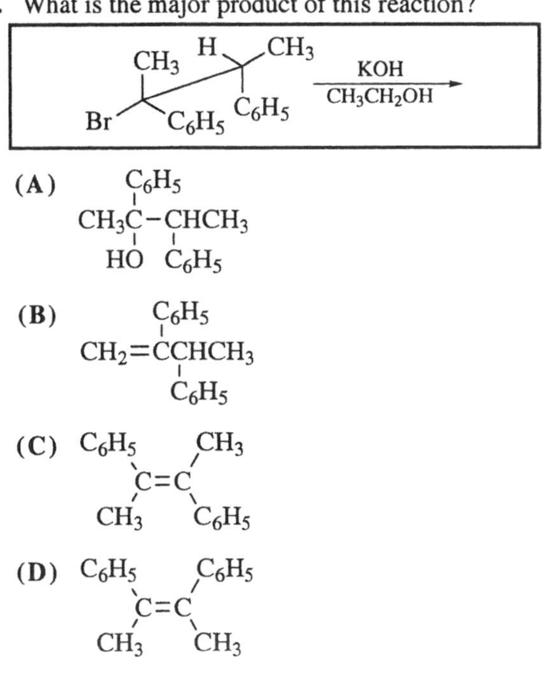

33. What are the expected products from the reaction shown?

$$C_6H_{11}-O-CH_2CH_2CH_2OH \xrightarrow{HBr\ (excess)}$$

(A) $C_6H_{11}-Br + BrCH_2CH_2CH_2OH$

(B) $C_6H_{11}-Br + BrCH_2CH_2CH_2Br$

(C) $C_6H_{11}-OH + BrCH_2CH_2CH_2Br$

(D) $C_6H_{11}-O-CH_2CH_2CH_2OH$

Answers to Study Questions

1. D
2. C
3. C
4. C
5. A
6. C
7. A
8. A
9. B
10. D
11. B
12. B
13. D
14. A
15. D
16. B

Answers to Practice Questions

1. A
2. C
3. D
4. D
5. B
6. D
7. B
8. B
9. A
10. C
11. B
12. C
13. A
14. D
15. A
16. B
17. D
18. C
19. B
20. C
21. A
22. B
23. C
24. C
25. B
26. C
27. A
28. B
29. C
30. C
31. A
32. C
33. B

Electrophilic Additions

The carbon–carbon double bond is an electron-rich region that is attractive to electrophiles. In the generalized mechanism shown in equation *1*, the electrophile bonds to one of the carbon atoms using two electrons of the double bond. This forms a carbocation intermediate. The electrophilic attack is usually regiospecific and the most stable carbocation predominates. Once formed, the carbocation may rearrange to a more stable carbocation by alkyl group or hydride migration. The carbocation intermediate is susceptible to nucleophilic attack, which completes the addition to the double bond (equation *1*).

$$\xrightarrow{E^+} \ {}^{+}\ E^+ \ \xrightarrow{:Nu^-} \ \text{Nu} \ \text{E} \qquad (1)$$

Conjugated dienes allow both 1,2- and 1,4-addition to occur via an allylic carbocation. The product is often a mixture of 1,2- and 1,4-addition products. The 1,2-addition product (often called the kinetic product) forms faster than the 1,4-addition product and dominates if the reaction is stopped before equilibrium is reached. The 1,4-addition product (often called the thermodynamic product) is usually more stable than the 1,2-addition product and dominates if the reaction is allowed to reach equilibrium.

The reactions of alkynes are similar to those of alkenes. Hydration of the triple bond and partial reduction to alkenes are common examples.

Some addition reactions often display stereoselectivity for *syn-* or *anti*-addition of the electrophile and nucleophile. Such observations are not consistent with the simple mechanism presented in equation *1*. For example, the *anti*-addition of halogens is rationalized by the formation of a halonium ion that blocks the subsequent addition of the nucleophile. The *syn*-addition of hydroboranes are rationalized by the more or less concerted nature of the bond makings and breakings as the addition occurs. Stereoselectivity is difficult to predict, and must become part of a student's empirical knowledge base.

Several addition reactions that are not necessarily electrophilic additions are included in this section. These reactions include epoxidation, carbene additions, and the Diels–Alder reaction. All are concerted addition reactions that do not involve the formation of intermediates. These reactions are recognized by structural features of reactants and reaction conditions. In each case, it is necessary to know the stereochemical result of that type of reaction.

Study Questions

EA-1. Predict the major product from the reaction shown.

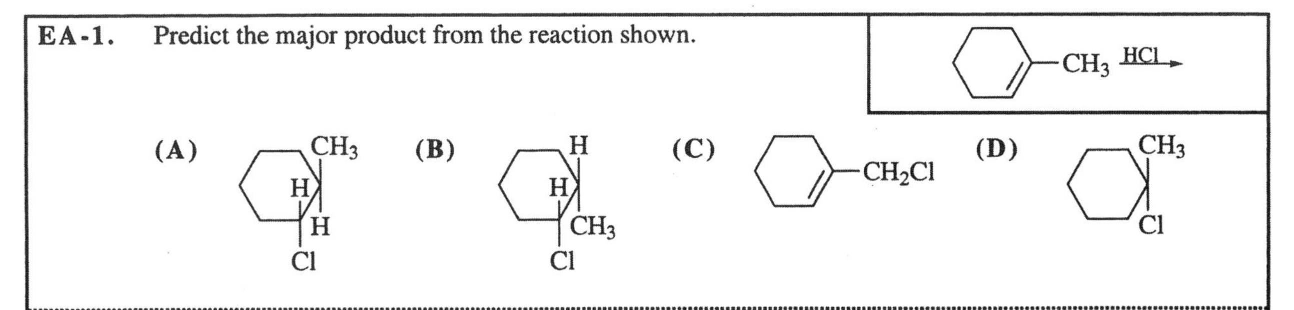

Knowledge Required: (1) Properties of nucleophiles and electrophiles. (2) Factors affecting carbocation stability.

Thinking it Through: The hydrogen–chlorine bond is polar, with the hydrogen atom slightly positive relative to the electronegative chlorine atom. The hydrogen ion is the electrophile, and the chloride ion is the nucleophile. The hydrogen ion adds to the double bond to give the tertiary carbocation rather than the less stable secondary carbocation. Choices **(A)** and **(B)** would form by addition of the nucleophile to the secondary carbocation, and are eliminated. Choice **(C)** does not involve addition to the double bond. Choice **(D)** forms by addition of the nucleophile to the tertiary carbocation, and is the correct answer.

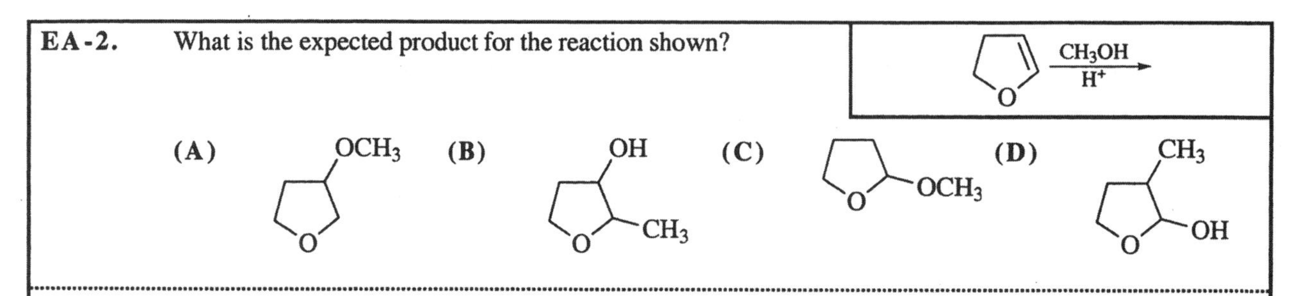

EA-2. What is the expected product for the reaction shown?

Knowledge Required: (1) Properties of nucleophiles and electrophiles. (2) Factors affecting carbocation stability. (3) Electron-releasing substituent effect of alkoxy groups.

Thinking it Through: The initial electrophilic attack is by the H^+ electrophile. The oxygen atom of methanol is the nucleophile. Choices **(B)** and **(D)** are eliminated because they require the electrophile to come from cleavage of the carbon–oxygen bond of the methanol. To choose between choices **(A)** and **(C)**, the relative stability of the two possible carbocations must be considered. In this case, the hydrogen ion becomes bound to the carbon atom that is further from the oxygen atom, forming a carbocation that is stabilized by resonance. Resonance stabilization results from electron donation from the oxygen atom that is attached to the double bond.

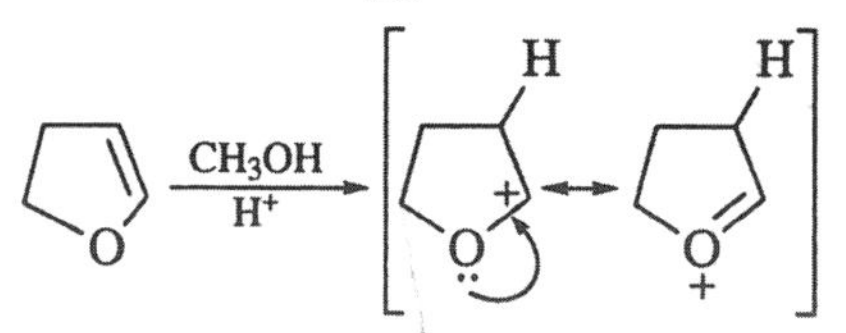

The methanol nucleophile then attacks the carbocation to yield a protonated ether, which then loses the proton to produce product **(C)**, the correct answer.

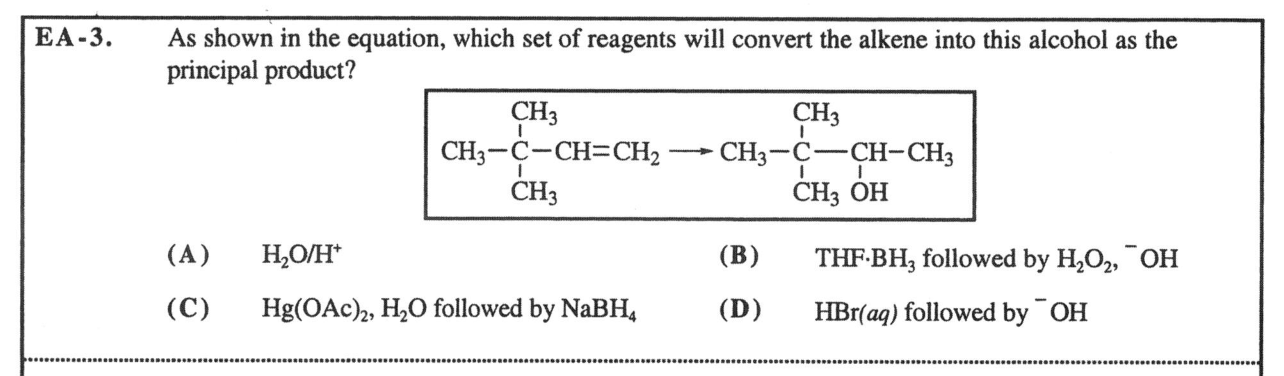

EA-3. As shown in the equation, which set of reagents will convert the alkene into this alcohol as the principal product?

$$CH_3-C(CH_3)_2-CH=CH_2 \longrightarrow CH_3-C(CH_3)_2-CH(OH)-CH_3$$

(A) H_2O/H^+

(B) $THF{\cdot}BH_3$ followed by H_2O_2, ^-OH

(C) $Hg(OAc)_2$, H_2O followed by $NaBH_4$

(D) $HBr(aq)$ followed by ^-OH

Knowledge Required: (1) Regiochemistry of electrophilic additions, specifically hydroboration–oxidation and oxymercuration–demercuration. (2) Carbocation stability and tendency to rearrange.

Thinking it Through: The desired transformation is the addition of water to the double bond in what is called "Markovnikov" regiochemistry. The hydroxyl group becomes bonded to what would have been the most stable carbocation. In addition, the formation of a secondary carbocation must be avoided since it can rearrange to a more stable tertiary carbocation by a methyl shift. All of the possible answers can result in the addition of water to a double bond. The hydroboration–oxidation sequence in choice **(B)** would result in "anti-Markovnikov" addition of water. The acidic conditions in choices **(A)** and **(D)** would result in free carbocations that would rearrange rapidly and result in very little of the desired regiochemistry. The oxymercuration–demercuration sequence in choice **(C)** is the correct answer because it avoids a free carbocation and gives the overall "Markovnikov" regiochemistry.

EA-4. What is the product of the reaction sequence shown?

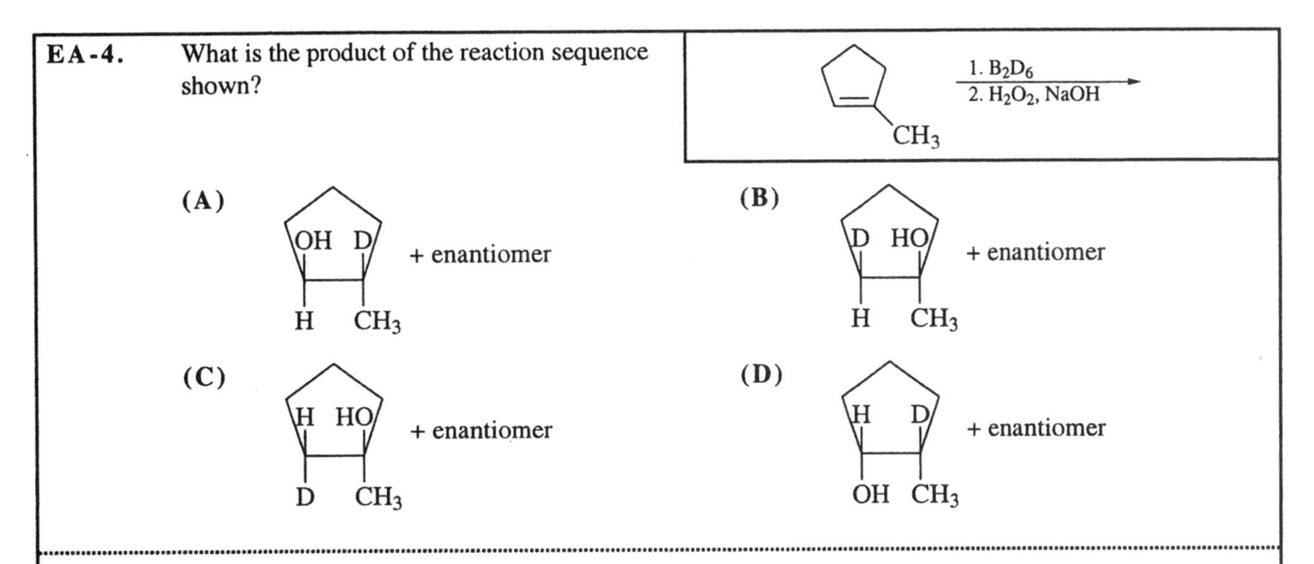

Knowledge Required: Regiochemistry and stereochemistry of the hydroboration–oxidation sequence.

Thinking it Through: This is a classic problem probing the regiochemistry and stereochemistry of a "standard" reaction sequence. Hydroboration–oxidation results in the "anti-Markovnikov" addition of water to the double bond. That means that the hydroxyl group will become bonded to the carbon atom that would be the *least* stable carbocation. The two steps are stereospecific and result in the *syn*-addition of water to the double bond. The reagent used for the hydroboration step in this problem is an isotopically labeled diborane, where the normal hydrogen atom has been replaced with a deuterium atom. In the hydroboration step, the boron becomes bonded to the carbon atom that would form the *least* stable carbocation and the deuterium atom becomes bonded to the carbon atom that would form the *most* stable carbocation. The choices provided for this question represent all possible combinations of regiochemistry and stereochemistry. Choices **(A)** and **(B)** have the correct stereochemistry for *syn*-addition, whereas choices **(C)** and **(D)** have the stereochemistry for *anti*-addition. Thus, choices **(C)** and **(D)** can be eliminated. Of choices **(A)** and **(B)**, only choice **(A)** has both the stereochemistry for *syn*-addition *and* the regiochemistry for anti-Markovnikov addition of water. Choice **(A)** is the correct answer. The addition can occur from either side of the double bond, producing enantiomers.

EA-5. Which intermediate is involved in the reaction shown?

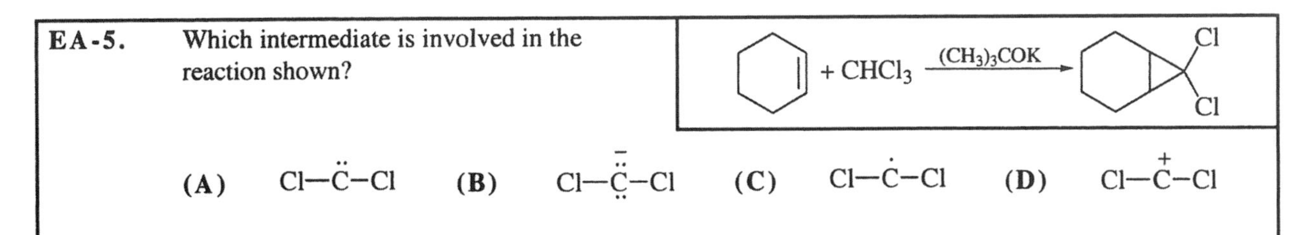

Knowledge Required: Formation and reactions of carbenes with alkenes.

Thinking it Through: The reaction conditions are rather specific for this reaction and probably must be taken into account. The potassium *tert*-butoxide is a sterically hindered strong base. When trihalomethanes are treated with the strong base, deprotonation gives the trihalomethyl anion, which loses a halide ion to give the neutral carbene. The dihalocarbene adds to the alkene in a cycloaddition reaction to give the dihalocyclopropane. The cycloaddition reaction is usually stereospecific, with the stereochemistry of the alkene maintained in the cyclopropane. The acidic nature of the hydrogen atom in the $CHCl_3$ can be predicted from the stabilizing effect of the three chlorine atoms on the negative charge of the trichloromethyl anion. The loss of a chloride ion leads to choice **(A)**, dichlorocarbene. The other choices are structurally implausible.

EA-6. The addition of bromine to the alkene shown will give

H, CH_2CH_3 / C=C / CH_3, H $\xrightarrow{Br_2}$

(A) one *meso* compound (B) one pair of enantiomers

(C) two *meso* compounds (D) four stereoisomers

Knowledge Required: (1) Stereochemistry of bromine addition. (2) Definitions of enantiomers and *meso* compounds

Thinking it Through: The addition of bromine occurs in an *anti* fashion via a bromonium ion that can form from either the top or the bottom of the double bond. This problem requires drawing the 3-dimensional representation of the product followed by analysis of the stereochemical possibilities. The *anti*-addition of Br_2 to the alkene shown gives the two stereoisomers shown below, which eliminates choice **(D)**. The chiral (stereogenic) centers are not mirror images of one another in either isomer, so they cannot be *meso* isomers. This eliminates choices **(A)** and **(C)**. Comparison shows them to be a pair of enantiomers. Thus, choice **(B)** is the correct answer.

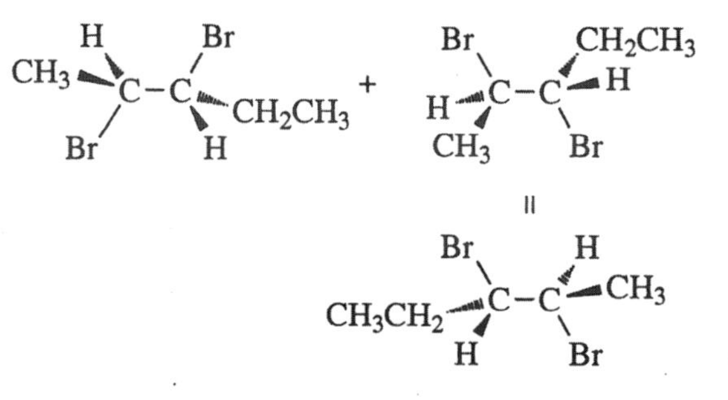

The addition of bromine to alkenes is one of a number of stereospecific addition reactions. Questions regarding the stereochemical outcome of addition reactions should always be worked out rather than relying on previously observed equations.

EA-7. Which product(s), shown below, would be formed in the reaction shown at right?

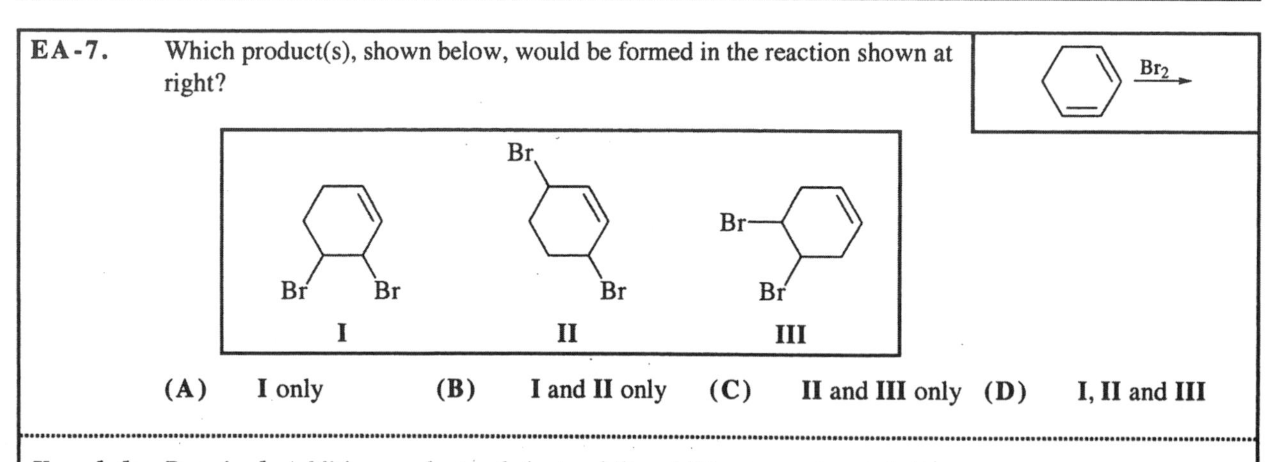

(A) I only (B) I and II only (C) II and III only (D) I, II and III

Knowledge Required: Addition products of electrophilic addition to conjugated dienes.

Thinking it Through: The electrophilic addition to conjugated dienes occurs at a terminal carbon atom of the four-carbon-atom diene system. The resonance-stabilized allylic carbocation intermediate has a partial positive charge at both ends, and therefore, the nucleophile attacks at both carbon atoms. The result is a mixture of 1,2- and 1,4-addition products. Product **I** is the result of 1,2-addition. Product **II** is the result of 1,4-addition. Product **III** is not possible from electrophilic addition. Therefore, choices **(C)** and **(D)**, which contain product **III**, are eliminated. Choice **(A)** fails to recognize the mixture of products that would be obtained. This leaves choice **(B)** with both 1,2- and 1,4-addition products as the correct answer.

EA-8. What is the product of the reaction of (*E*)-1-phenylpropene with peroxybenzoic acid?

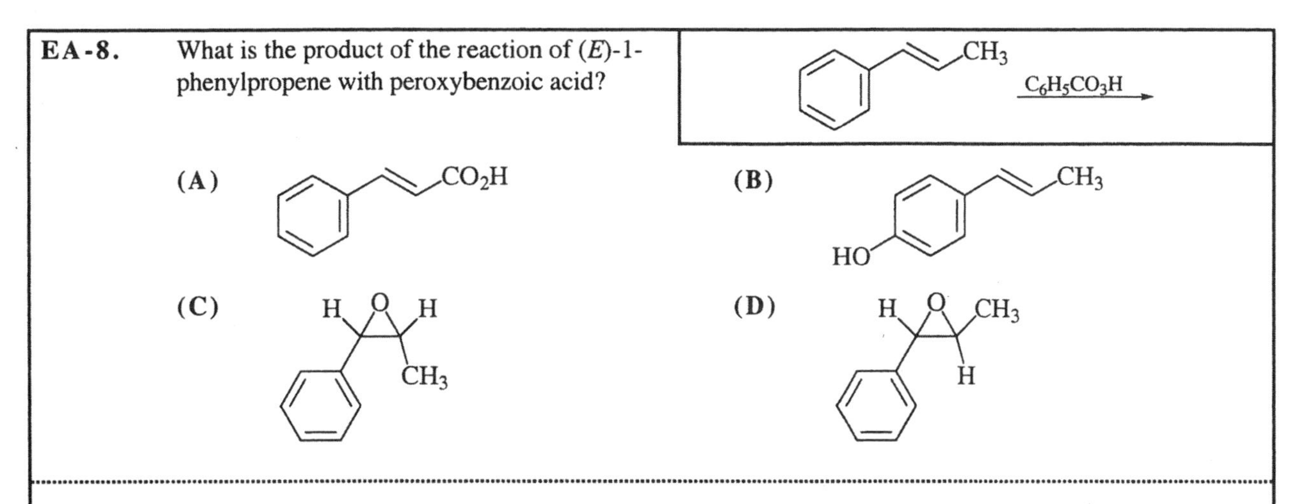

Knowledge Required: Reaction products of peroxyacids with alkenes.

Thinking it Through: Peroxyacids contain the unique functional group, $-CO_3H$, and are reagents that convert alkenes to epoxides (three-membered cyclic ethers). The epoxidation reaction is stereospecific in that the stereochemistry of the alkene is maintained in the epoxide. Choices **(A)** and **(B)** are not epoxides and are eliminated. Choices **(C)** and **(D)** are epoxides, but with different stereochemistry. The substituents are *cis* to each other in choice **(C)**, but *trans* to each other in choice **(D)**. Since the substituents are *trans* in the starting alkene, choice **(D)** is the correct answer.

EA-9. What is the major product isolated from the reaction shown?

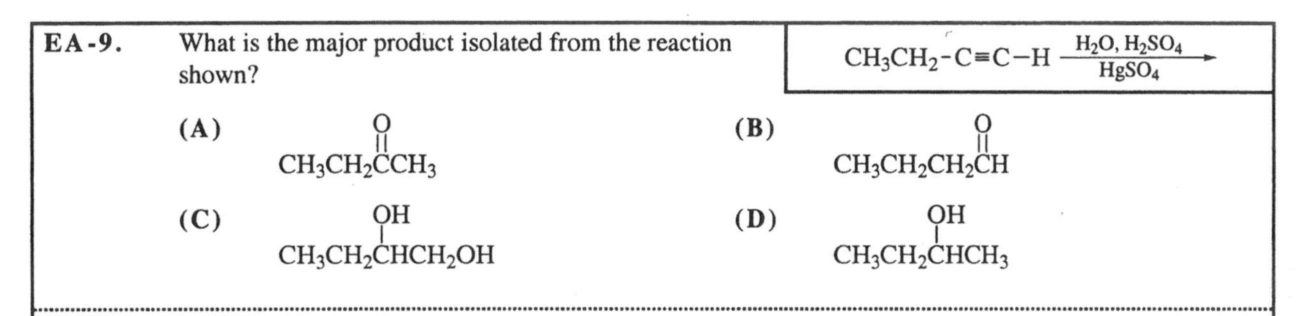

Knowledge Required: (1) Properties of nucleophiles and electrophiles. (2) Factors affecting carbocation stability (Markovnikov regiochemistry). (3) Products of keto–enol isomerism.

Thinking it Through: Mercuric sulfate is a catalyst that makes the hydration of an alkyne proceed smoothly, and the Hg^{2+} ion is the electrophile that initiates the reaction. However, the reaction outcome is the same, and much easier to understand, if H^+ ion is considered to be the electrophile. The oxygen atom of water is the nucleophile. The regiochemistry of the protonation of an alkyne is the same as for an alkene; the more 2° stable vinyl carbocation preferentially forms. Water then attacks the carbocation to give an enol, with the hydroxyl group attached to the more substituted carbon atom of the triple bond (Markovnikov regiochemistry). The enol rapidly rearranges to the corresponding ketone.

$$\underset{\text{enol}}{CH_3CH_2-\overset{\overset{OH}{|}}{C}=CH_2} \rightleftharpoons \underset{\text{ketone}}{CH_3CH_2-\overset{\overset{O}{\|}}{C}-CH_3}$$

Choices **(C)** and **(D)** are eliminated because they are neither an enol nor a ketone. Choices **(A)** and **(B)** could both be formed from isomerization of an initial enol, but the enol necessary for choice **(B)** would require the formation of the less stable 1° vinyl carbocation. Thus, choice **(A)** is the correct answer.

EA-10. What is the major product of the Diels–Alder reaction shown?

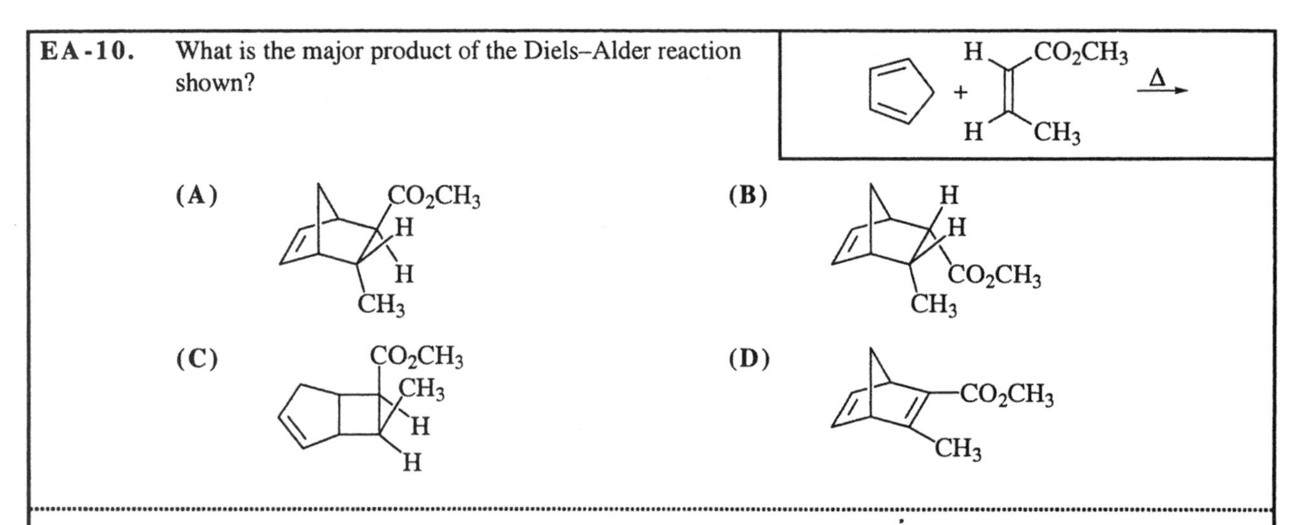

Knowledge Required: (1) The electron reorganization that occurs in the Diels–Alder reaction. (2) The stereochemical preferences observed in the Diels–Alder reaction.

Thinking it Through: The Diels–Alder reaction involves the formation of a six-membered ring via a reorganization of the π electrons from a conjugated diene and an alkene (a dienophile). The reaction is considered to be a concerted 4+2 cycloaddition reaction, and occurs simply by heating the reactants together in solution. The stereochemistry of the alkene is maintained in the resulting six-membered ring. When cyclopentadiene is the diene, a bicyclo[2.2.1]heptene results. The preferred transition state leads to the stereoisomer in which an unsaturated substituent is oriented *endo* (*trans* to the one-carbon-atom bridge). The *exo* orientation (*cis* to the one-carbon-atom bridge) is more thermodynamically stable, but it has a higher transition state energy. Consequently, the *exo* product forms more slowly. Choice **(C)** is 2+2 cycloaddition that does not occur thermally, and can be eliminated. Choice **(D)** can also be eliminated because the π electrons of the dienophile have not taken part in the reaction; the double bond of the dienophile still exists. Choice **(A)** is incorrect because the two substituents are *trans* to each other, whereas they were *cis* in the dienophile. Choice **(B)** has the substituents *cis* to each other with the preferred *endo* orientation, and is the correct answer.

EA-11. What is the major product from the reaction of 1-butene with bromine in methanol?

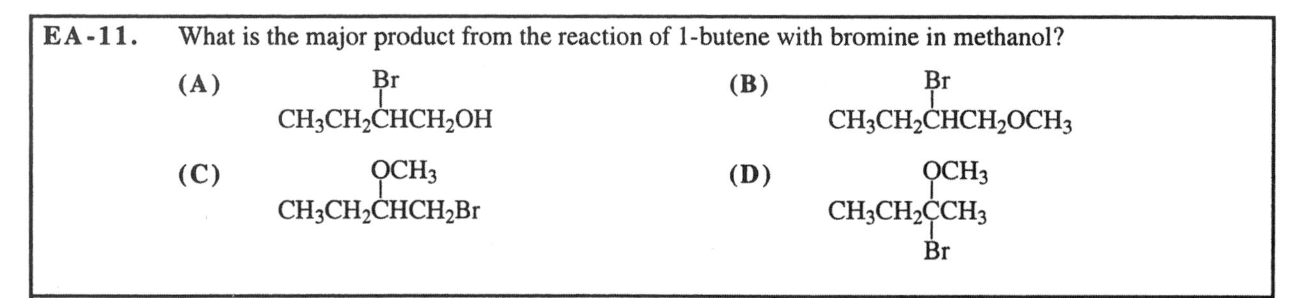

Knowledge Required: (1) Mechanism for halogenation of alkenes. (2) Regiochemistry for nucleophilic attack on cyclic halonium ions.

Thinking it Through: Bromine addition to alkenes proceeds through an intermediate bromonium ion. When the solvent is non-nucleophilic (chloroform or methylene chloride, for example), the only nucleophile present is the bromide counter ion. Bromide ion attacks the bromonium ion to give a vicinal dibromide. However, when the solvent is nucleophilic (water or an alcohol, for example), the solvent nucleophile can attack the bromonium ion to give a β-bromoalcohol (a bromohydrin) or a β-bromoether. Since the solvent is in large excess over the bromide ion, nucleophilic attack by the solvent predominates. Here, methanol is the solvent, and the product must be a β-bromoether. Choice **(A)**, a bromoalcohol, and choice **(D)**, an α–bromoether, are eliminated. Choices **(B)** and **(C)** differ only in the regiochemistry of the nucleophilic opening of the intermediate bromonium ion. The solvent preferentially attacks the carbon atom with the greatest amount of partial positive charge, which is the more highly substituted one. Therefore, choice **(C)** is the correct answer.

Practice Questions

1. What is the major product from the reaction shown?

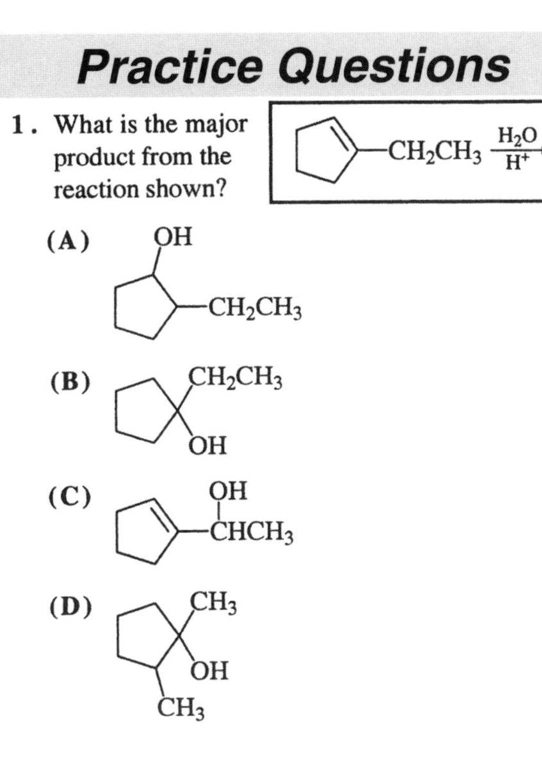

2. Predict the major product of the reaction shown.

CH_3 $\xrightarrow[\text{2. } H_2O_2\text{, NaOH}]{\text{1. } B_2H_6}$

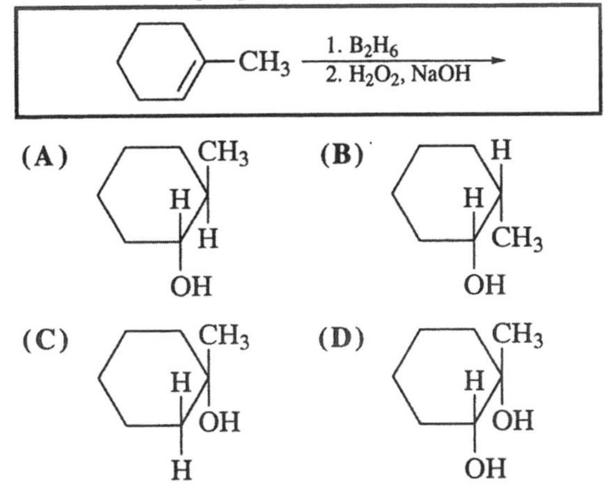

3. Which would be the best method to carry out the transformation shown?

$(CH_3)_2CHCH{=}CH_2 \longrightarrow (CH_3)_2CHCH(OH)CH_3$

(A) H_2O, H^+

(B) HBr, ROOR followed by NaOH

(C) BH_3 followed by H_2O_2, NaOH

(D) $Hg(OAc)_2$, H_2O followed by $NaBH_4$

4. Predict the product from the reaction shown.

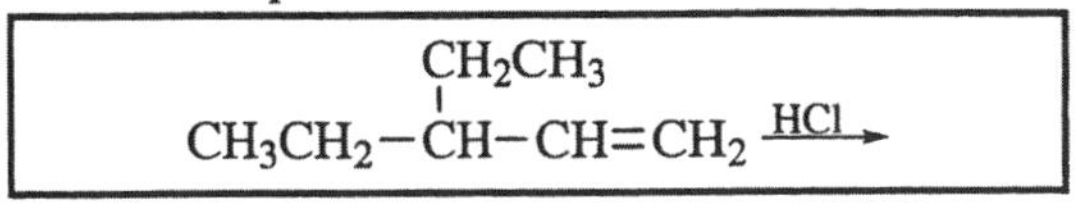

$CH_3CH_2{-}CH(CH_2CH_3){-}CH{=}CH_2 \xrightarrow{HCl}$

(A) $CH_3CH_2{-}C(CH_2CH_3)(Cl){-}CH_2CH_3$

(B) $CH_3CH_2{-}CH(CH_2CH_3){-}CH(Cl)CH_3$

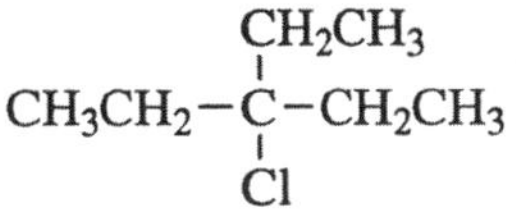

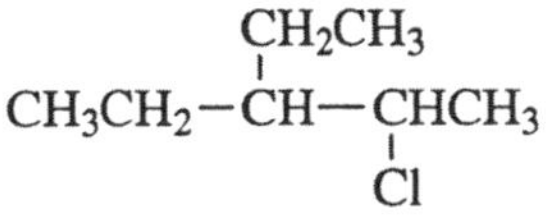

(C) $CH_3CH_2{-}C(CH_3)(Cl){-}CH(CH_3){-}CH_3$

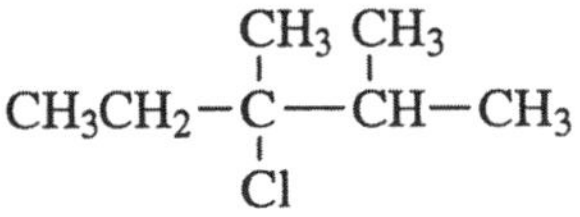

(D) $CH_3CH_2{-}CH(Cl){-}CH(CH_2CH_3)CH_3$

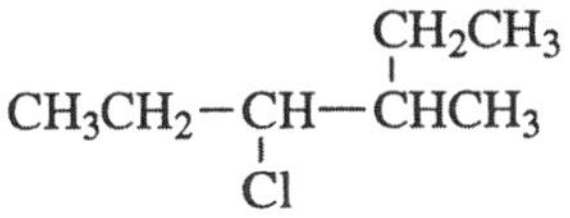

5. Identify the nucleophile that attacks the carbocation intermediate in the acid-catalyzed hydration shown.

$(CH_3)_2C{=}CHCH_3 \longrightarrow (CH_3)_2C(OH)CH_2CH_3$

(A) HO^- (B) H_2O

(C) H^+ (D) H_3O^+

6. What is the product of the reaction shown?

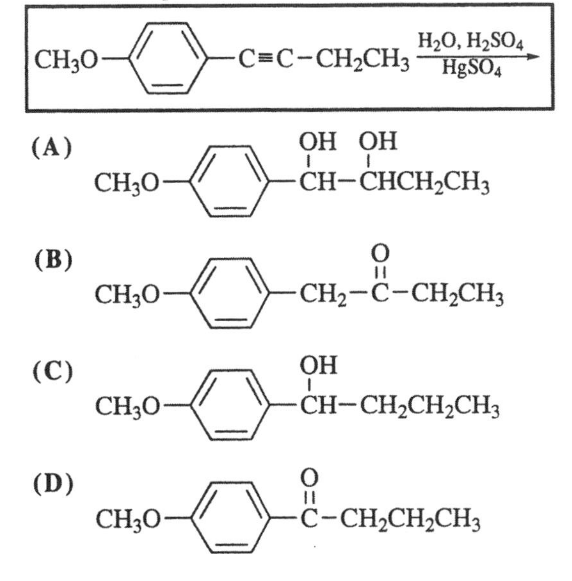

[reaction: CH₃O-C₆H₄-C≡C-CH₂CH₃ with H₂O, H₂SO₄, HgSO₄]

(A) CH₃O-C₆H₄-CH(OH)-CH(OH)CH₂CH₃

(B) CH₃O-C₆H₄-CH₂-C(=O)-CH₂CH₃

(C) CH₃O-C₆H₄-CH(OH)-CH₂CH₂CH₃

(D) CH₃O-C₆H₄-C(=O)-CH₂CH₂CH₃

7. Which set of reagents will carry out the conversion shown?

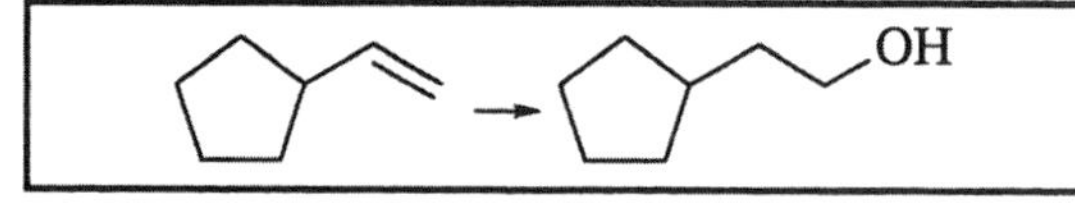

(A) H_2O, peroxides

(B) H_2O, H^+

(C) B_2H_6 followed by H_2O_2, NaOH

(D) $Hg(OAc)_2$, H_2O followed by $NaBH_4$

8. The addition of bromine to (*E*)-3-hexene will give

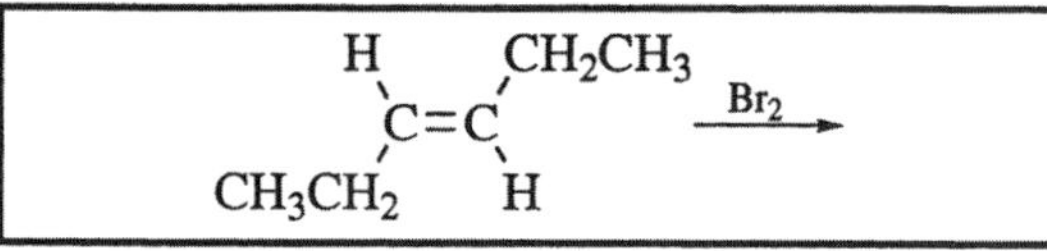

(A) one *meso* compound.

(B) one pair of enantiomers.

(C) two *meso* compounds.

(D) four stereoisomers.

9. What is the major product from this reaction?

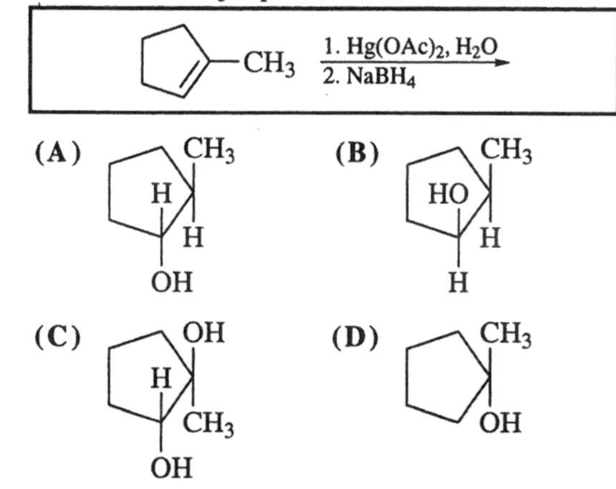

10. What is the product of the reaction shown?

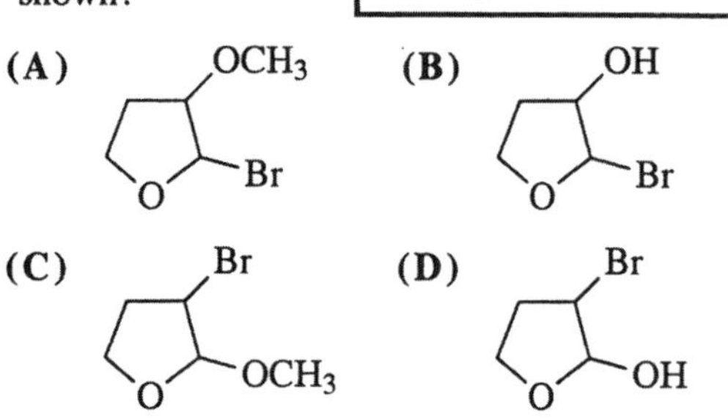

11. Which is the expected product of this reaction?

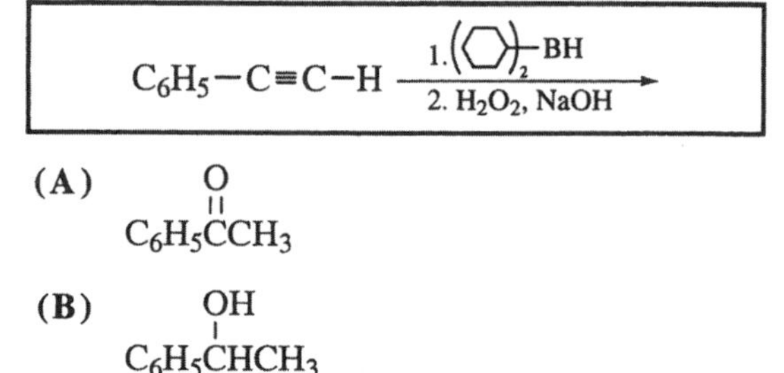

(A) $C_6H_5C(=O)CH_3$

(B) $C_6H_5CH(OH)CH_3$

(C) $C_6H_5CH_2C(=O)H$

(D) $C_6H_5CH_2CH_2OH$

12. Which molecule will correctly complete the reaction shown?

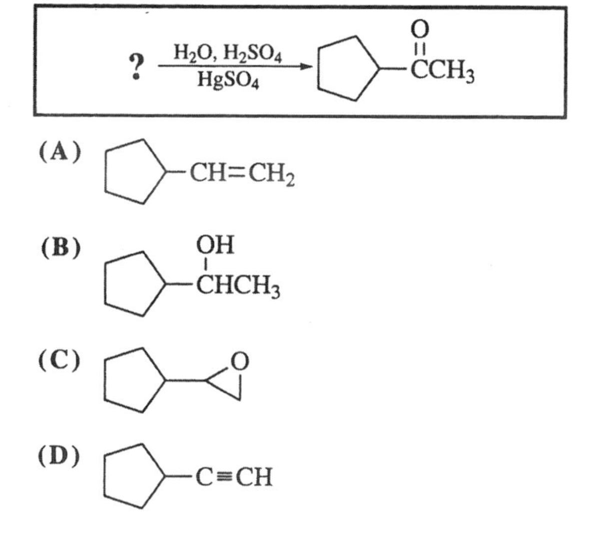

13. What is the principal product of the reaction sequence shown?

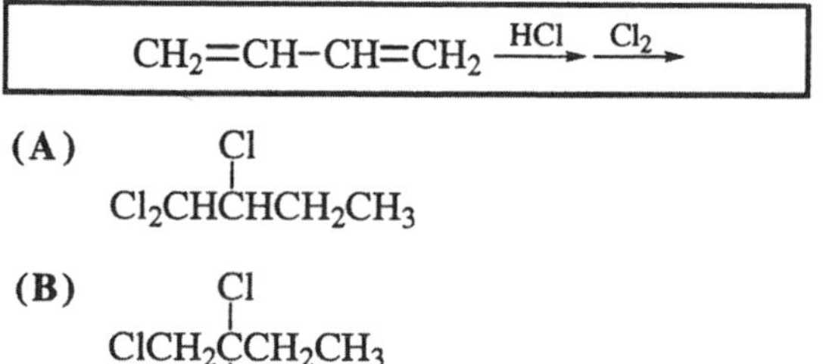

(A) $Cl_2CHCH(Cl)CH_2CH_3$

(B) $ClCH_2C(Cl)_2CH_2CH_3$

(C) $ClCH_2CH(Cl)CH(Cl)CH_3$

(D) $ClCH_2CH(Cl)CH_2CH_2Cl$

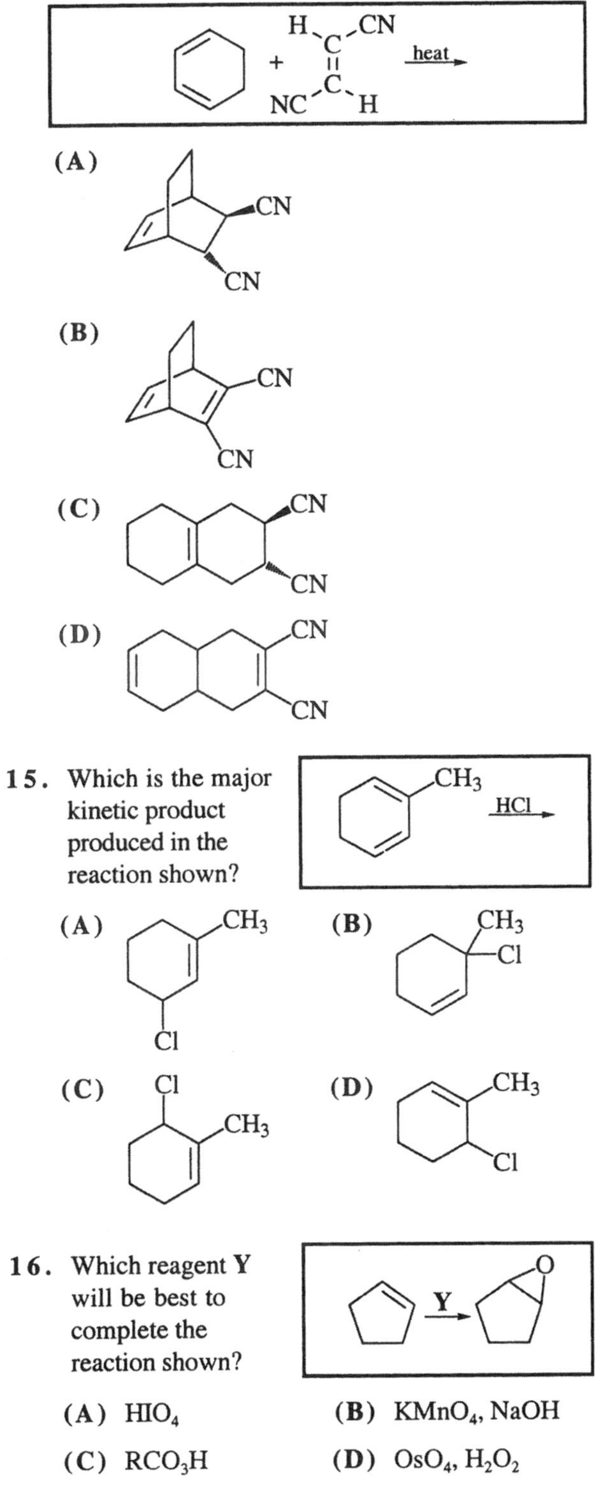

14. What is the major product of the Diels–Alder reaction shown?

(A)

(B)

(C)

(D)

15. Which is the major kinetic product produced in the reaction shown?

(A) (B)

(C) (D)

16. Which reagent **Y** will be best to complete the reaction shown?

(A) HIO_4 **(B)** $KMnO_4$, NaOH

(C) RCO_3H **(D)** OsO_4, H_2O_2

17. With which reagents can the ketone shown be prepared *via* a Diels–Alder reaction?

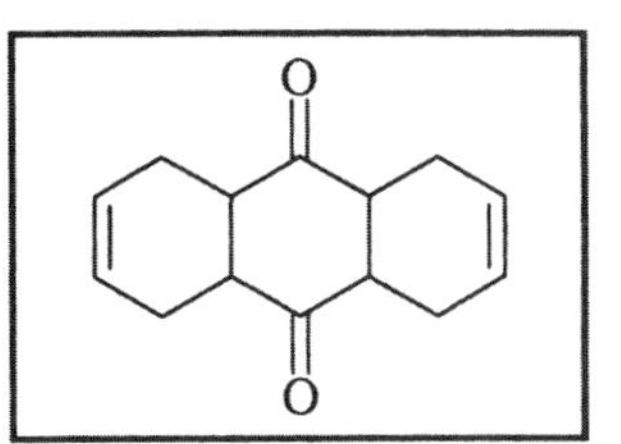

(A)

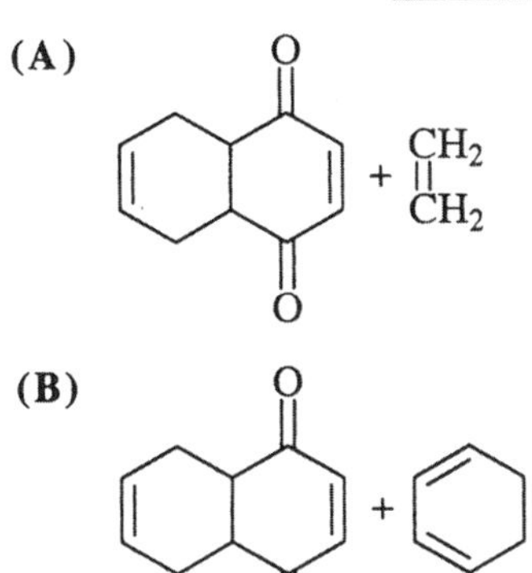

(B)

(C)

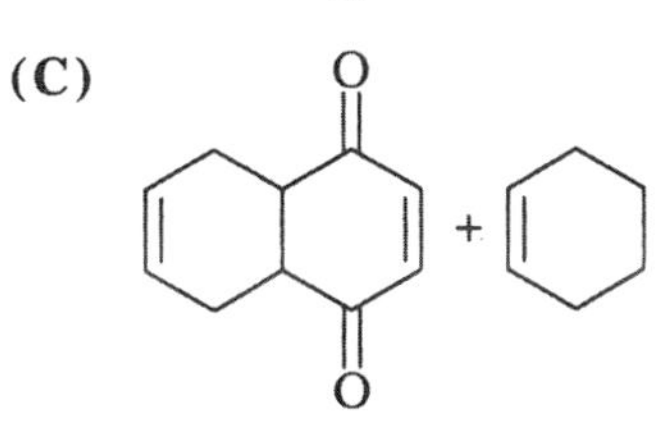

(D)

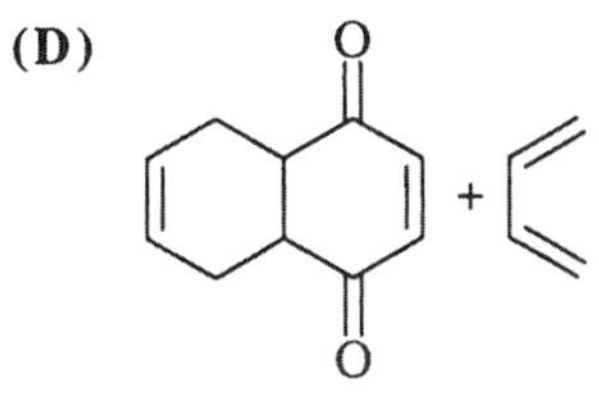

18. What is the product of the reaction shown?

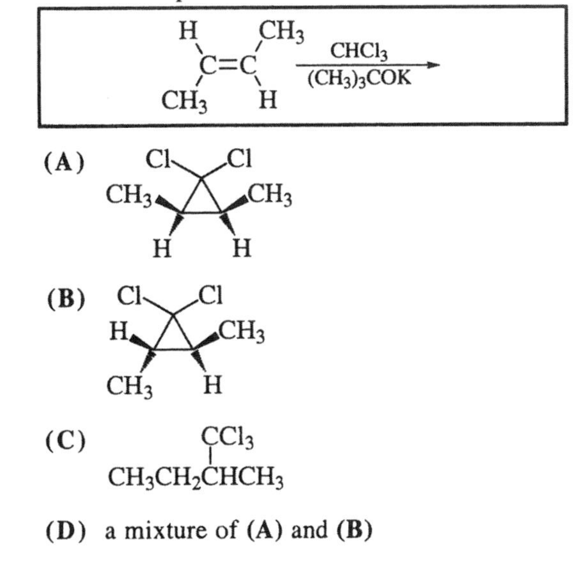

(A)

(B)

(C) $CH_3CH_2CH(CCl_3)CH_3$

(D) a mixture of **(A)** and **(B)**

19. Which is the expected product of the reaction shown?

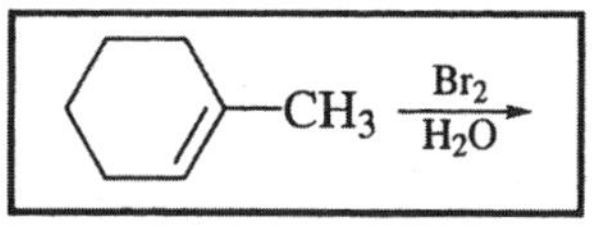

(A)
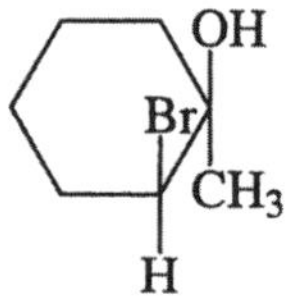

(B)
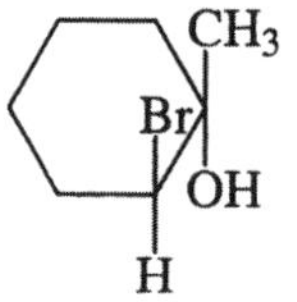

(C)
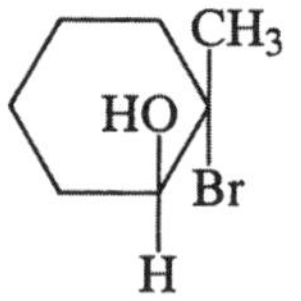

(D)
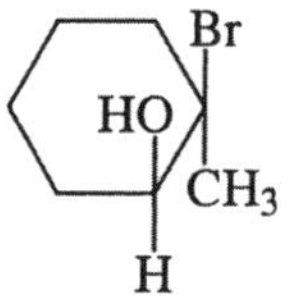

20. Under which set of conditions is the reaction shown best carried out?

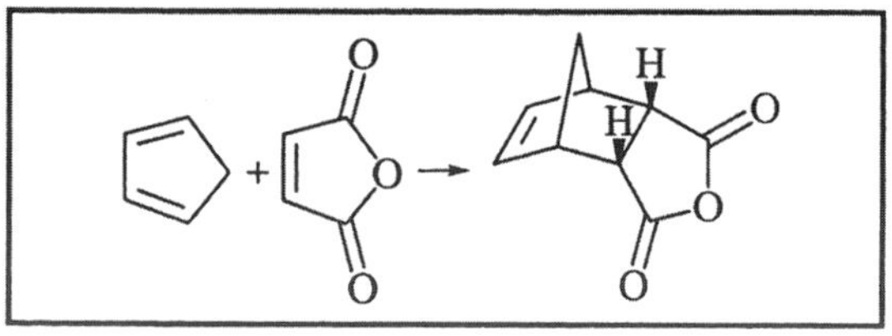

(A) 6 M H_2SO_4

(B) 6 M NaOH

(C) heating in hexane

(D) exposing to UV light in hexane

21. Which Newman projection represents the product of the reaction shown?

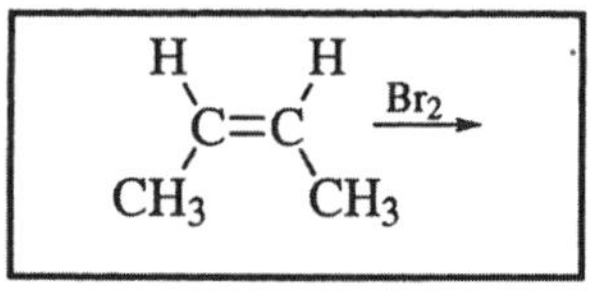

(A)
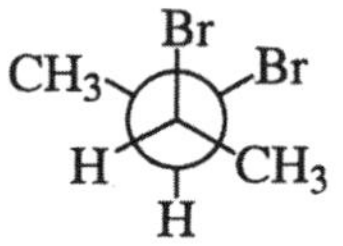

(B)
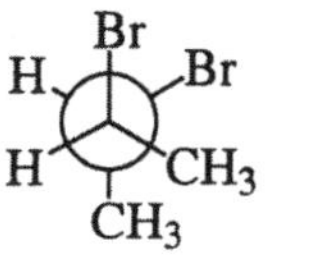

(C)
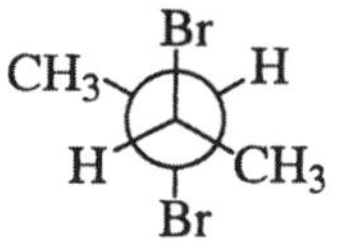

(D)
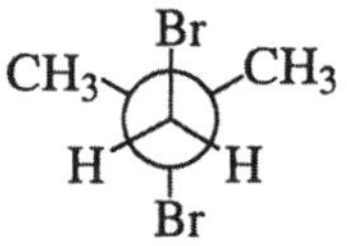

Answers to Study Questions

1. D
2. C
3. C
4. A
5. A
6. B
7. B
8. D
9. A
10. B
11. C

Answers to Practice Questions

1. B
2. A
3. D
4. A
5. B
6. D
7. C
8. A
9. D
10. C
11. C
12. D
13. C
14. A
15. B
16. C
17. D
18. B
19. B
20. C
21. A

Nucleophilic Addition at Carbonyl Groups

The carbonyl group is polar with the electronegative oxygen atom creating a partial positive charge on the carbon atom. Nucleophiles are attracted to the electrophilic carbon atom. The formation of a sigma bond between the nucleophilic atom and the carbonyl carbon atom breaks the carbon–oxygen π bond, which localizes the π-bonding electrons as nonbonding electrons on the oxygen atom. The resulting alkoxide species is often referred to as a tetrahedral intermediate (equation *1*). In protic environments, the alkoxide becomes protonated and the addition to the carbonyl group is complete. The ease of nucleophilic attack at the carbonyl carbon atom is subject to steric hindrance and electronic considerations.

$$\overset{\delta-}{O}=\overset{\delta+}{C}< \;+\; Nu^- \longrightarrow -\overset{O^-}{\underset{Nu}{C}}- \xrightarrow{H^+} -\overset{OH}{\underset{Nu}{C}}- \qquad (1)$$

The nucleophilic addition is catalyzed by acid. Protonation of the carbonyl oxygen atom results in an even more electrophilic carbon atom as suggested by the resonance structure in equation 2. Nucleophilic attack forms the addition product directly.

$$\overset{\delta-}{O}=\overset{\delta+}{C}< \xrightarrow{H^+} \left[>C=\overset{+}{O}H \longleftrightarrow >\overset{+}{C}-OH \right] \xrightarrow[-H^+]{NuH} -\overset{OH}{\underset{Nu}{C}}- \qquad (2)$$

If the nucleophile is a weak base, and therefore a good leaving group, the nucleophilic addition is reversible. An equilibrium is established that reflects the relative stabilities of the carbonyl compound and the adduct. Since the substituents are closer together in the tetrahedral intermediate than in the starting material, steric effects may be important in assessing relative stabilities. In addition, the stability of the carbonyl compound is greatly affected by the electronic nature of the substituent attached to the carbonyl carbon atom. Electron-releasing groups diminish the positive charge on the carbonyl group, which is a stabilizing factor. Electron-withdrawing groups intensify the positive charge on the carbon atom, and destabilize the carbonyl group.

If the nucleophile is a primary amine or a so-called derivative of ammonia ($Y{-}NH_2$), the tetrahedral intermediate can lose water to form the carbon–nitrogen double bond of an **imine**. A secondary amine can lose water from the carbon–carbon double bond giving an **enamine**.

The formation of acetals and imines can be favored by the removal of the water, which is also a product of the reactions. Reversing the process, acetals, imines, and enamines can be hydrolyzed back to the carbonyl compound using an aqueous acid solution.

Aldehydes and ketones that are α,β-unsaturated may undergo either 1,2-additon or 1,4-addition with nucleophiles. The *kinetic* product results from 1,2-nucleophilic addition, while the *thermodynamic* product results from 1,4-nucleophilic addition. Strongly basic nucleophiles such as Grignard and organolithium reagents and complex metal hydrides generally give 1,2-additions. Weakly basic nucleophiles and organocuprates (R_2CuLi) generally give 1,4-additions.

Study Questions

NA-1. Which compound exists to the greatest extent as its hydrate when dissolved in aqueous solution?

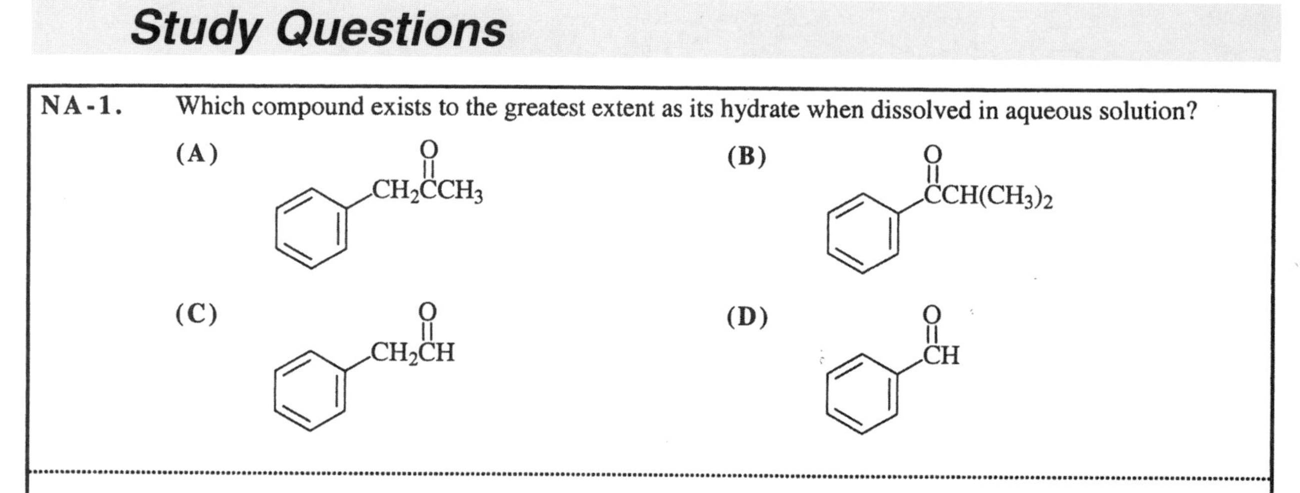

Knowledge Required: (1) Mechanism of nucleophilic addition to a carbonyl group. (2) Electron-releasing characteristic of alkyl groups. (3) Steric factors in formation of reaction intermediate. (4) Stabilizing effect of conjugation.

Thinking it Through: The nucleophilic attack by water at the carbonyl carbon atom to form the tetrahedral intermediate is reversible. The relative concentrations of reactants and products reflect their relative stabilities. Ketones have two electron-releasing groups attached to the carbonyl carbon atom and therefore, that carbon atom is less electrophilic than the carbonyl carbon atom of an aldehyde. In addition, the two alkyl or aryl groups of the ketone must come closer to each other as the tetrahedral intermediate forms. Steric repulsion between these groups is greater than the repulsion between one alkyl (or aryl) group and a hydrogen atom in an aldehyde. Both the electronic effect and the steric effect suggest that the carbonyl group of ketones is more stable than the carbonyl group of aldehydes, so less of the ketone adduct would be present at equilibrium. On this basis, choices **(A)** and **(B)** are eliminated. Of the two aldehydes shown, choice **(D)** is conjugated, and therefore more stable than choice **(C)**, which is the correct answer.

NA-2. Arrange the ketones in order of decreasing reactivity toward cyanohydrin formation with HCN/KCN.

$CH_3\overset{O}{\overset{\|}{C}}CH_3$	$(CH_3)_2CH\overset{O}{\overset{\|}{C}}CH_3$	$CF_3\overset{O}{\overset{\|}{C}}CF_3$
1	**2**	**3**

(A) 1 > 2 > 3 **(B)** 2 > 1 > 3 **(C)** 3 > 1 > 2 **(D)** 3 > 2 > 1

Knowledge Required: (1) Mechanism of nucleophilic addition to a carbonyl group. (2) Electron-withdrawing characteristic of fluoroalkyl groups. (3) Steric factors in formation of reaction intermediate.

Thinking it Through: The three structures shown are all ketones, and the reaction involves nucleophilic addition of HCN to the carbonyl group. Nucleophilic attack at the carbonyl carbon atom is a consequence of the electrophilic character of the carbonyl carbon atom. The electron-withdrawing trifluoromethyl group greatly increases the electrophilic character of the carbonyl carbon atom in ketone **3** relative to that of the other two ketones. Therefore, choices **(A)** and **(B)** are eliminated. The intermediate for ketone **2** is more sterically hindered than the intermediate for ketone **1**, so ketone **2** should be less reactive. This gives the sequence **3 > 1 > 2** as the correct answer, which is choice **(C)**.

NA-3. What is the principal product of the acid-catalyzed reaction of cyclopentanone with ethylene glycol ($HOCH_2CH_2OH$)?

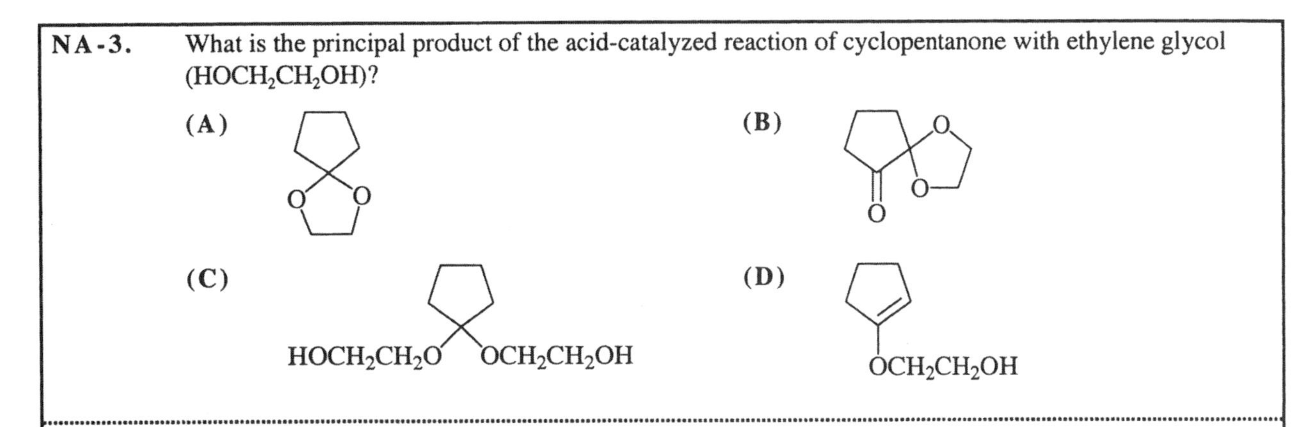

Knowledge Required: (1) Mechanism of nucleophilic addition to a carbonyl group. (2) Relative rates of intramolecular and intermolecular reactions.

Thinking it Through: The general mechanism of nucleophilic addition under acidic conditions begins with protonation of the carbonyl group, which activates it toward nucleophilic attack by the alcohol, forming a hemiacetal. Under acidic conditions, the hydroxyl group of the hemiacetal is protonated, and loses water to give a resonance-stabilized carbocation. *Intramolecular* nucleophilic attack by the remaining hydroxyl would lead to the acetal shown in choice **(A)**. Choice **(C)** is an acetal, but results from *intermolecular* reaction of a second ethylene glycol with the carbocation. The intramolecular reaction proceeds much more rapidly than the intermolecular reaction, which eliminates choice **(C)**. Choice **(D)** is not an acetal, but an enol ether that would react to form the resonance stabilized carbocation, then choice **(A)**. Therefore, choice **(D)** is eliminated. Choice **(B)** contains both ketone and acetal functional groups, and it would not form under any conceivable reaction conditions. This leaves choice **(A)** as the only correct answer.

NA-4. Which is the product that can be isolated from the reaction shown?

$$CH_3OCH(CH_3)CH_2C(CH_3)(OCH_3)_2 \xrightarrow[\Delta]{H_2O,\ H^+}$$

(A) $CH_3OCH(CH_3)CH_2C(=O)CH_3$

(B) $HOCH(CH_3)CH_2C(=O)CH_3$

(C) $HOCH(CH_3)CH_2C(CH_3)(OCH_3)_2$

(D) $HOCH(CH_3)CH_2C(CH_3)(OH)_2$

Knowledge Required: (1) Structure of acetal functional group. (2) Mechanism of acid-catalyzed hydrolysis of acetals. (3) Stability of simple ethers to mild acidic conditions.

Thinking it Through: Many nucleophilic addition reactions are reversible. This question deals with the reverse reaction of acetal formation. The reactant molecule contains both an acetyl functional group and a simple ether functional group. Only the acetal group is reactive under mild acidic conditions. Simple ethers require extreme conditions for hydrolysis. Choices **(B)**, **(C)**, and **(D)** are eliminated as possible products because each shows the simple ether as having been hydrolyzed. Only choice **(A)** is left as a possible answer. Choice **(A)** correctly shows the product of the acetal hydrolysis to be a ketone group as well as showing the simple ether as having not reacted.

NA-5. What are the products of the hydrolysis reaction shown?

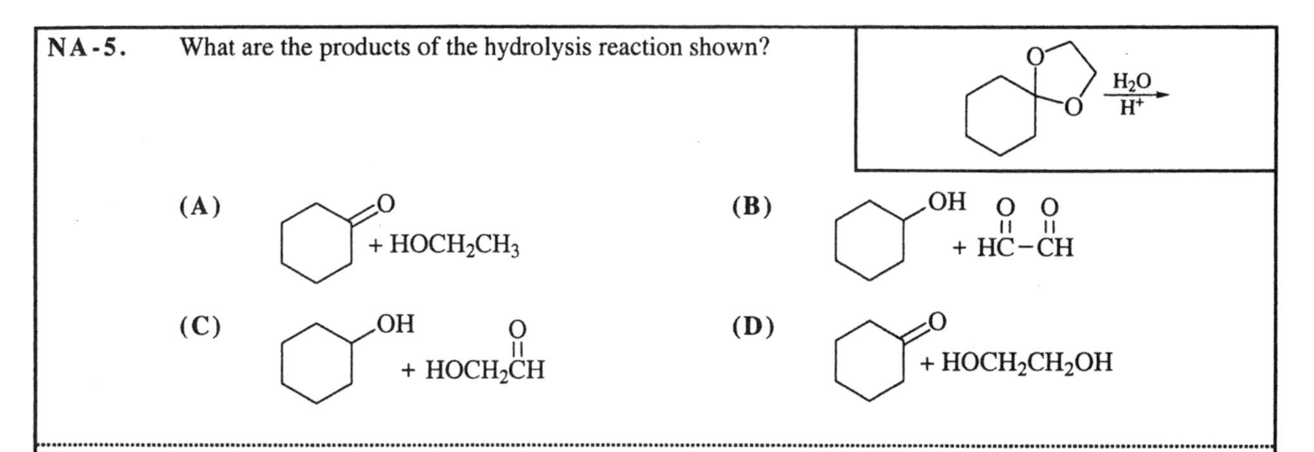

Knowledge Required: (1) Mechanism of nucleophilic addition to a carbonyl group. (2) Mechanism of acid-catalyzed hydrolysis of acetals.

Thinking it Through: This is another example of the reverse reaction for nucleophilic addition. This question considers the reverse of the reaction in which an acetal is formed from a ketone. The first thing is to recognize that the cyclic compound is an acetal that could only have formed from a diol. In the hydrolysis reaction, protonation on one of the oxygen atoms of the acetal produces a good leaving group (an alcohol), which can dissociate to a resonance-stabilized carbocation.

Nucleophilic attack on the carbocation by water then produces a hemiacetal. Subsequent protonation on the alkoxyl oxygen atom again produces a good leaving group, which then dissociates to the protonated carbonyl group. The overall result is that hydrolysis of the carbon atom bearing the two alkoxyl groups converts it to a carbonyl group, and the alkoxyl groups become alcohol groups. Either two alcohol molecules or a diol must be counted among the products. Choice **(A)** has the correct ketone, but only one alcohol group among the products. Choices **(B)** and **(C)** have the carbonyl group appearing in the wrong fragment. Choice **(D)** has the correct ketone, and there are two alcohol groups in the diol. It is the correct answer.

NA-6. Which is the product that can be isolated from the reaction shown?

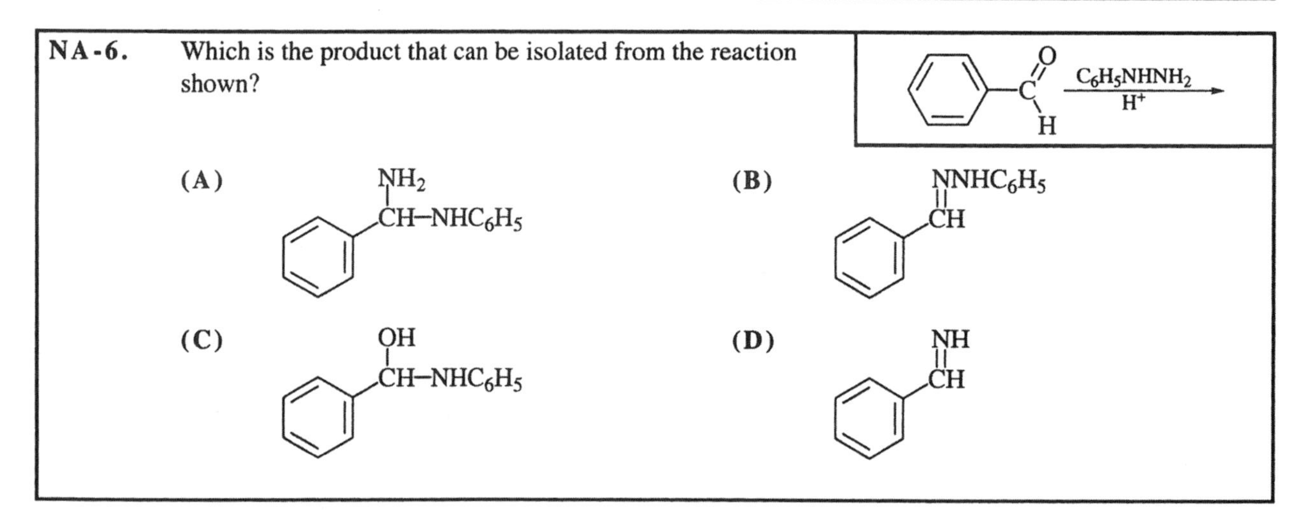

Knowledge Required: Mechanism of nucleophilic addition to a carbonyl group when the nucleophile is a primary amine (Y–NH_2).

Thinking it Through: The reaction shows an aldehyde reacting with phenylhydrazine, a primary amine, in the presence of an acid. Acid-catalyzed nucleophilic addition would give a species with a hydroxyl and an amine attached to what was the carbonyl carbon atom. Protonation of the hydroxyl and loss of water forms a resonance-stabilized carbocation that can lose a proton from the nitrogen atom to give a carbon–nitrogen double bond. Choices **(A)** and **(C)** are immediately eliminated because they do not contain a carbon–nitrogen double bond. Choice **(B)** results if the terminal N–H bond is broken during the elimination of water, and is the correct answer. Choice **(D)** would form if the N–N bond were broken, which cannot happen under these conditions.

NA-7. What is the most reasonable structure for product of the reaction shown?

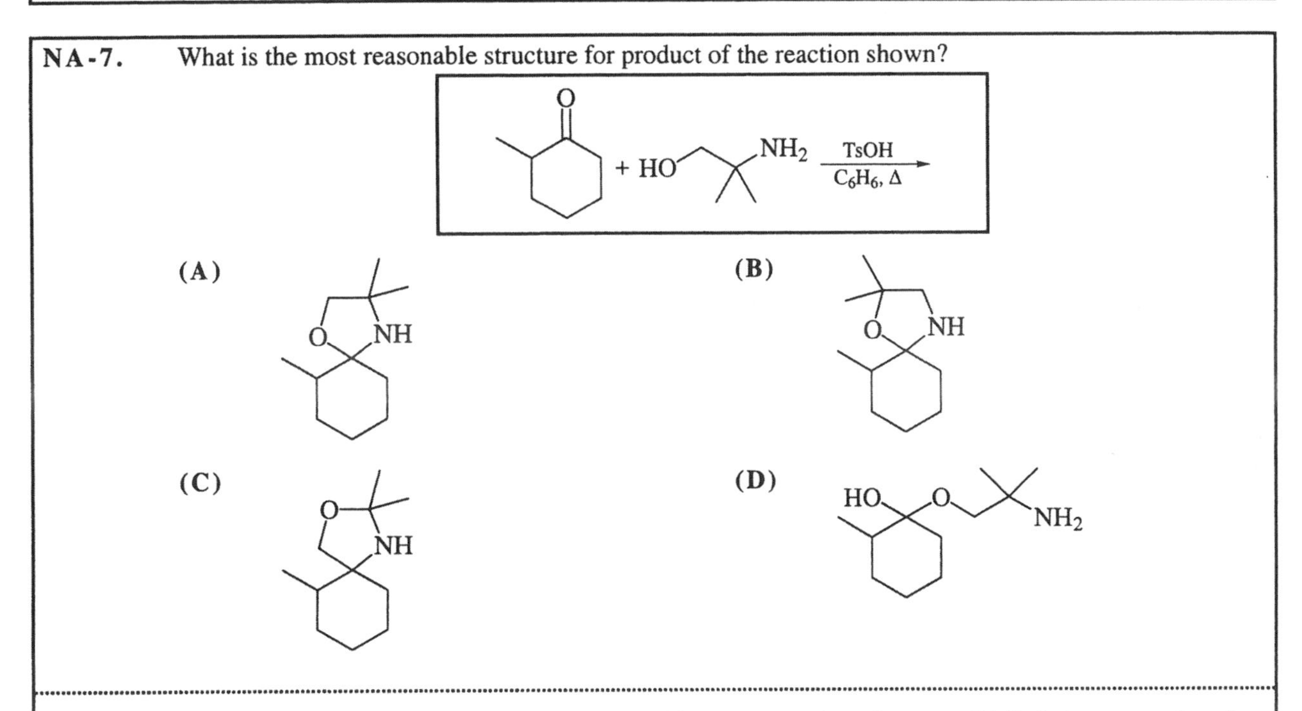

Knowledge Required: (1) Mechanism of nucleophilic addition to a carbonyl group. (2) Relative strengths of alcohol oxygen atoms and amine nitrogen atoms as nucleophiles.

Thinking it Through: Under acidic reaction conditions, the carbonyl group of the ketone becomes protonated. The nitrogen atom is an intrinsically better nucleophile (stronger base) than the oxygen atom, and preferentially attacks the carbonyl carbon atom, producing a hemiaminal (analogous to a hemiacetal). In analogy to the mechanism of acetal formation with a diol, the hemiaminal undergoes protonation of the hydroxyl group followed by rapid *intramolecular* nucleophilic attack at the carbocation by the hydroxyl oxygen atom. The product is the mononitrogen analogue of an acetal, and must have both a nitrogen atom and an oxygen atom directly attached to the carbon atom. Choice **(C)** is eliminated because a methylene group is shown between the carbonyl carbon atom and the oxygen atom. Choice **(D)** is eliminated because it shows initial nucleophilic attack by the hydroxyl oxygen atom, with no further reaction. Considering choices **(A)** and **(B)**, the difference is in the linkage of the nitrogen atom and oxygen atom to the tertiary carbon atom. Choice **(A)** correctly shows the nitrogen atom directly connected to the tertiary carbon atom, while **(B)** shows the nitrogen atom connected through a methylene group.

NA-8. What is the product of the $NaBH_4$ reduction shown?

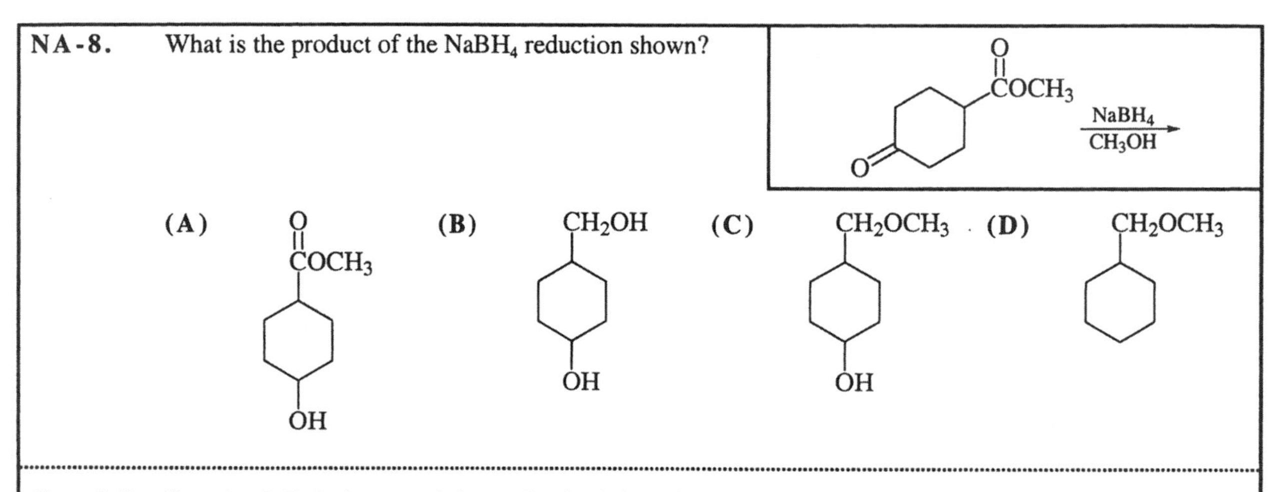

Knowledge Required: Relative reactivity and selectivity of hydride reagents.

Thinking it Through: Sodium borohydride, $NaBH_4$, is a source of hydride ion and reduces aldehydes and ketones by hydride attack at the carbonyl group. In general, $NaBH_4$ does not readily reduce esters. Choice **(A)** shows the reduction of the ketone without reaction with the ester and is the correct answer. Choice **(B)** shows reduction of both the ketone and ester groups. The reaction shown in choices **(C)** and **(D)** indicates the ester reduction product to be an ether, which would not occur. If the reduction did occur, an alcohol would result, as in choice **(B)**. Note that $LiAlH_4$ is a more reactive source of hydride and readily reduces aldehydes, ketones, esters, and carboxylic acids.

NA-9. Which combination would *not* produce the alcohol shown?

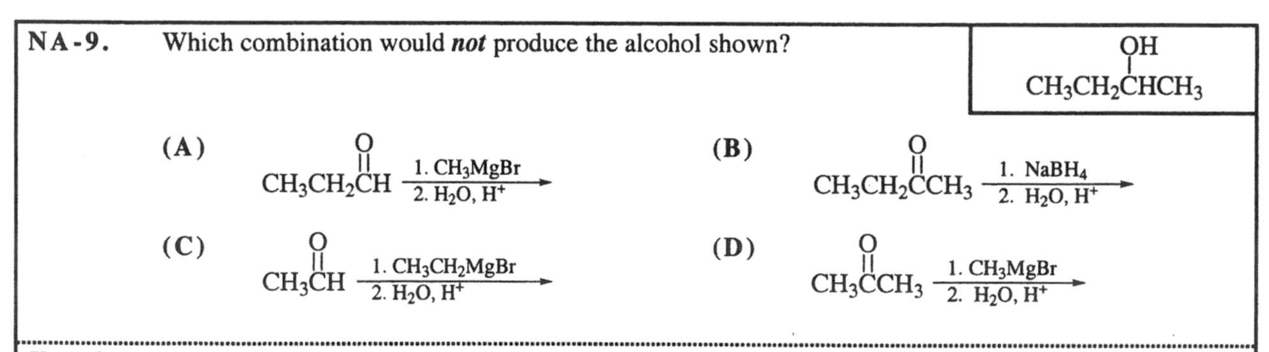

Knowledge Required: (1) Mechanism of nucleophilic addition to a carbonyl group. (2) Organomagnesium compounds (Grignard reagents) as sources of alkyl nucleophiles. (3) $NaBH_4$ as a source of nucleophilic hydride.

Thinking it Through: A systematic approach to the synthesis of alcohols is to focus on transformation of a carbonyl group into the carbon atom bearing the hydroxyl. The third substituent attached to the carbon atom bearing the hydroxyl is introduced as the nucleophile reacting with the carbonyl group. Carbon nucleophiles can derive from Grignard reagents or organolithium reagents. The nucleophilic hydrogen atom can come from $NaBH_4$ or $LiAlH_4$. This notion of "thinking backwards" is called **retrosynthetic analysis**. Mentally disconnect one of the substituents attached to the carbon atom bearing the hydroxyl. That substituent is the nucleophile and the carbon atom bearing the hydroxyl becomes the carbonyl carbon atom. In this question, however, it is necessary to examine each of the combinations to determine which is incorrect. The final secondary alcohol contains a methyl group and an ethyl group attached to the original carbonyl carbon atom. Choice **(A)** begins with propanal and adds a methyl group, so it will produce the desired 2-butanol. Choice **(C)** produces the same result by beginning with ethanal and adding an ethyl group. Choice **(B)** already has the correct substituents on the carbonyl carbon atom, and the hydride reduction converts the ketone group to an alcohol group. Choice **(D)** forms a tertiary alcohol (2-methyl-2-propanol) rather than a secondary alcohol, and is the correct answer.

NA-10. What is the product of the reaction sequence shown?

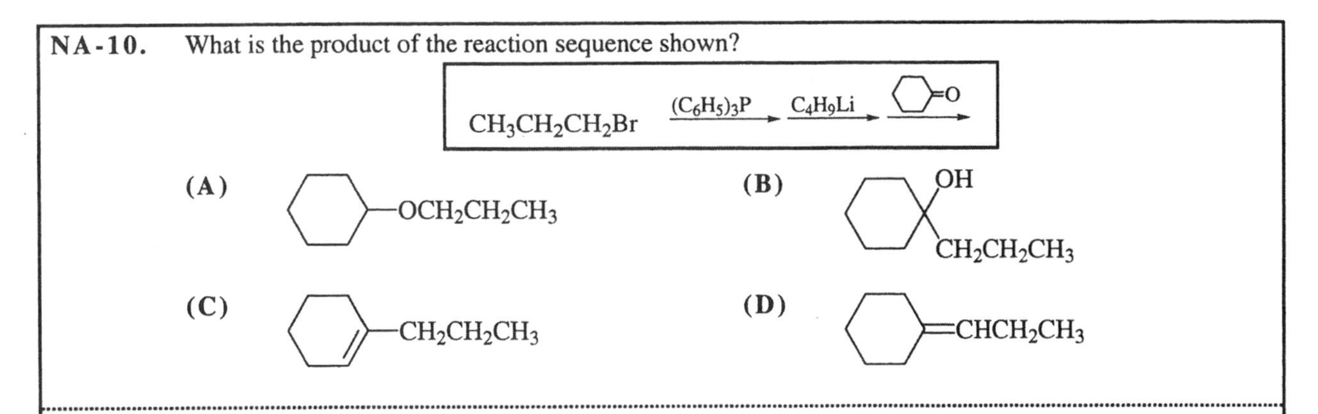

Knowledge Required: The formation of Wittig reagents and their reaction with aldehydes and ketones.

Thinking it Through: The Wittig reaction is an important method for the synthesis of alkenes. The overall transformation converts a carbon–oxygen double bond into a carbon–carbon double bond. In a Wittig reaction, an aldehyde or ketone reacts with a phosphonium ylide. An ylide is a species with positive and negative charge on adjacent atoms. In a phosphonium ylide, a phosphorus atom is positive and a carbon atom is negative, and therefore nucleophilic. The ylide is formed by the deprotonation of a phosphonium salt, which is first formed by the reaction of a trivalent phosphorus compound with an alkyl halide. This sequence is shown below for the reagents in this question.

$$CH_3CH_2CH_2Br \xrightarrow{(C_6H_5)_3P} \underset{\text{phosphonium salt}}{CH_3CH_2CH_2\overset{+}{P}(C_6H_5)_3\ Br^-} \xrightarrow{C_4H_9Li} \underset{\text{phosphonium ylide}}{CH_3CH_2\overset{..}{C}H\overset{+}{P}(C_6H_5)_3}$$

The ylide is the nucleophile that attacks the carbonyl carbon atom of aldehydes and ketones in the key step of the Wittig reaction. A new carbon–carbon bond is formed that eventually becomes a carbon–carbon double bond with the elimination of oxygen and phosphorus. Formally, one carbon atom of the carbon–carbon double bond derives from the carbonyl carbon atom and the other carbon atom is from the alkyl halide to which the halogen had been attached.

Choice **(A)** is immediately eliminated because it would require a nucleophilic attack on the carbonyl oxygen atom instead of the carbonyl carbon atom. Choice **(B)** suggests a nucleophilic attack at the carbonyl carbon atom, but there is no logical basis for the elimination of the phosphorus atom. Choice **(C)** indicates formation of an alkene after a nucleophilic attack, but the double bond involves the wrong carbon atom. Choice **(D)** would result from the nucleophilic attack of the carbon atom from the phosphonium ylide on the carbonyl carbon atom, followed by removal of the oxygen atom by triphenylphosphine, and it is the correct answer.

NA-11. Which nucleophilic reagent, Z, will give the most 1,4-addition product with methyl vinyl ketone?

$$CH_3-\overset{O}{\overset{\|}{C}}-CH=CH_2 \xrightarrow{Z}$$

(A) $LiAlH_4$ **(B)** NaCN **(C)** CH_3MgBr **(D)** $(C_6H_5)_3\overset{+}{P}\overset{..}{C}H_2$

Knowledge Required: The factors that control whether 1,2-addition or 1,4-addition is the major mode of reaction of nucleophiles with α,β-unsaturated carbonyl compounds.

Thinking it Through: Methyl vinyl ketone is a conjugated ketone and may undergo either 1,2-addition or 1,4-addition with nucleophiles. Kinetically, 1,2-additions are favored over 1,4-additions. When the nucleophile is strongly basic, the initial1,2-addition is irreversible and the1,2-addition product is the dominant product. However, when the nucleophile is less basic, the initial 1,2-addition is reversible and the thermodynamically more stable 1,4-addition product is the dominant product. The reagents in choices **(A)**, **(C)** and **(D)** are strongly basic and give mostly 1,2-addition. The cyanide ion, in choice **(B)**, is a much weaker base ($pK_a = 9$ for HCN) and gives predominantly 1,4-addition. Choice **(B)** is the correct answer.

Practice Questions

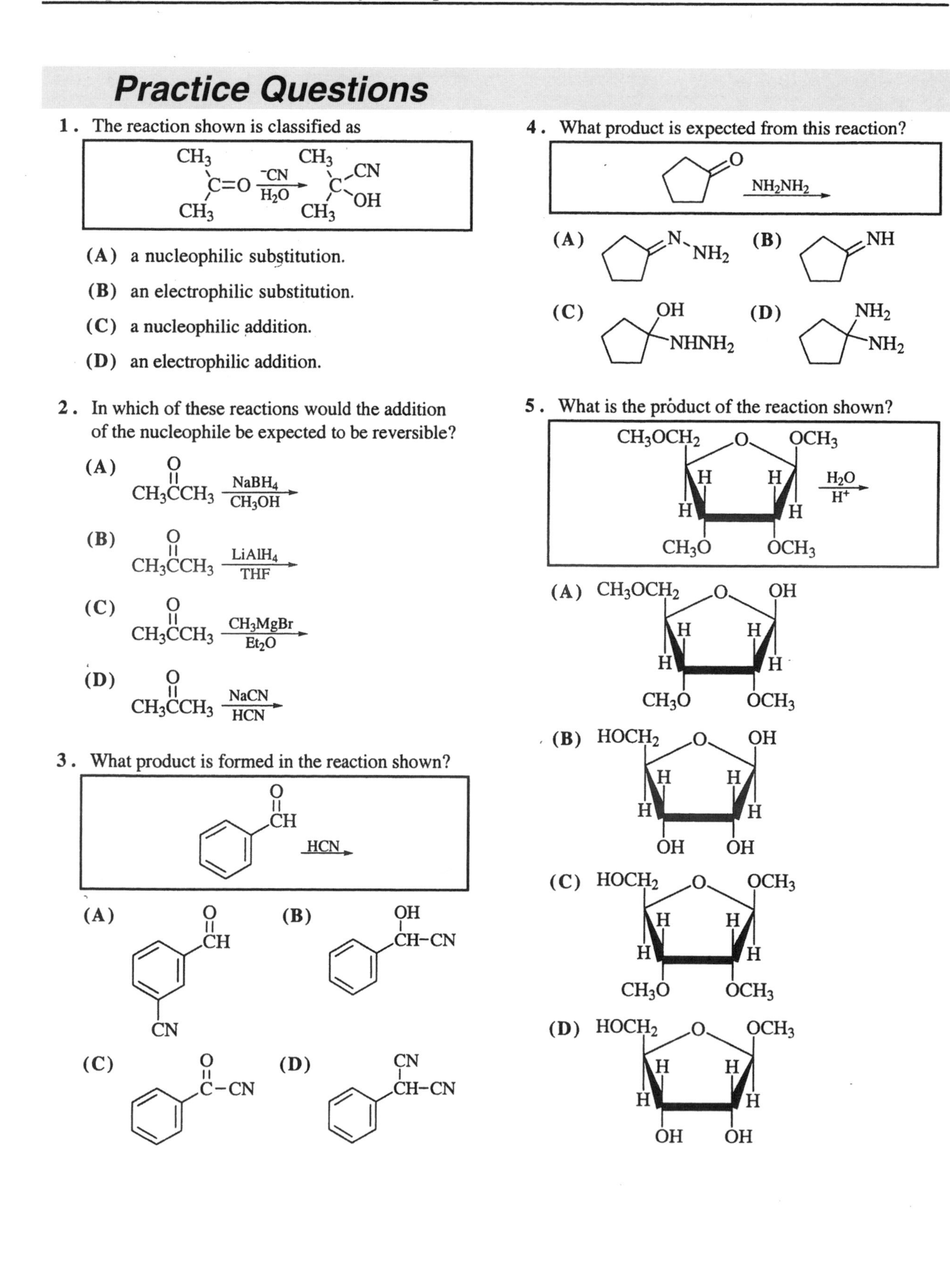

1. The reaction shown is classified as

$(CH_3)_2C{=}O \xrightarrow[H_2O]{^-CN} (CH_3)_2C(CN)(OH)$

(A) a nucleophilic substitution.
(B) an electrophilic substitution.
(C) a nucleophilic addition.
(D) an electrophilic addition.

2. In which of these reactions would the addition of the nucleophile be expected to be reversible?

(A) $CH_3\overset{O}{\overset{\|}{C}}CH_3 \xrightarrow[CH_3OH]{NaBH_4}$

(B) $CH_3\overset{O}{\overset{\|}{C}}CH_3 \xrightarrow[THF]{LiAlH_4}$

(C) $CH_3\overset{O}{\overset{\|}{C}}CH_3 \xrightarrow[Et_2O]{CH_3MgBr}$

(D) $CH_3\overset{O}{\overset{\|}{C}}CH_3 \xrightarrow[HCN]{NaCN}$

3. What product is formed in the reaction shown?

4. What product is expected from this reaction?

5. What is the product of the reaction shown?

6. Which is the ***main*** product that can be isolated from the reaction shown?

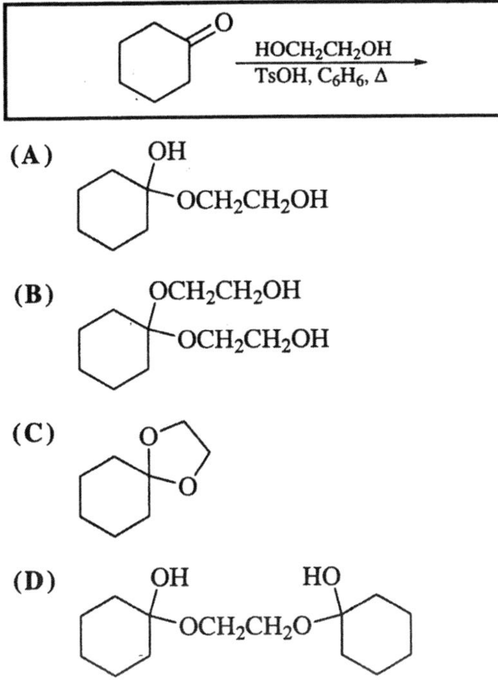

7. The initial stable product from the reaction of 4-methyl-2-pentanone and excess methanol with H_2SO_4 as a catalyst would be

$$CH_3-\overset{O}{\overset{\|}{C}}-CH_2-\overset{CH_3}{\overset{|}{C}H}-CH_3 \xrightarrow[H_2SO_4]{CH_3OH}$$

(A) $CH_3-\underset{OCH_3}{\underset{|}{\overset{OCH_3}{\overset{|}{C}}}}-CH_2-\overset{CH_3}{\overset{|}{C}H}-CH_3$

(B) $CH_3OCH_2-\overset{O}{\overset{\|}{C}}-CH_2-\overset{CH_3}{\overset{|}{C}H}-CH_3$

(C) $CH_3-\overset{OCH_3}{\overset{|}{C}H}-CH_2-\overset{CH_3}{\overset{|}{C}H}-CH_3$

(D) $CH_3-\overset{OCH_3}{\overset{|}{C}}=CH-\overset{CH_3}{\overset{|}{C}H}-CH_3$

8. Which compound would be most rapidly hydrolyzed by aqueous HCl to give methanol as one of the products?

(A) $CH_3OCH_2CH_2CH_3$

(B) $CH_3OCH_2CH_2OCH_3$

(C) $CH_3OCH_2CH_2OH$

(D) $CH_3CH_2CH(OCH_3)_2$

9. Table sugar is (+)-sucrose. Which of these products forms as the result of enzymatic or acid-catalyzed test-tube hydrolysis?

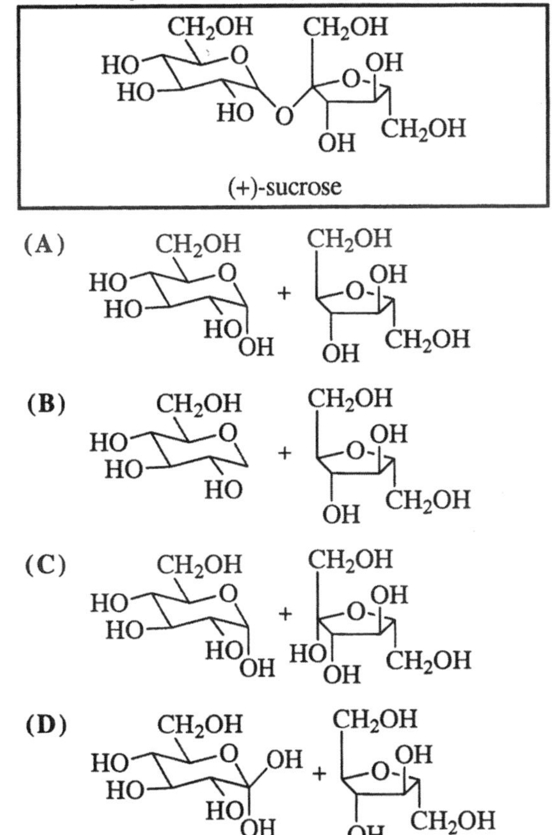

10. Which is a major product of the reaction shown?

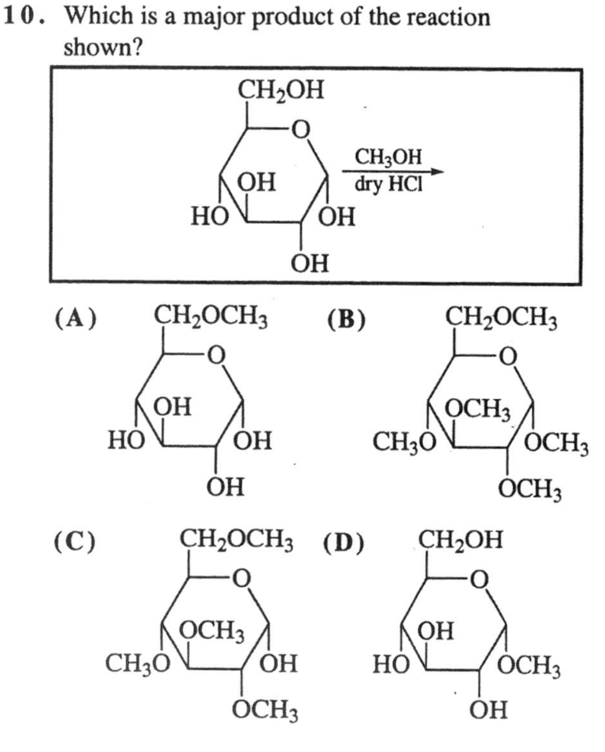

11. What is the product for the reaction shown?

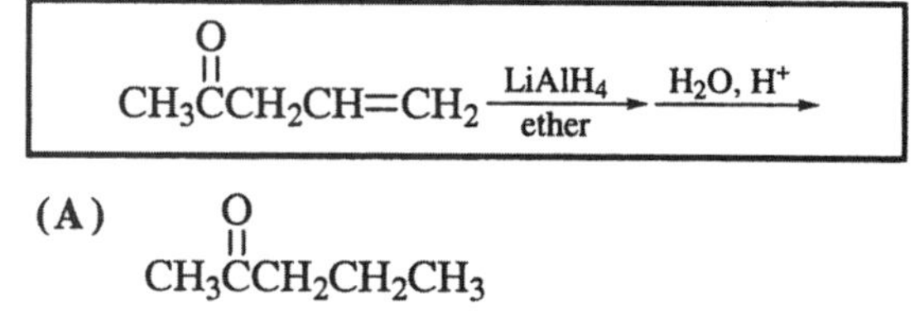

(A) $CH_3\overset{O}{\overset{\|}{C}}CH_2CH_2CH_3$

(B) $CH_3\overset{OH}{\overset{|}{C}}HCH_2CH{=}CH_2$

(C) $CH_3\overset{OH}{\overset{|}{C}}HCH_2CH_2CH_3$

(D) $CH_3CH_2OH + CH_3CH_2CH_3$

12. Which is the best reagent for this conversion?

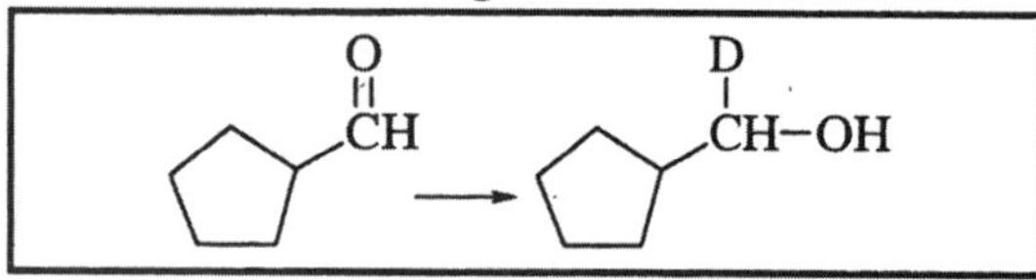

(A) D_2O, containing catalytic amounts of HCl

(B) $NaBD_4$ in CH_3CH_2OH (and an aqueous workup)

(C) NaOD in CH_3CH_2OD (and an aqueous workup)

(D) D_2O_2 in CH_3CO_2H

13. Predict the products of the reaction shown.

14. What combination of reagents would *not* form the alcohol shown after an aqueous workup?

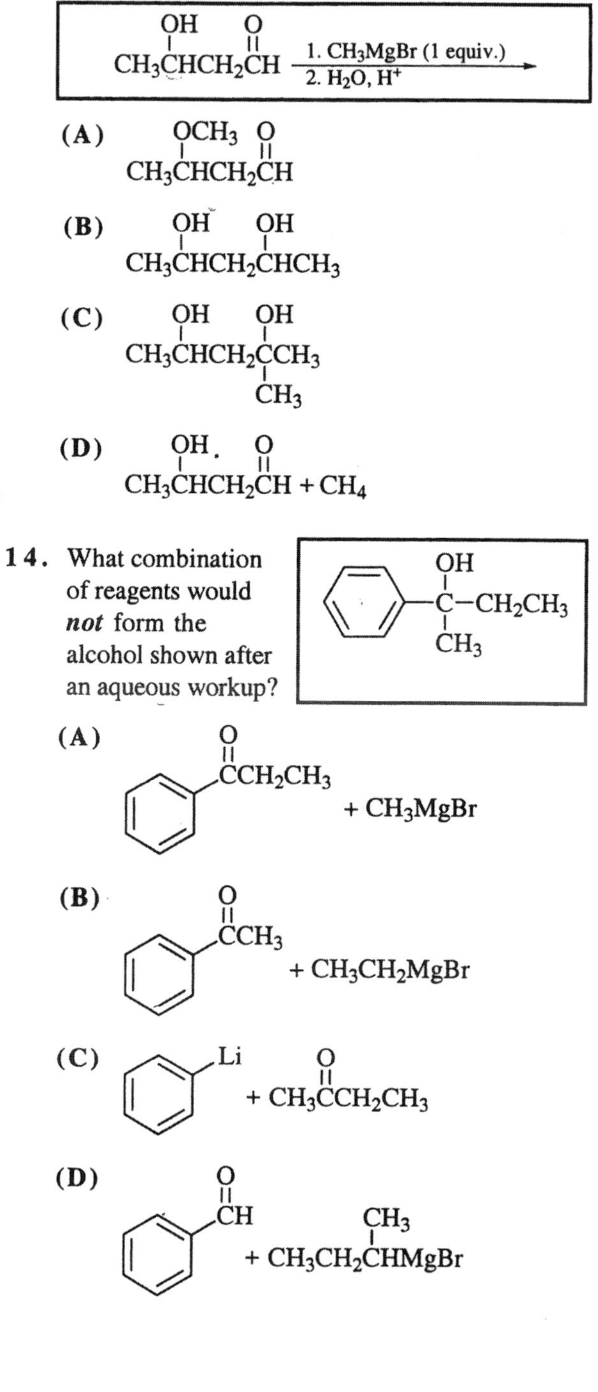

15. What is the major product of the reaction shown?

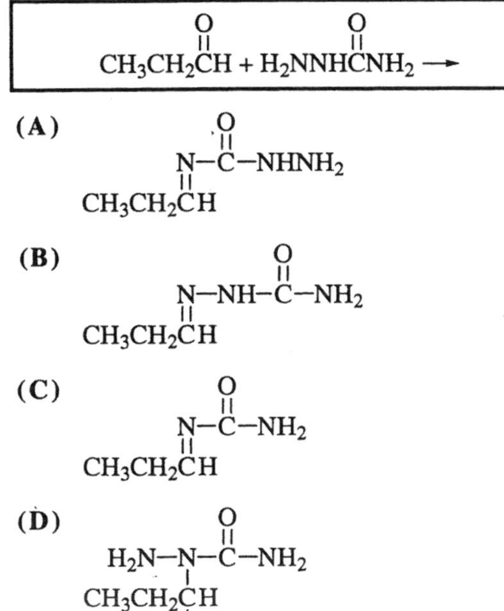

16. What would be the product from this reaction?

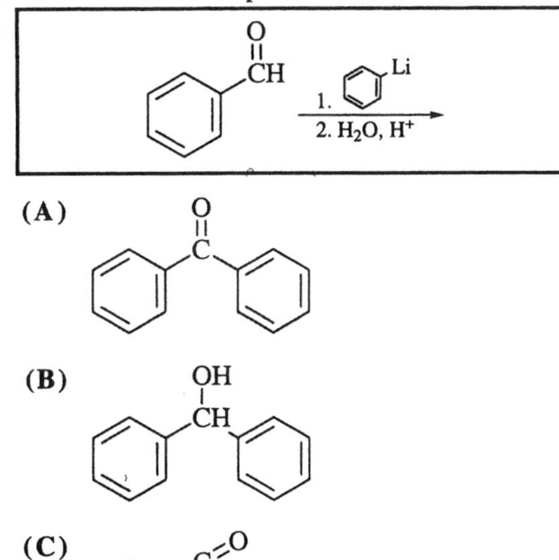

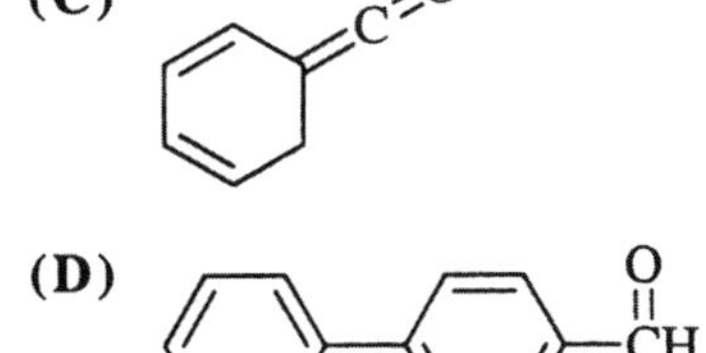

17. Which reagent will accomplish the conversion shown?

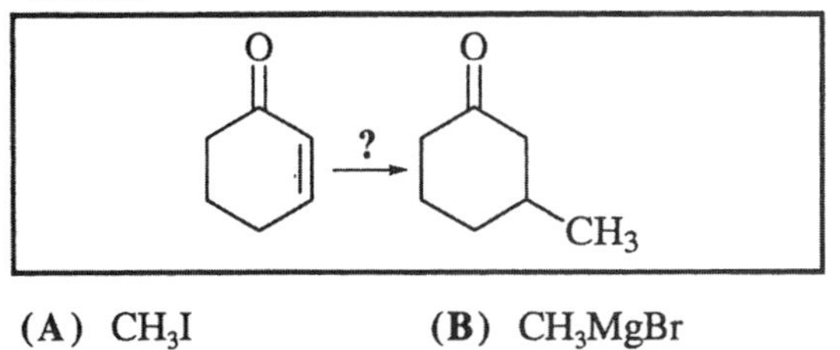

(A) CH_3I (B) CH_3MgBr

(C) CH_3Li (D) $(CH_3)_2CuLi$

18. Which reagent sequence will effect the transformation shown?

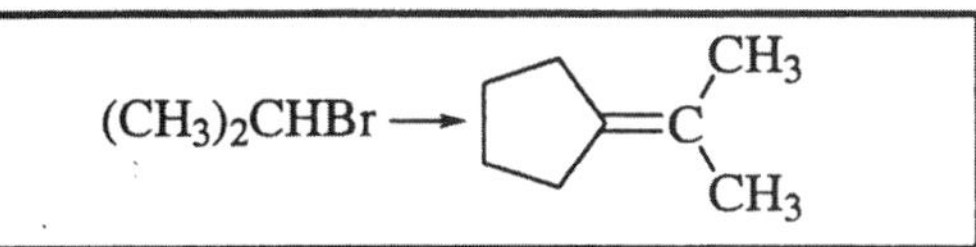

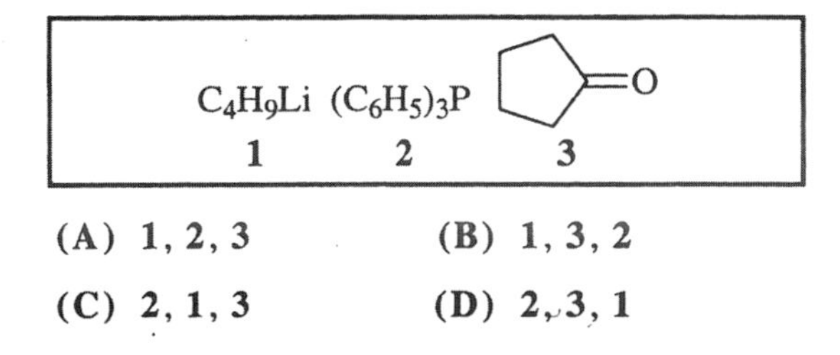

(A) 1, 2, 3 (B) 1, 3, 2

(C) 2, 1, 3 (D) 2, 3, 1

19. What set of ketone (1) and aldehyde (2) will provide the same alcohol product (3) when submitted to the reaction conditions shown?

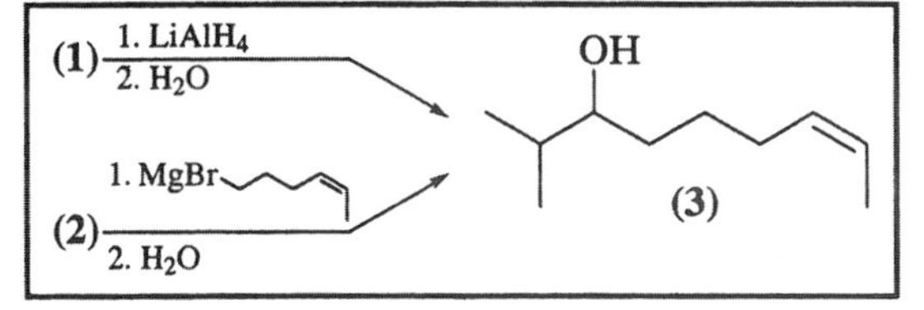

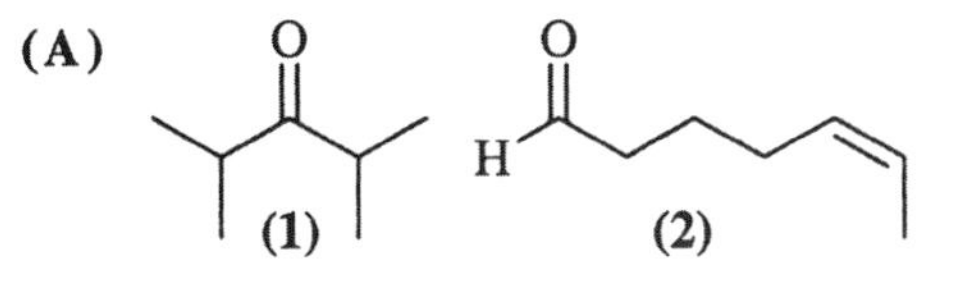

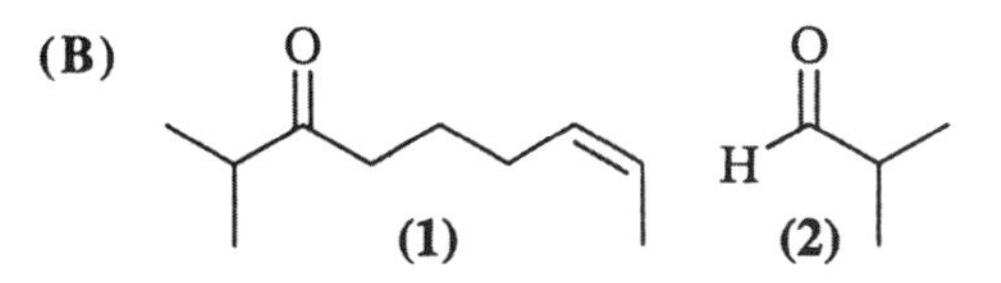

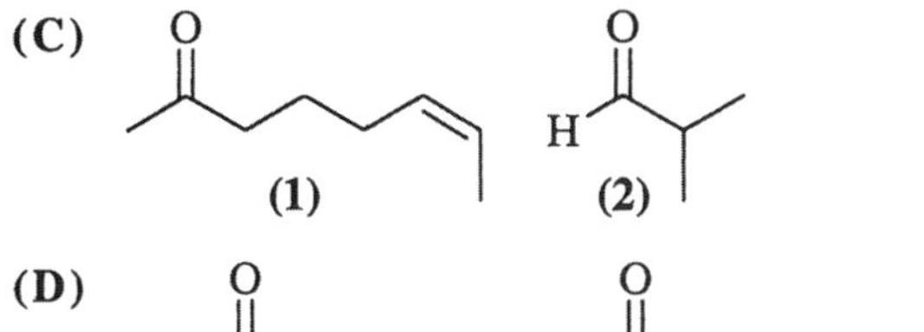

20. What is the product of this reaction sequence?

$$C_6H_5CH_2Cl \xrightarrow{(C_6H_5)_3P} \xrightarrow{(CH_3)_3COK} \xrightarrow{(C_6H_5)_2C=O}$$

(A) $(C_6H_5)_3COH$

(B) 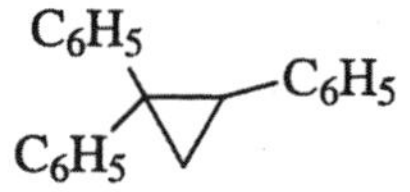 [cyclopropane bearing C_6H_5, C_6H_5 on one carbon and C_6H_5 on another]

(C) $(C_6H_5)_2C=CHC_6H_5$

(D) $(C_6H_5)_3COCH_2C_6H_5$

21. What is the product of this reaction sequence?

$$(C_6H_5)_3\overset{+}{P}CH_3\ I^- \xrightarrow{C_4H_9Li} \xrightarrow{CH_3\overset{O}{\overset{\|}{C}}CH=CHCH_3}$$

(A) $CH_2=C(CH_3)-CH=CHCH_3$

(B) $CH_3-\overset{O}{\overset{\|}{C}}-C_6H_5$

(C) $CH_3-C(OH)(CH_3)-CH=CHCH_3$

(D) $C_4H_9-C(OH)(CH_3)-CH=CHCH_3$

22. Which is the ***major*** product of this reaction?

$$CH_2=CH-\overset{O}{\overset{\|}{C}}-C_6H_5 \xrightarrow[CH_3OH]{CH_3ONa}$$

(A) $CH_3-CH(OCH_3)-\overset{O}{\overset{\|}{C}}-C_6H_5$

(B) $CH_2=CH-C(OCH_3)_2-C_6H_5$

(C) $CH_2=CH-C(OH)(OCH_3)-C_6H_5$

(D) $CH_3O-CH_2-CH_2-\overset{O}{\overset{\|}{C}}-C_6H_5$

Answers to Study Questions

1. C
2. C
3. A
4. A
5. D
6. B
7. A
8. A
9. D
10. D
11. B

Answers to Practice Questions

1. C
2. D
3. B
4. A
5. A
6. C
7. A
8. D
9. C
10. D
11. B
12. B
13. D
14. D
15. B
16. B
17. D
18. C
19. B
20. C
21. A
22. D

Nucleophilic Substitution at Carbonyl Groups

Carboxylic acids and their derivatives, acyl halides, anhydrides, esters, and amides, undergo the general reaction termed nucleophilic acyl substitution. The carbonyl group is polarized with the carbon atom partially positive and therefore, electrophilic. Nucleophiles can bond to the carbonyl carbon atom forming what is often referred to as a tetrahedral intermediate, because of the sp^3 hybridized carbon atom. The tetrahedral intermediate can reform the strong carbon–oxygen double bond with the loss of the leaving group $Y:^-$. The nucleophile has thus substituted for the leaving group at the acyl carbon atom.

$$R-\overset{\overset{\delta-}{:O:}}{\underset{\delta+}{C}}-Y + :Nu^- \longrightarrow R-\underset{Nu}{\overset{:\ddot{O}:^-}{C}}-Y \longrightarrow R-\overset{:O:}{\overset{\|}{C}}-Nu + :Y^- \qquad (1)$$

tetrahedral intermediate

Nucleophilic acyl substitution is subject to acid catalysis. Protonation of the carbonyl oxygen atom increases the electrophilic character of the carbonyl carbon atom as indicated by the resonance below:

$$R-\overset{:O:}{\overset{\|}{C}}-Y \xrightarrow{H^+} \left[R-\overset{:\overset{+}{O}-H}{\overset{\|}{C}}-Y \longleftrightarrow R-\overset{:\ddot{O}-H}{\overset{|}{\overset{+}{C}}}-Y \right] \qquad (2)$$

Nucleophilic attack on the protonated carbonyl compound forms the tetrahedral intermediate directly, which can then collapse with the loss of a leaving group. Under acidic conditions, protonation of a poor leaving group increases its tendency to dissociate.

Although they do not involve nucleophilic addition to carbonyl groups, the similarities of nitrile chemistry and condensation polymers formed by repeated nucleophilic acyl substitution justify their inclusion in the problems in this section.

Study Questions

NC-1. Which would be a reasonable intermediate in the mechanism for this reaction?

$$CH_3-\overset{O}{\overset{\|}{C}}-OCH_3 + H_2O \xrightarrow{H^+} CH_3-\overset{O}{\overset{\|}{C}}-OH + CH_3OH$$

(A) $CH_3-\overset{O^-}{\overset{|}{\underset{OH}{\underset{|}{C}}}}-OCH_3$

(B) $CH_3-\overset{OH}{\overset{|}{\underset{+\underset{H}{OH}}{\underset{|}{C}}}}-OCH_3$

(C) $CH_3-\overset{O}{\overset{\|}{C}}-\overset{+}{\underset{H}{O}}CH_3$

(D) $CH_3-\overset{O^-}{\overset{|}{\underset{+\underset{H}{OH}}{\underset{|}{C}}}}-OCH_3$

Knowledge Required: Mechanism of the acid-catalyzed nucleophilic acyl substitution reaction that is characteristic of carboxylic acids and carboxylic acid derivatives.

Thinking it Through: Acidic reaction conditions catalyze nucleophilic acyl substitution through protonation of the carbonyl oxygen atom. This makes the carbonyl carbon atom more electrophilic, and more susceptible to attack by the water nucleophile. Choice **(B)** would result from water attacking the protonated ester, and is the correct answer. Choices **(A)** and **(D)** would result from the nonprotonated ester being attacked by hydroxide ion and water, respectively. Choice **(C)** has the ester protonated on the wrong oxygen atom.

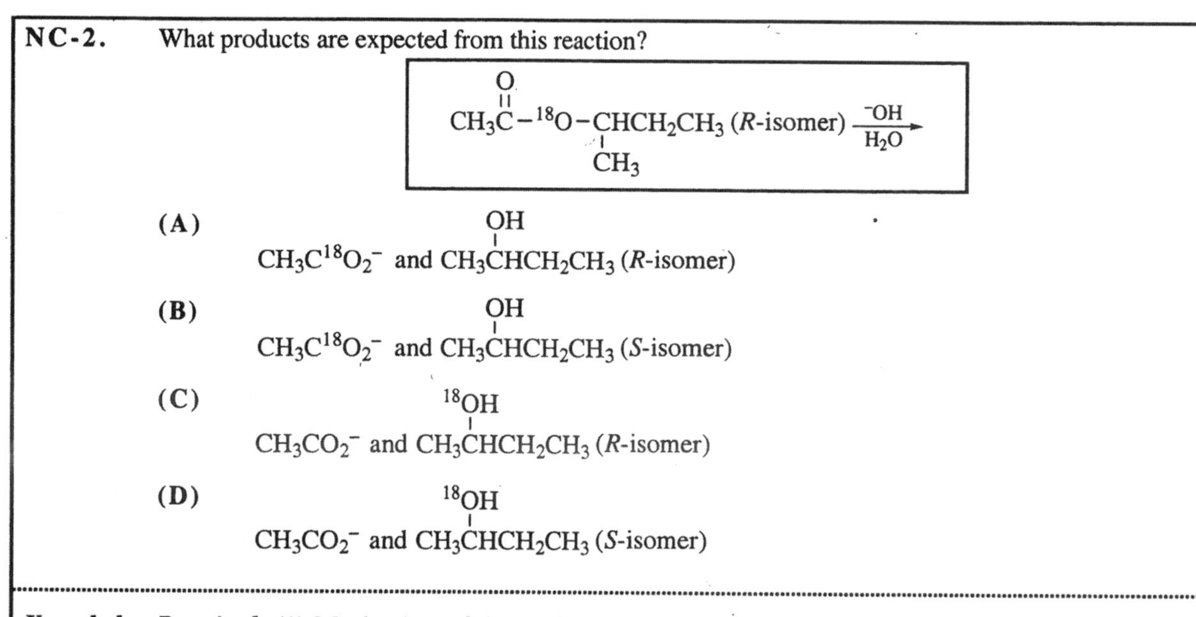

NC-2. What products are expected from this reaction?

$$CH_3C(=O)-{}^{18}O-CH(CH_3)CH_2CH_3 \ (R\text{-isomer}) \xrightarrow[H_2O]{^-OH}$$

(A) $CH_3C^{18}O_2^-$ and $CH_3CH(OH)CH_2CH_3$ (R-isomer)

(B) $CH_3C^{18}O_2^-$ and $CH_3CH(OH)CH_2CH_3$ (S-isomer)

(C) $CH_3CO_2^-$ and $CH_3CH({}^{18}OH)CH_2CH_3$ (R-isomer)

(D) $CH_3CO_2^-$ and $CH_3CH({}^{18}OH)CH_2CH_3$ (S-isomer)

Knowledge Required: (1) Mechanism of the nucleophilic acyl substitution under basic (or neutral) conditions. (2) Configuration at a stereogenic center.

Thinking it Through: The nucleophile is the hydroxide ion, which attacks the carbonyl carbon atom of the ester, forming a tetrahedral intermediate. The tetrahedral intermediate collapses with the dissociation of the alkoxide ion. Strongly basic conditions result in immediate deprotonation of the resulting carboxylic acid. The bond is broken between the carbonyl carbon atom and the oxygen atom of the alkoxide. The labeled oxygen atom remains with the alkoxide, and the configuration at the stereogenic center (chiral center) remains the same.

Choice **(A)** is eliminated because the alkyl oxygen bond has been broken instead of the bond between the oxygen atom and the carbonyl carbon atom. Choice **(D)** shows the cleavage of the correct carbon–oxygen bond, but the configuration has been inverted. Choice **(B)** is chemically plausible, but would require an S_N2 attack by hydroxide ion on the alkyl carbon atom. This is not an observed mechanism for basic hydrolysis of esters. Only choice **(C)** shows cleavage of the correct bond and retention of configuration. It is the correct answer.

NC-3. The order from most reactive to least reactive with aqueous ammonia is

(A) $CH_3-C(=O)-O-C(=O)-CH_3 > CH_3-C(=O)-OCH_2CH_3 > CH_3-C(=O)-Cl$

(B) $CH_3-C(=O)-OCH_2CH_3 > CH_3-C(=O)-Cl > CH_3-C(=O)-O-C(=O)-CH_3$

(C) $CH_3-C(=O)-Cl > CH_3-C(=O)-OCH_2CH_3 > CH_3-C(=O)-O-C(=O)-CH_3$

(D) $CH_3-C(=O)-Cl > CH_3-C(=O)-O-C(=O)-CH_3 > CH_3-C(=O)-OCH_2CH_3$

Knowledge Required: The relative reactivity of carboxylic acid derivatives toward nucleophilic acyl substitution.

Thinking it Through: In general, the relative reactivity of the carboxylic acid derivatives follows the order: acyl chloride > acid anhydride > ester > amide. This trend may be predicted from two considerations: (1) the extent of resonance delocalization and stabilization of the starting material; and (2) the basicity of the leaving group from the tetrahedral intermediate. For example, resonance is least important for acyl chlorides, and the Cl^- is the weakest base and thus the best leaving group. Therefore choices **(A)** and **(B)** may be eliminated because the acyl chloride is not listed as the most reactive. Choice **(D)** must be the correct answer, since it has the anhydride listed as being more reactive than the ester.

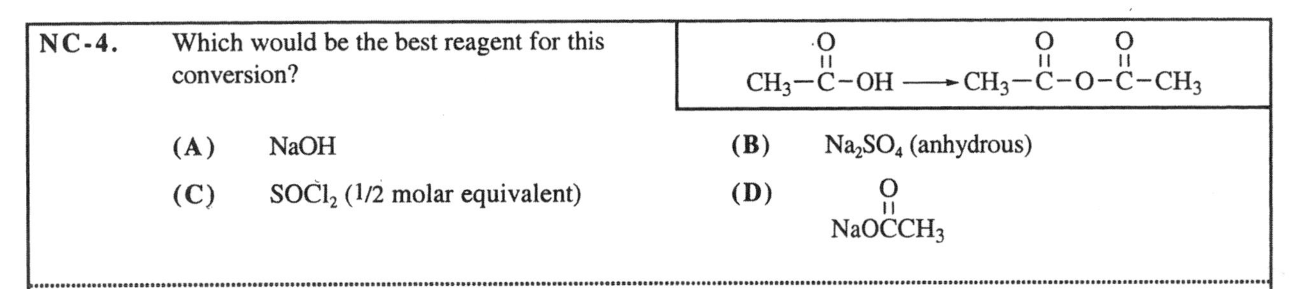

NC-4. Which would be the best reagent for this conversion?

$CH_3-C(=O)-OH \longrightarrow CH_3-C(=O)-O-C(=O)-CH_3$

(A) NaOH

(B) Na_2SO_4 (anhydrous)

(C) $SOCl_2$ (1/2 molar equivalent)

(D) $NaOC(=O)CH_3$

Knowledge Required: (1) The relative reactivity of carboxylic acid derivatives. (2) Products of the reaction of carboxylic acids with thionyl chloride, $SOCl_2$.

Thinking it Through: Primarily because of the reactivity of acyl chlorides, their preparation from carboxylic acids and $SOCl_2$ is a necessary precursor to deal with many synthesis problems. Choice **(A)** is a strong base and would simply deprotonate the acid. The carboxylate anion does not undergo nucleophilic substitution, so **(A)** is eliminated. Choice **(B)** is a drying agent for small amounts of water, but cannot "dehydrate" the carboxylic acid. This eliminates **(B)**. In choice **(D)**, the acetate ion cannot displace the stronger base, OH^-, from the carboxylic acid. This leaves choice **(C)** as the correct answer. In choice **(C)**, conversion of half of the acid to the acid chloride allows the other half to provide the acetate nucleophile, and this is a reasonable synthesis of the anhydride.

NC-5. What is the best way to perform this transformation?

$CH_3-C(=O)-OCH_3 \xrightarrow{?} CH_3-C(=O)-OCH_2CH_3$

(A) $\xrightarrow[CH_3CH_2OH]{CH_3CH_2OK}$

(B) $\xrightarrow[CH_3OH]{CH_3OK} \xrightarrow{CH_3I}$

(C) $\xrightarrow{CH_3CH_2MgBr} \xrightarrow{H_2O,\ H^+}$

(D) $\xrightarrow{Br_2,\ h\nu} \xrightarrow[ether]{Mg} \xrightarrow{CH_3I}$

Knowledge Required: (1) The mechanism of nucleophilic acyl substitution. (2) The reaction of esters with Grignard reagents. (3) The acidic nature of hydrogen atoms on a carbon atom alpha to a carbonyl group.

Thinking it Through: The transesterification is the result of nucleophilic acyl substitution of the CH_3O^- by a $CH_3CH_2O^-$. This is exactly the reagent encountered first in **(A)**. The other choices are eliminated. Choice **(B)** treats the ester with the strong base, CH_3O^-. Deprotonation of the α–carbon atom would lead to a Claisen condensation. The Grignard reagent in choice **(C)** provides a $CH_3CH_2^-$ nucleophile, which would attack the carbonyl carbon atom. Choice **(D)** is an imaginative sequence with several flaws. The free radical bromination would give a mixture of products. Formation of a Grignard would be futile since Grignard reagents react with esters. And finally, Grignard reagents do not couple efficiently with alkyl halides. Choice **(B)** suggests that methoxide ion will deprotonate the methoxy methyl group and the resulting carbanion will then be alkylated with CH_3I. However, any deprotonation would occur at the much more acidic methyl group that is bonded to the carbonyl group.

NC-6. Other than ethanol, what is the major product of this reaction sequence?

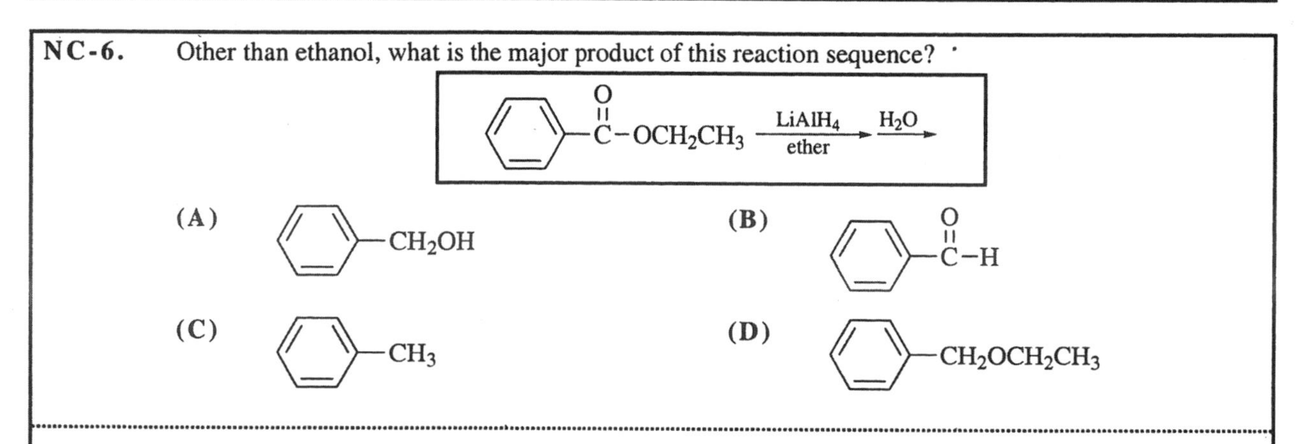

Knowledge Required: Hydride reduction of carboxylic acids and derivatives.

Thinking it Through: Recognizing $LiAlH_4$ as a source of the nucleophilic hydride ion, H:⁻, allows this reaction to be analyzed. Nucleophilic attack by hydride ion produces a tetrahedral intermediate, which can then collapse with the dissociation of the less basic group originally on the carbonyl carbon atom. The resulting aldehyde is then reduced further to a primary alcohol. The correct answer is **(A)**. The initially formed aldehyde is choice **(B)**. Choices **(C)** and **(D)** would require an S_N2 displacement of an ^-OH group, which is not observed.

NC-7. What is the product of this reaction?

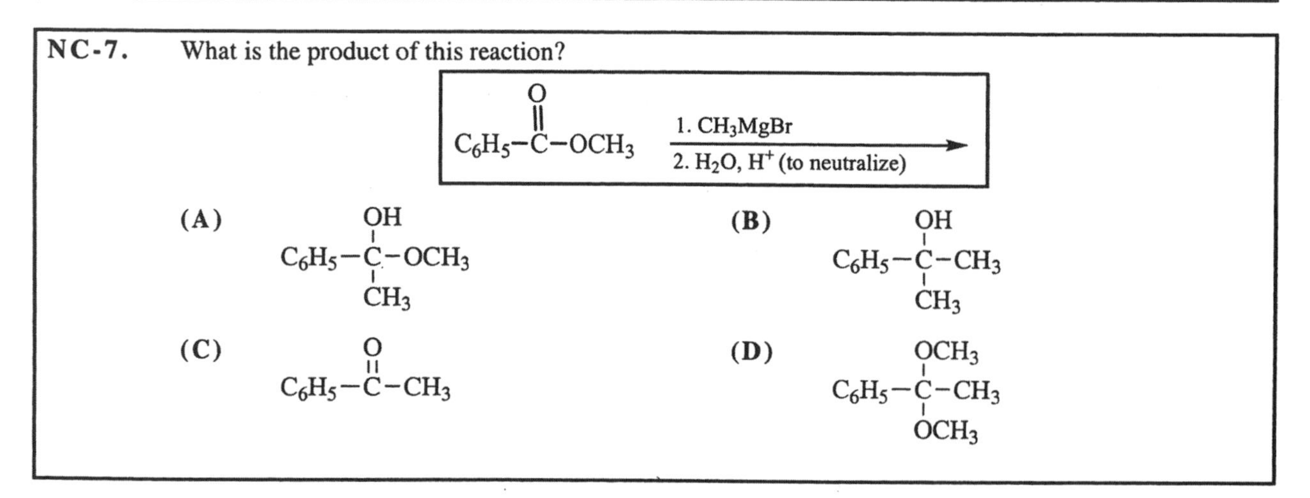

Knowledge Required: (1) Reactions of organometallic reagents with carboxylic acids and derivatives. (2) Relative reactivity of organometallic reagents.

Thinking it Through: The common organometallic reagents can be viewed as containing a nucleophilic carbon atom. Nucleophilic attack on the carbonyl group produces a tetrahedral intermediate that can collapse with the dissociation of the less basic group originally on the carbonyl carbon atom. Depending on the reactivity of the organometallic reagent, the resulting carbonyl group may then be attacked again, leading to an alcohol after hydrolysis. The reaction of esters with Grignard reagents and organolithium reagents are particularly useful. The product is a tertiary alcohol, with two identical alkyl groups attached to the carbon atom bearing the hydroxyl group. A tertiary alcohol is produced because the initially formed ketone is more reactive than the ester and reacts with additional reagent as fast as it is formed. Regardless of the amount of Grignard or organolithium reagent used, a tertiary alcohol is produced; however, two moles of reagent per mole of ester must be used to obtain good yields. The less reactive organocuprate reagents, R_2CuLi, react only with acyl halides, and not with the resulting ketone. The reaction of acid chlorides with organocuprate reagents is thus a useful synthesis for ketones. Choices **(A)**, **(B)**, and **(C)** could all be produced by the reaction of a Grignard reagent with an ester, but choices **(A)** and **(C)** are intermediates that will react with additional Grignard reagent to produce the tertiary alcohol in choice **(B)**, which is the correct answer. Choice **(D)** cannot be synthesized with these reactants.

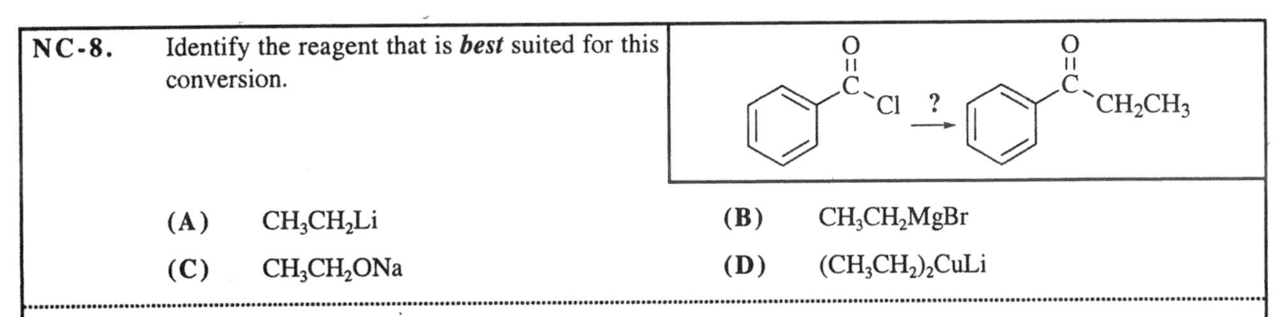

NC-8. Identify the reagent that is ***best*** suited for this conversion.

(A) CH_3CH_2Li **(B)** CH_3CH_2MgBr

(C) CH_3CH_2ONa **(D)** $(CH_3CH_2)_2CuLi$

Knowledge Required: Relative reactivities of organometallic reagents toward carbonyl compounds.

Thinking it Through: A new carbon–carbon bond is being formed in this conversion of an acyl chloride to a ketone, which requires an organometallic reagent, Choice **(C)** may be eliminated because it is an alkoxide, not an organometallic reagent, and would produce an ester. The remaining three choices will all react to produce the ketone. However, the alkyllithium reagent in choice **(A)** and the Grignard reagent in choice **(B)** will subsequently react with the ketone group to yield the corresponding tertiary alcohol. This leaves choice **(D)** as the correct answer. Lithium dialkylcuprates are less reactive than alkyllithium and Grignard reagents, and will not continue to react with the ketone product.

NC-9. What is the order from most reactive to least reactive for these esters towards aqueous hydroxide?

$CH_3-C(=O)-OC(CH_3)_3$	$CH_3-C(=O)-OCH_3$	$CF_3-C(=O)-OCH_3$
1	2	3

(A) 3 > 2 > 1 **(B)** 1 > 2 > 3 **(C)** 3 > 1 > 2 **(D)** 2 > 3 > 1

Knowledge Required: (1) Mechanism of ester hydrolysis. (2) Electronic and steric effects on hydrolysis rates.

Thinking it Through: All three esters undergo hydrolysis by the same mechanism: addition of hydroxide ion to the carbonyl carbon atom followed by expulsion of the alkoxide. Electron-withdrawing groups in the ester will increase the partial positive charge on the carbonyl carbon atom, making it more susceptible to nucleophilic attack. On the other hand, bulky groups in the ester will hinder nucleophilic attack on the carbonyl carbon atom, because the large groups are forced closer together in the tetrahedral intermediate. Inspection of the three esters reveals that the main difference between esters **2** and **3** is that ester **3** has three electron-withdrawing fluorine atoms. The fact that ester **3** should hydrolyze faster than ester **2** eliminates choices **(B)** and **(D)**. Ester **1** contains a bulky tertiary butyl group and should hydrolyze more slowly than ester **2**. This leaves choice **(A)** as the correct answer.

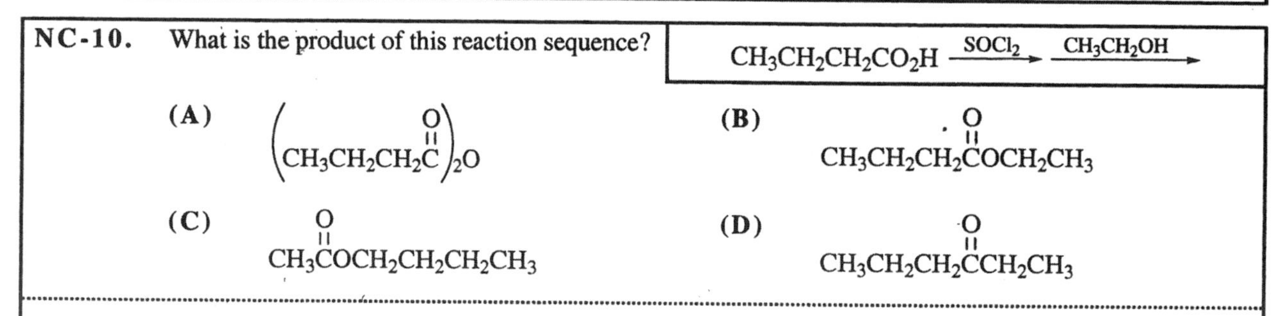

Knowledge Required: (1) Preparation of acyl chlorides from carboxylic acids. (2) Reactions of acyl chlorides with nucleophiles.

Thinking it Through: Thionyl chloride converts the carboxylic acids into the corresponding acyl chloride. The acyl chloride then reacts with the ethanol to form the corresponding ester. Choices **(A)** and **(D)**, which are not esters, may be eliminated. Both choices **(B)** and **(C)** are esters, but choice **(C)** has the wrong number of carbon atoms in the alcohol and carboxylic acid portions. Choice **(B)** is the correct answer.

Practice Questions

1. This reaction that is typical of carboxylic acids, esters, acid halides, anhydrides, and amides is called

$$R-\overset{O}{\overset{\|}{C}}-Cl + H_2O \longrightarrow R-\overset{O}{\overset{\|}{C}}-OH + HCl$$

(A) nucleophilic non-acyl substitution

(B) nucleophilic addition

(C) nucleophilic acyl substitution

(D) electrophilic substitution

2. Which structure is a reasonable intermediate in the acid-catalyzed hydrolysis of ethyl acetate in dilute aqueous acid?

(A) $CH_3-C(O^-)(OH)-OCH_2CH_3$

(B) $CH_3-C(OH)(^+OH_2)-OCH_2CH_3$

(C) $CH_3-C(O^+)(H)-OCH_2CH_3$

(D) $CH_3-\overset{O}{\overset{\|}{C}}{}^+$

3. Which would be hydrolyzed ***most slowly*** with aqueous NaOH?

(A) $CH_3-\overset{O}{\overset{||}{C}}-NHCH_3$

(B) $CH_3-\overset{O}{\overset{||}{C}}-OCH_3$

(C) $CH_3-\overset{O}{\overset{||}{C}}-O-\overset{O}{\overset{||}{C}}-CH_3$

(D) $CH_3-\overset{O}{\overset{||}{C}}-Cl$

4. Using your knowledge of base strengths, predict which of these reactions would be most likely to occur.

(A) $R\overset{O}{\overset{||}{C}}OCH_3 + R\overset{O}{\overset{||}{C}}O^- \rightarrow R\overset{O}{\overset{||}{C}}O\overset{O}{\overset{||}{C}}R + {}^-OCH_3$

(B) $R\overset{O}{\overset{||}{C}}OH + Cl^- \rightarrow R\overset{O}{\overset{||}{C}}Cl + {}^-OH$

(C) $R\overset{O}{\overset{||}{C}}O-C_6H_5 + {}^-OCH_3 \rightarrow R\overset{O}{\overset{||}{C}}OCH_3 + {}^-O-C_6H_5$

(D) $R\overset{O}{\overset{||}{C}}NH_2 + CH_3SH \rightarrow R\overset{O}{\overset{||}{C}}SCH_3 + NH_3$

5. Which intermediate is involved in the mechanism of this base-promoted hydrolysis?

$C_6H_5-\overset{O}{\overset{||}{C}}-OCH_2CH_3 \xrightarrow[H_2O]{^-OH}$

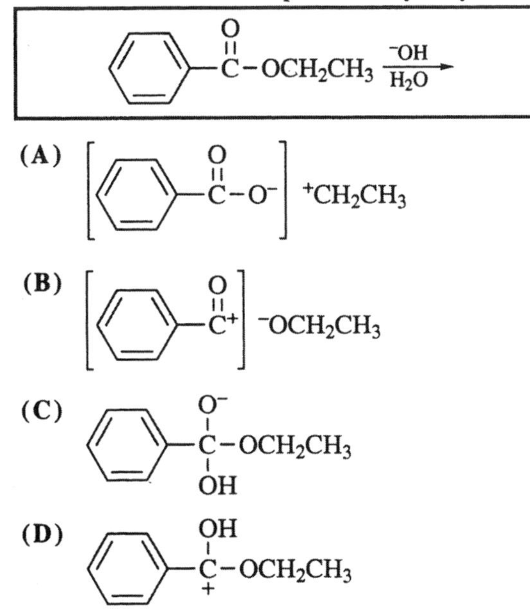

6. With which reagent is benzoic acid readily converted into benzoyl chloride in high yield?

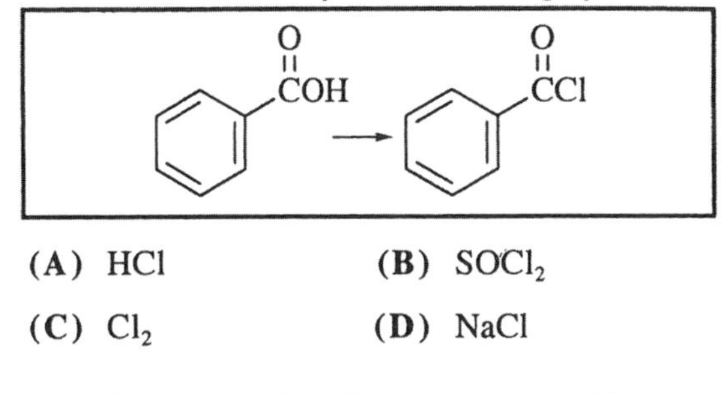

(A) HCl (B) $SOCl_2$

(C) Cl_2 (D) NaCl

7. Which set of products is obtained from this acid hydrolysis?

$CH_3\underset{NH_2}{\underset{|}{C}}H\overset{O}{\overset{||}{C}}-NHCH_2\overset{O}{\overset{||}{C}}OH \xrightarrow[H^+]{H_2O}$

(A) $CH_3CH_2\overset{+}{N}H_3 + HO\overset{O}{\overset{||}{C}}NH_2 + CH_3\overset{O}{\overset{||}{C}}OH$

(B) $CH_3\underset{^+NH_3}{\underset{|}{C}}H\overset{O}{\overset{||}{C}}NH_2 + CH_3\overset{O}{\overset{||}{C}}OH$

(C) $CH_3\underset{^+NH_3}{\underset{|}{C}}H\overset{O}{\overset{||}{C}}OH + H_3\overset{+}{N}CH_2\overset{O}{\overset{||}{C}}OH$

(D) $CH_3CH_2\overset{O}{\overset{||}{C}}OH + CH_3\overset{O}{\overset{||}{C}}OH + 2NH_4^+$

8. What product would you isolate from these reactants?

$CH_3-\overset{O}{\overset{||}{C}}-Cl \xrightarrow{CH_3CH_2NH_2}$

(A) $CH_3-\overset{O}{\overset{||}{C}}-\overset{CH_3}{\overset{|}{C}}HNH_2$

(B) $CH_3-\underset{Cl}{\underset{|}{\overset{OH}{\overset{|}{C}}}}-NHCH_2CH_3$

(C) $CH_3-\overset{NCH_2CH_3}{\overset{||}{C}}-Cl$

(D) $CH_3-\overset{O}{\overset{||}{C}}-NHCH_2CH_3$

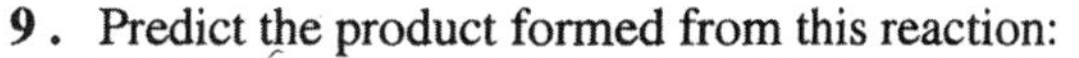

9. Predict the product formed from this reaction:

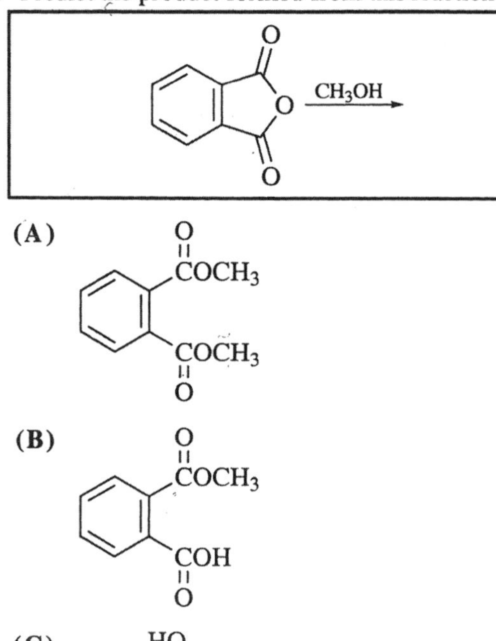

(A)

(B)

(C)

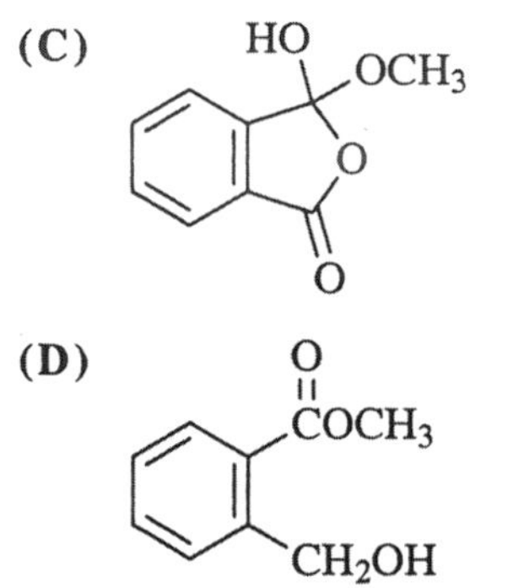

(D)

10. What are the products of this reaction?

$CH_3C(=O)NHCH_2CH_3 \xrightarrow[H_2O]{^-OH}$

(A) $CH_3C(=O)NH_2 + CH_3CH_2OH$

(B) $CH_3OH + NH_3 + CH_3CH_2OH$

(C) $CH_3C(=O)O^- + CH_3CH_2NH_2$

(D) $CH_3C(=O)NHO^- + CH_3CH_3$

11. Which of these reactions would produce this product?

$CH_3-C(=O)-NHCH_3$

(A) $CH_3-C(=O)-OCH_3 \xrightarrow{NH_3}$

(B) $CH_3-C(=O)-O-C(=O)-CH_3 \xrightarrow{CH_3NH_2}$

(C) $CH_3-C(=O)-NH_2 \xrightarrow[H^+]{CH_3OH}$

(D) $CH_3-C(=O)-ONa \xrightarrow{CH_3NH_2}$

12. What is the product of this reaction?

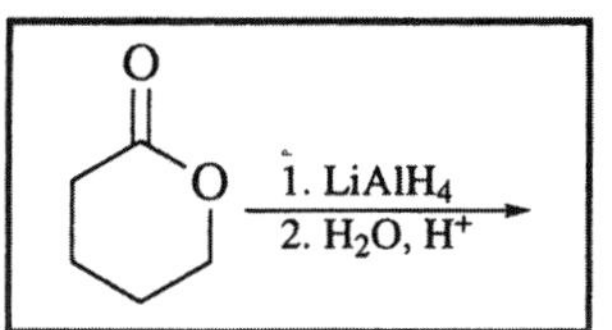

(A)

(B) $HOCH_2CH_2CH_2CH_2CH_3$

(C) $HC(=O)CH_2CH_2CH_2CH_2OH$

(D) $HOCH_2CH_2CH_2CH_2CH_2OH$

13. Which reaction sequence is preferred for this conversion?

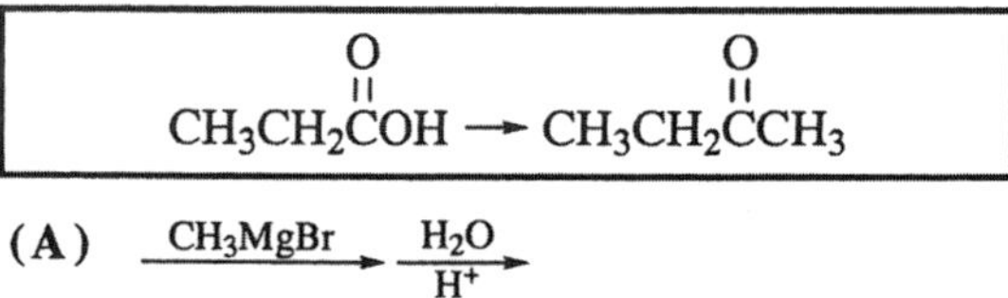

$CH_3CH_2C(=O)OH \rightarrow CH_3CH_2C(=O)CH_3$

(A) $\xrightarrow{CH_3MgBr} \xrightarrow[H^+]{H_2O}$

(B) $\xrightarrow{SOCl_2} \xrightarrow{(CH_3)_2CuLi}$

(C) $\xrightarrow{SOCl_2} \xrightarrow{CH_3Li} \xrightarrow[H^+]{H_2O}$

(D) $\xrightarrow{SOCl_2} \xrightarrow{CH_3MgBr} \xrightarrow[H^+]{H_2O}$

14. What is the product of this Grignard reaction?

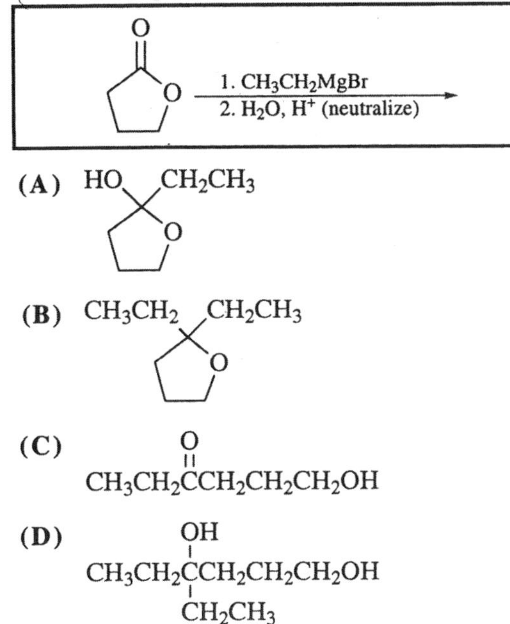

15. Which reaction sequence would ***best*** accomplish this transformation?

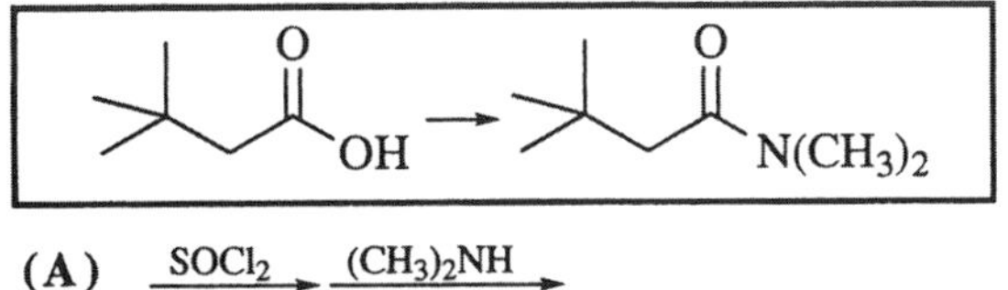

(A) $\xrightarrow{SOCl_2}\ \xrightarrow{(CH_3)_2NH}$

(B) $\xrightarrow{NH_3}\ \xrightarrow{CH_3I\ (excess)}$

(C) $\xrightarrow{NaBH_4}\ \xrightarrow{(CH_3)_2NH}$

(D) $\xrightarrow{LiN(CH_3)_2}$

16. What is the product of this reduction?

$$CH_3CH_2CH(CH_3){-}C(=O){-}NHCH_3 \xrightarrow[ether]{LiAlH_4} \xrightarrow{H_2O}$$

(A) $CH_3CH_2CH(CH_3)NHCH_3$

(B) $CH_3CH_2CH(CH_3)CH_2OH$

(C) $CH_3CH_2CH(CH_3)CH{=}NCH_3$

(D) $CH_3CH_2CH(CH_3)CH_2NHCH_3$

17. What product is formed from this reaction sequence?

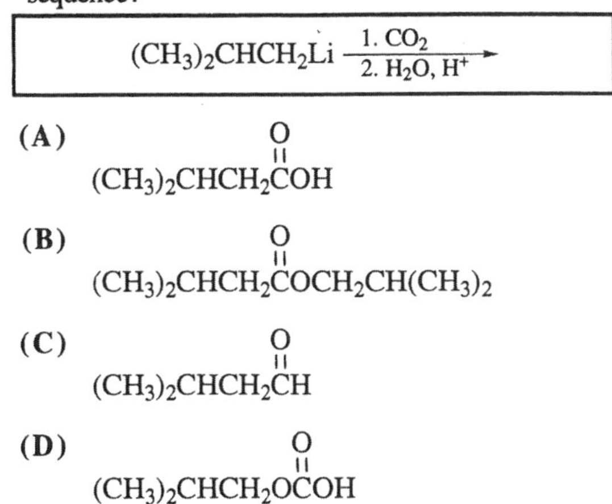

18. Which of these polymers could be readily prepared by condensation polymerization?

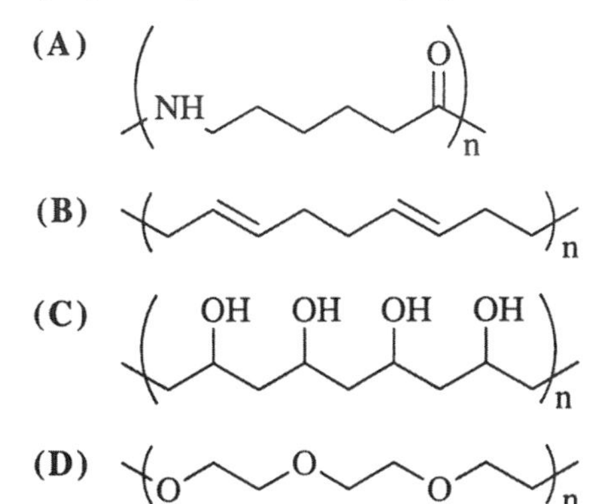

19. Predict the product from this reaction sequence.

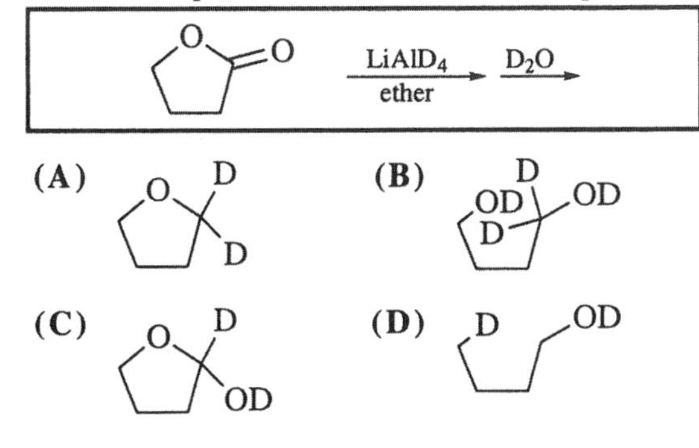

20. What monomers are needed to produce this polymer?

$$\left(\overset{O}{\overset{\|}{C}}CH_2CH_2\overset{O}{\overset{\|}{C}}OCH_2CH_2O \right)_n$$

(A) $HOCH_2CH_2\overset{O}{\overset{\|}{C}}OH + HOCH_2CH_2\overset{O}{\overset{\|}{C}}OH$

(B) $ClCH_2COCH_2Cl + HOCH_2CH_2\overset{O}{\overset{\|}{C}}OCH_2CH_2OH$

(C) $HO\overset{O}{\overset{\|}{C}}-\overset{O}{\overset{\|}{C}}OH + HOCH_2CH_2CH_2CH_2OH$

(D) $HO\overset{O}{\overset{\|}{C}}CH_2CH_2\overset{O}{\overset{\|}{C}}OH + HOCH_2CH_2OH$

21. What is the product of this reaction?

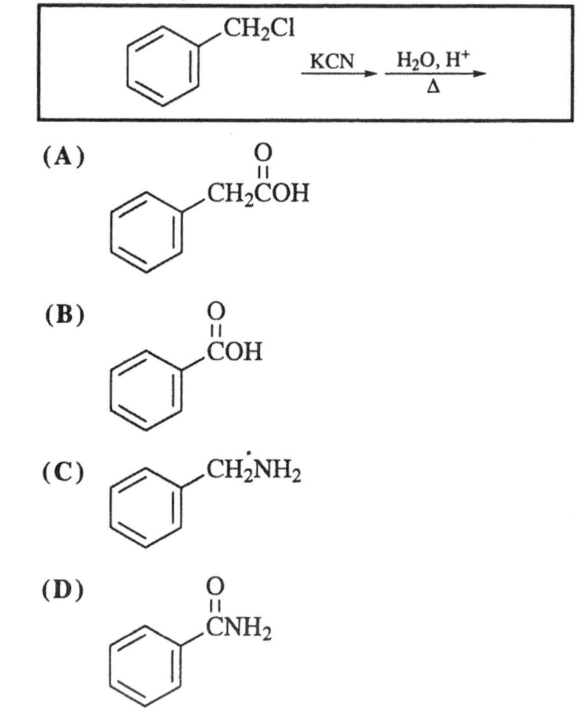

Answers to Study Questions

1. B
2. C
3. D
4. C
5. A
6. A
7. B
8. D
9. A
10. B

Answers to Practice Questions

1. C
2. B
3. A
4. C
5. C
6. B
7. C
8. D
9. B
10. C
11. B
12. D
13. B
14. D
15. A
16. D
17. A
18. A
19. B
20. D
21. A

Enols and Enolate Ion Reactions

All carbonyl compounds with at least one α–hydrogen atom are in equilibrium with their respective enol tautomers. At equilibrium the amount of enol tautomer is vanishingly small with the exception of β–dicarbonyl compounds, for which the enol tautomer may even be the dominant tautomer. The rate of interconversion of the tautomers is greatly accelerated by the presence of a trace of either acid or base.

There are two "hot spots" for reactivity in aldehydes and ketones: (1) The carbonyl carbon atom is subject to addition reactions by nucleophiles (equation *1*); (2) the α–carbon atom can be deprotonated, converting it into a nucleophile. A hydrogen atom on an sp^3 carbon atom α to a carbonyl group is thirty orders of magnitude more acidic than a hydrogen atom on an sp^3 carbon atom of an ordinary hydrocarbon! Resonance, of course, provides the explanation (equation 2).

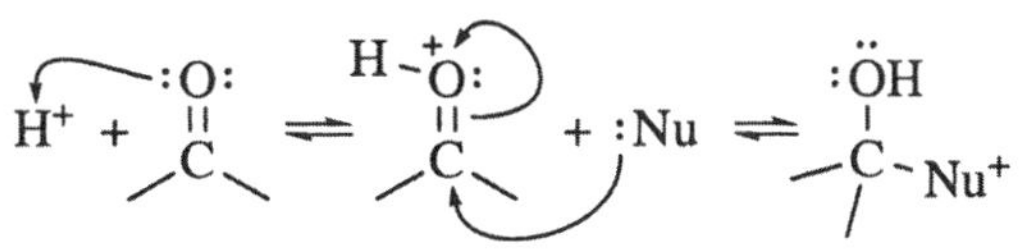

$$H^+ + \mathrm{{>}C{=}\ddot{O}\colon} \rightleftharpoons \mathrm{{>}C{=}\overset{+}{O}{-}H} + \mathrm{\colon Nu} \rightleftharpoons \mathrm{{-}\overset{|}{\underset{|}{C}}(\ddot{O}H){-}Nu^+} \qquad (1)$$

$$\mathrm{{>}C({=}O){-}\overset{|}{C}{-}H} + {}^-OH \rightleftharpoons \left[\mathrm{{>}C({=}O){-}\ddot{C}^-{<}} \longleftrightarrow \mathrm{{>}C(O^-){=}C{<}}\right] \qquad (2)$$

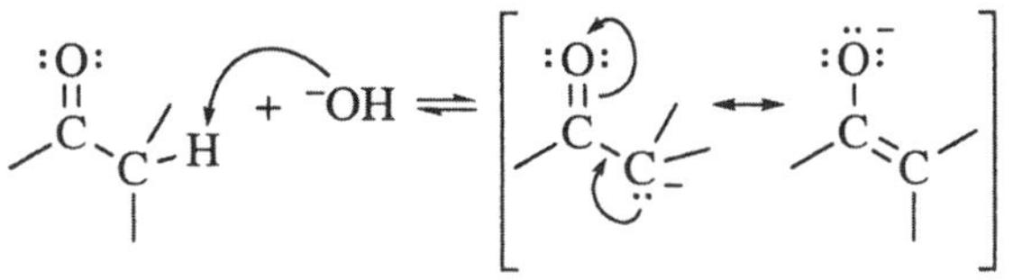

Bases such as alkoxide ions produce only small amounts of the enolates of simple carbonyl compounds, but bases that are significantly stronger than alkoxide ions quantitatively convert simple carbonyl compounds into their enolates. LDA, $[(CH_3)_2CH]_2N^- Li^+$, and sodium hydride, NaH, are examples of strong bases that produce enolates.

A nucleophilic α–carbon atom can be halogenated or alkylated. A halogenation is shown in equation *3*.

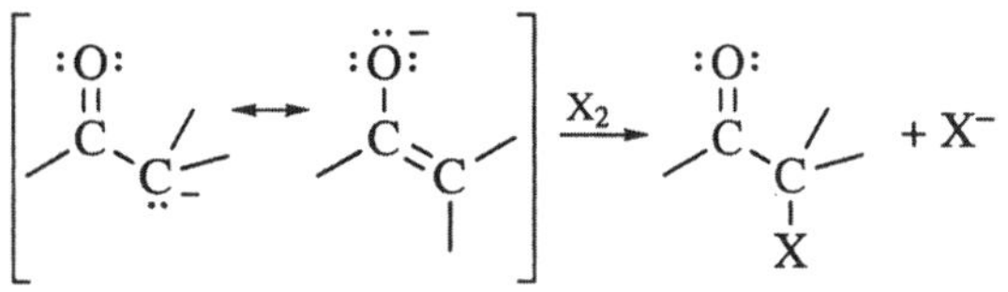

$$\left[\mathrm{{>}C({=}O){-}\ddot{C}^-{<}} \longleftrightarrow \mathrm{{>}C(O^-){=}C{<}}\right] \xrightarrow{X_2} \mathrm{{>}C({=}O){-}\overset{|}{\underset{X}{C}}{<}} + X^- \qquad (3)$$

Many other examples of nucleophilic addition to the carbonyl group are provided in the *Nucleophilic Addition at Carbonyl Groups* chapter. The nucleophilic α–carbon atom of one aldehyde can attack the carbonyl carbon atom of another aldehyde molecule in a dimerization reaction called an aldol reaction (or aldol condensation if an α,β–unsaturated carbonyl compound is the product). The mechanism of the aldol condensation is shown in equations *4* through *6*. As in the general nucleophilic additions to the carbonyl group, aldehydes are more reactive than ketones for both steric and electronic reasons.

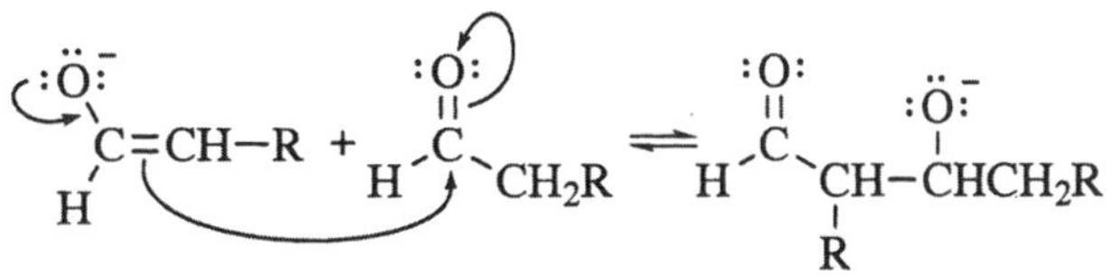

$$\mathrm{H(^-\colon\ddot{O}\colon)C{=}CH{-}R} + \mathrm{H{-}C({=}O){-}CH_2R} \rightleftharpoons \mathrm{H{-}C({=}O){-}CH(R){-}CH(\ddot{O}^-)CH_2R} \qquad (4)$$

$$\mathrm{H{-}C({=}O){-}CH(R){-}CH(\ddot{O}^-)CH_2R} \xrightleftharpoons{H_2O} \mathrm{H{-}C({=}O){-}CH(R){-}CH(\ddot{O}H)CH_2R} + {}^-OH \qquad (5)$$

$$\mathrm{H{-}C({=}O){-}CH(R){-}CH(\ddot{O}H)CH_2R} \xrightleftharpoons{^-OH} \mathrm{H{-}C({=}O){-}C(R){=}CHCH_2R} + H_2O \qquad (6)$$

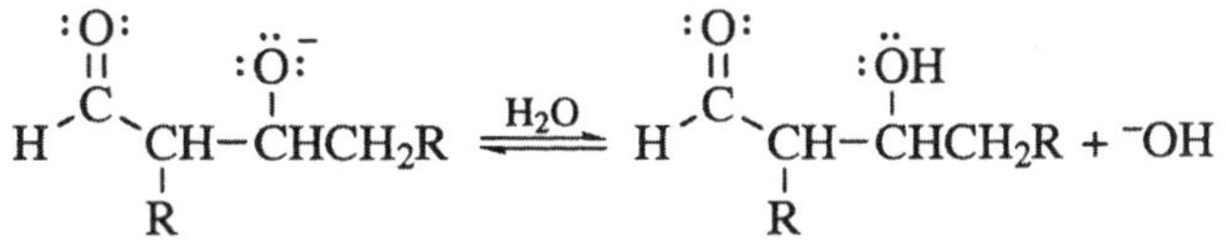

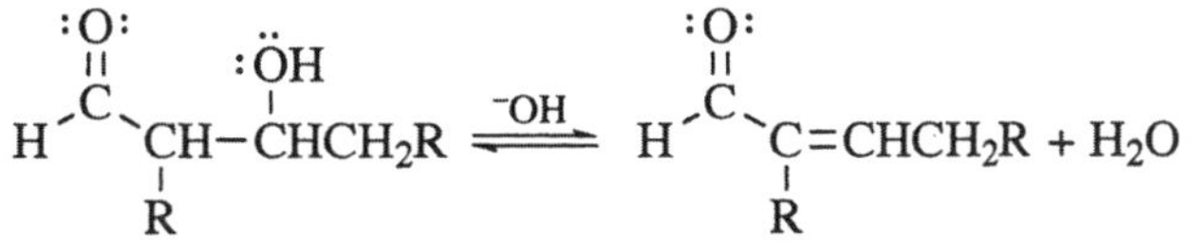

If the carbonyl-containing compound is an ester, the alcohol portion can be displaced in a nucleophilic substitution reaction by the α–carbon atom in a reaction called the Claisen condensation.

There are many different combinations of aldehydes, ketones, esters, nitriles, nitroalkanes, and β–dicarbonyl compounds that undergo aldol-type or Claisen-type reactions. When confronted with a combination not previously seen, one can usually predict the correct product this way: First, write the formula for the anion of the component with the most acidic α–hydrogen atom. Then, write the formula for the product that would form following attack on the carbonyl group of the other component. This second step can be either an addition reaction or a substitution reaction.

Although less nucleophilic than enolate ions, enols may act as nucleophiles towards very reactive electrophiles such as halogens or protonated carbonyl groups.

Study Questions

EE-1. Consider the hydrogen atoms attached to three different carbon atoms (labeled **1**, **2**, and **3**). Rank the attached hydrogen atoms in order from most acidic to least acidic.

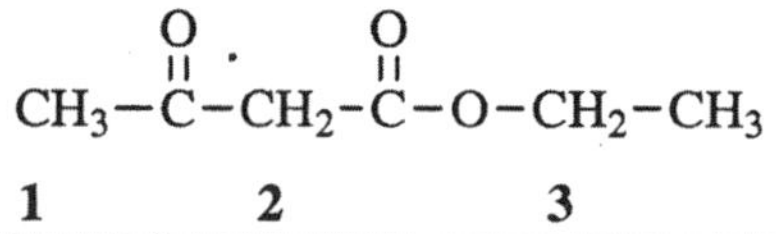

$$\underset{1}{CH_3}-\overset{O}{\overset{\|}{C}}-\underset{2}{CH_2}-\overset{O}{\overset{\|}{C}}-O-CH_2-\underset{3}{CH_3}$$

(A) 1 > 2 > 3 **(B)** 2 > 3 > 1 **(C)** 3 > 2 > 1 **(D)** 2 > 1 > 3

Knowledge Required: Resonance stabilization of enolate anions.

Thinking it Through: The acidities of each of the three types of hydrogen atoms depend on the relative stabilities of the conjugate bases that result from removal of the proton. The stability of each conjugate base depends on the number of the resonance structures that can be drawn for it, and on the stability of each resonance structure.

Removing a proton from carbon atom **1** yields a conjugate base with two resonance structures.

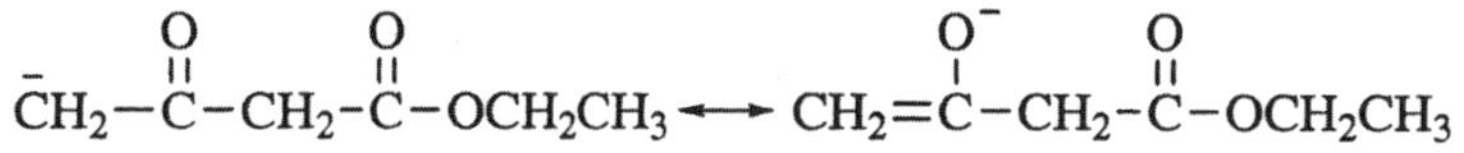

$$\overset{-}{C}H_2-\overset{O}{\overset{\|}{C}}-CH_2-\overset{O}{\overset{\|}{C}}-OCH_2CH_3 \longleftrightarrow CH_2=\overset{O^-}{\overset{|}{C}}-CH_2-\overset{O}{\overset{\|}{C}}-OCH_2CH_3$$

Removing a proton from carbon atom **2** yields a conjugate base with three resonance structures, so this hydrogen atom is more acidic than the hydrogen atom attached to carbon atom **1**. Thus, choice **(A)** may be eliminated.

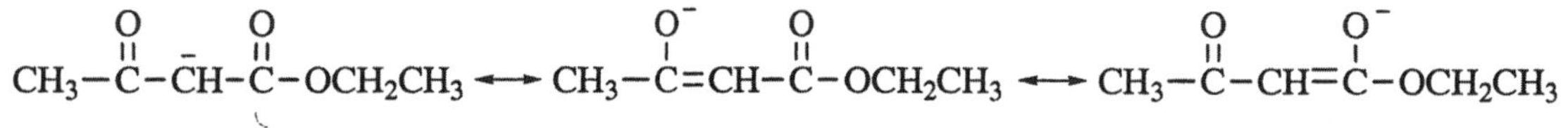

$$CH_3-\overset{O}{\overset{\|}{C}}-\overset{-}{C}H-\overset{O}{\overset{\|}{C}}-OCH_2CH_3 \longleftrightarrow CH_3-\overset{O^-}{\overset{|}{C}}=CH-\overset{O}{\overset{\|}{C}}-OCH_2CH_3 \longleftrightarrow CH_3-\overset{O}{\overset{\|}{C}}-CH=\overset{O^-}{\overset{|}{C}}-OCH_2CH_3$$

The conjugate base formed by removing a proton from carbon atom **3** has no resonance structures, so it must be the least acidic of the three. This eliminates choices **(B)** and **(C)**, leaving choice **(D)** as the correct answer.

EE-2. What would be the major product of this reaction?

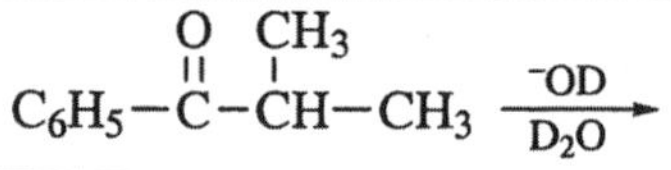

$$C_6H_5-\overset{O}{\overset{\|}{C}}-\overset{CH_3}{\overset{|}{C}H}-CH_3 \xrightarrow[D_2O]{^-OD}$$

(A) $C_6H_5-\overset{O}{\overset{\|}{C}}-\overset{CD_3}{\overset{|}{C}H}-CD_3$

(B) $C_6H_5-\overset{O}{\overset{\|}{C}}-\overset{CH_3}{\overset{|}{C}D}-CH_3$

(C) $C_6D_5-\overset{O}{\overset{\|}{C}}-\overset{CH_3}{\overset{|}{C}H}-CH_3$

(D) $C_6H_5-\overset{O}{\overset{\|}{C}}-\overset{CD_3}{\overset{|}{C}D}-CD_3$

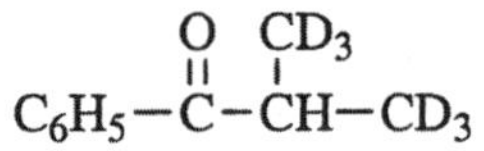

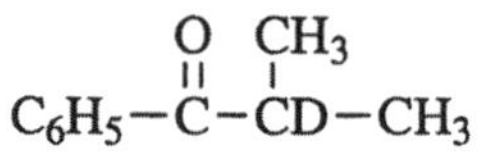

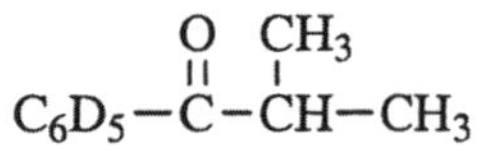

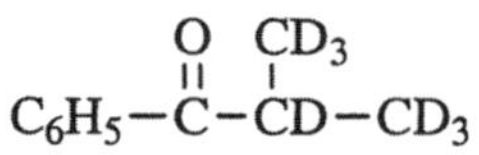

Knowledge Required: (1) The identity and acidity of hydrogen atoms α to a carbonyl group. (2) Ability of α–hydrogen atoms to undergo deuterium exchange with deuterated solvents.

Thinking it Through: In the substrate there is only one α–hydrogen atom. This proton can be removed by the base, ^{-}OD, leaving the resonance-stabilized enolate ion.

$$C_6H_5-\overset{O}{\overset{\|}{C}}-\overset{CH_3}{\overset{|}{C}H}-CH_3 \xrightarrow{^{-}OD} C_6H_5-\overset{O}{\overset{\|}{C}}-\overset{CH_3}{\overset{|}{\underset{-}{C}}}-CH_3 \xrightarrow{D_2O} C_6H_5-\overset{O}{\overset{\|}{C}}-\overset{CH_3}{\overset{|}{C}D}-CH_3$$

Deuteration of the negatively charged α–carbon atom can only be accomplished by deuterium in the D_2O medium. Choice **(B)** is the only possible correct answer, because it is the one choice that shows replacement of only the α–hydrogen atom with a deuterium atom.

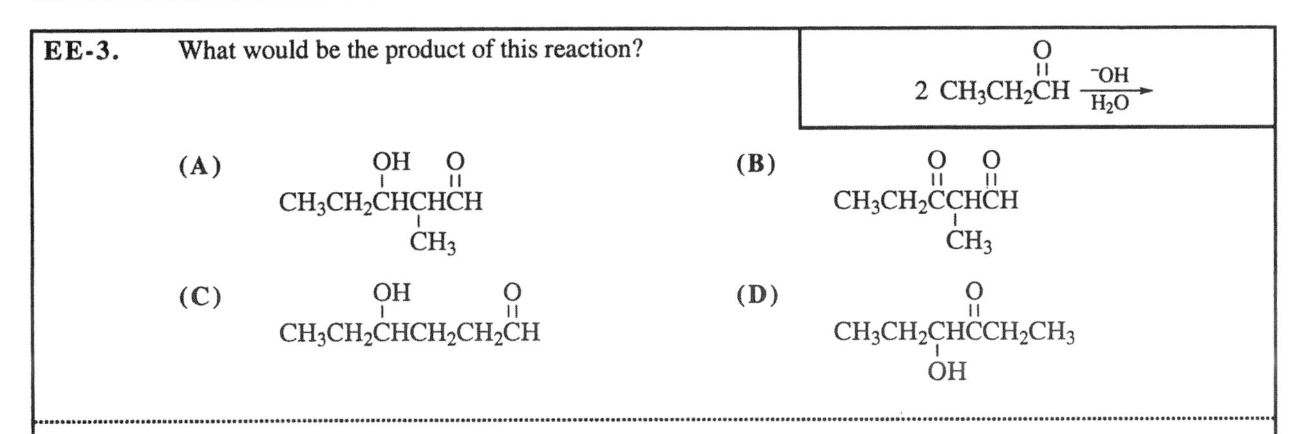

Knowledge Required: (1) How to identify the most acidic hydrogen atoms. (2) The mechanism of the aldol reaction.

Thinking it Through: First, an enolate ion is formed by abstraction an α–hydrogen atom, and then the enolate ion attacks the carbonyl carbon atom of another aldehyde to form a new carbon–carbon bond. Protonation of the new oxygen anion by water gives the aldol product. This sequence will yield the atom connectivity in choice **(A)**, which is the correct answer. Choice **(B)**, a β–keto aldehyde, is incorrect because it would require the initial enolate ion to displace the strongly basic hydride ion from the carbonyl carbon atom of another aldehyde in a substitution reaction. At first glance, Choice **(C)** may appear reasonable because it contains an aldehyde and an alcohol function; but, its formation would require the abstraction of a β–hydrogen atom instead of an α–hydrogen atom. Furthermore, an aldol condensation yields a β-hydroxy carbonyl compound. The structure shown is a γ-hydroxy carbonyl compound. Choice **(D)** would require the abstraction of the hydrogen atom attached to the carbonyl carbon atom to produce an anion, which would then react with another aldehyde molecule. Choices **(B)**, **(C)** and **(D)** are all commonly chosen wrong answers for this type of question.

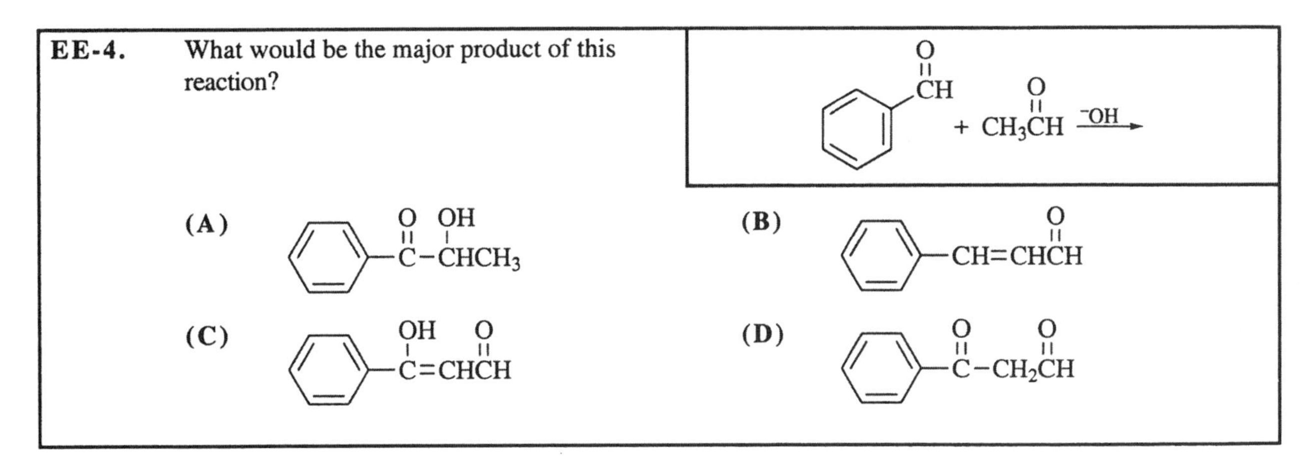

Knowledge Required: (1) The identity of the α–hydrogen atom. (2) The mechanism of the aldol reaction.

Thinking it Through: In the present case, only the acetaldehyde has an α–hydrogen atom, so it forms the enolate ion. Then the enolate ion adds to the carbonyl group of the benzaldehyde, producing the aldol product shown below. This aldol product does not match any of the choices, so something more must be involved. Recall that aldol products may undergo base-catalyzed dehydration to give α,β–unsaturated carbonyl compounds if left in contact with base, especially at elevated temperature.

$$C_6H_5CH(OH)CH_2CH{=}O \xrightarrow{^{-}OH} C_6H_5CH{=}CHCH{=}O$$

This dehydration occurs spontaneously even at normal temperatures when the new double bond extends the conjugation of another π system (a benzene ring in this instance) with the carbonyl group. The dehydrated aldol shown here matches the structure in choice **(B)**, and is the correct answer. Choice **(A)** would require the abstraction of the hydrogen atom attached to the carbonyl carbon atom of the benzaldehyde to produce an anion that would then add to the acetaldehyde carbonyl group. However, this hydrogen atom is much less acidic than an α–hydrogen atom, eliminating choice **(A)**. Choice **(D)** would require the enolate ion to displace the strongly basic hydride ion from the benzaldehyde carbonyl group in a substitution reaction, so it is eliminated. Choice **(C)** would require the loss of a molecule of hydrogen instead of a molecule of water from the initial aldol adduct, which is not logical.

EE-5. What is the product of this reaction sequence?

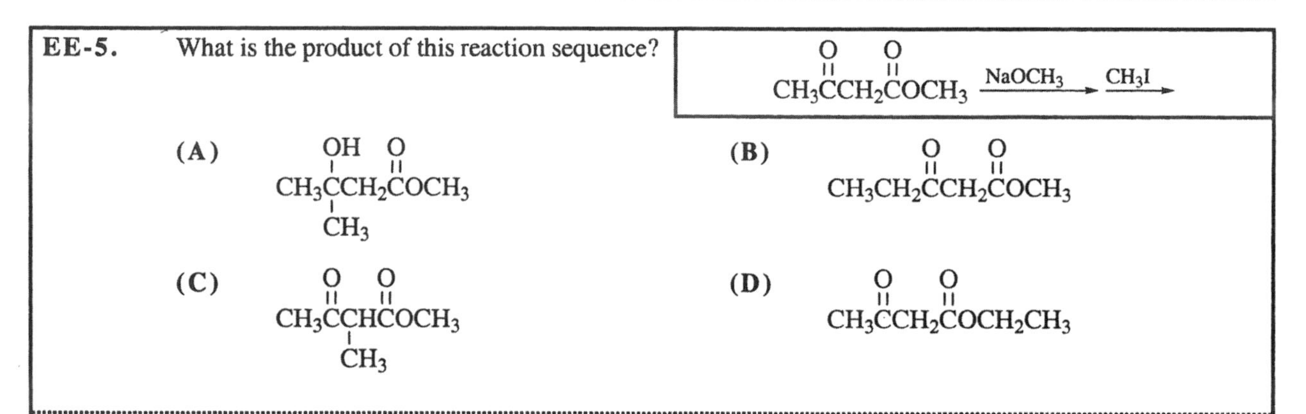

Knowledge Required: (1) Relative acidities of α–hydrogen atoms. (2) Products of alkylation of enolate carbon atoms.

Thinking it Through: As discussed in the first study question, the hydrogen atoms on a carbon atom flanked by two carbonyl groups are the most acidic. One is removed by sodium methoxide to yield the enolate ion shown below. In this resonance structure, the carbon atom is negatively charged and can act as a nucleophile toward the CH_3I added in the second step.

$$CH_3C(=O)\underline{C}HC(=O)OCH_3 \xrightarrow{CH_3I} CH_3C(=O)CH(CH_3)C(=O)OCH_3$$

Choice **(C)** matches the product formed by this sequence, and is the correct answer. Choices **(B)** and **(D)** are eliminated because they would require the formation of much less stable anions, followed by methylation with CH_3I. The product of Choice **(A)** is eliminated because it would require the addition of a nucleophilic methyl anion (some type of a methyl organometallic reagent).

EE-6. What would be the product of this reaction sequence?

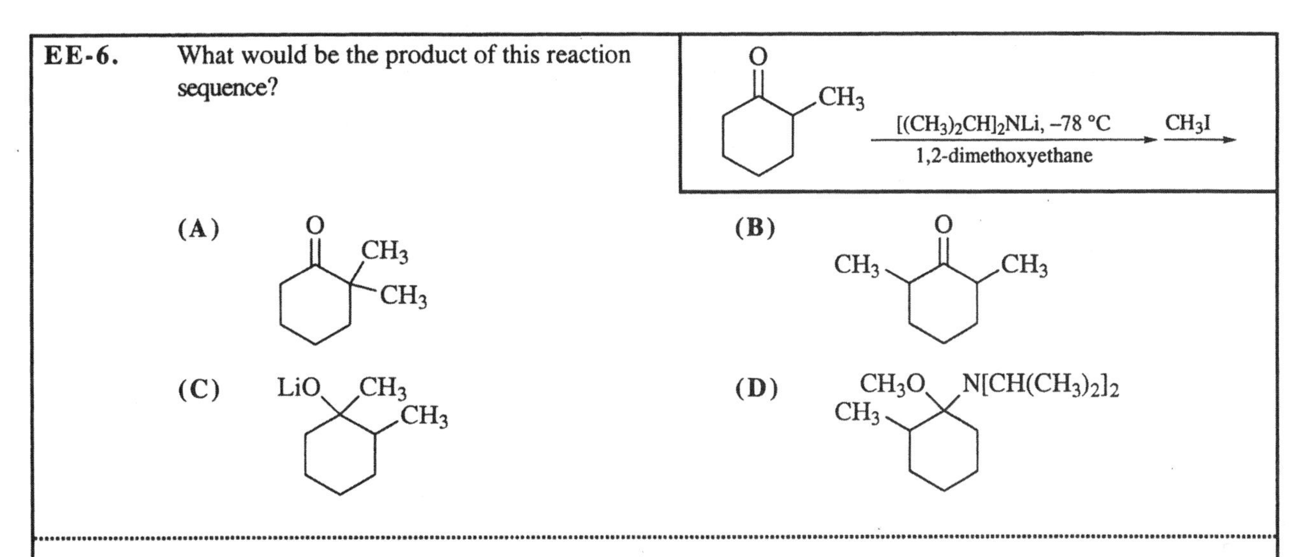

Knowledge Required: (1) Formation of kinetic *vs.* thermodynamic enolate ions. (2) Alkylation of enolate ions with alkyl halides.

Thinking it Through: The strong base lithium diisopropylamide, $[(CH_3)_2CH]_2NLi$, is used to form the enolate of the cyclohexanone reactant quantitatively. Two enolates ions are possible. The one that is formed in this reaction arises from removal of the secondary hydrogen atom, which is more accessible to the bulky diisopropylamide base. This enolate is formed more rapidly, and the reaction is not reversible. Consequently, it is referred to as the *kinetic* enolate ion. The *thermodynamic* enolate ion is more stable as a result of its more highly substituted double bond, and is formed with weaker bases (e.g. alkoxide ions) under conditions where the kinetic enolate is in equilibrium with the starting ketone. The nucleophilic α–carbon atom of the kinetic enolate ion reacts with the methyl group, displacing iodide ion, to form choice **(B)**. Choice **(A)** is eliminated because it would have been the product formed by alkylation of the thermodynamic enolate ion. Choice **(C)** is eliminated because its formation would require the organometallic reagent CH_3Li. Choice **(D)** is unlikely, and is eliminated, because the bulkiness of diisopropylamide makes it a poor nucleophile

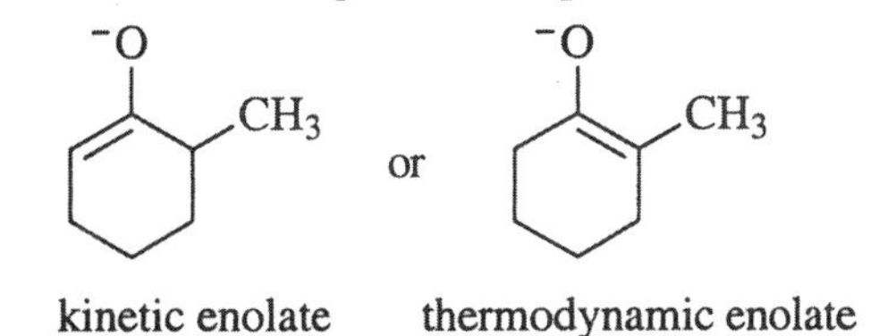

EE-7. What would be the major product of this reaction?

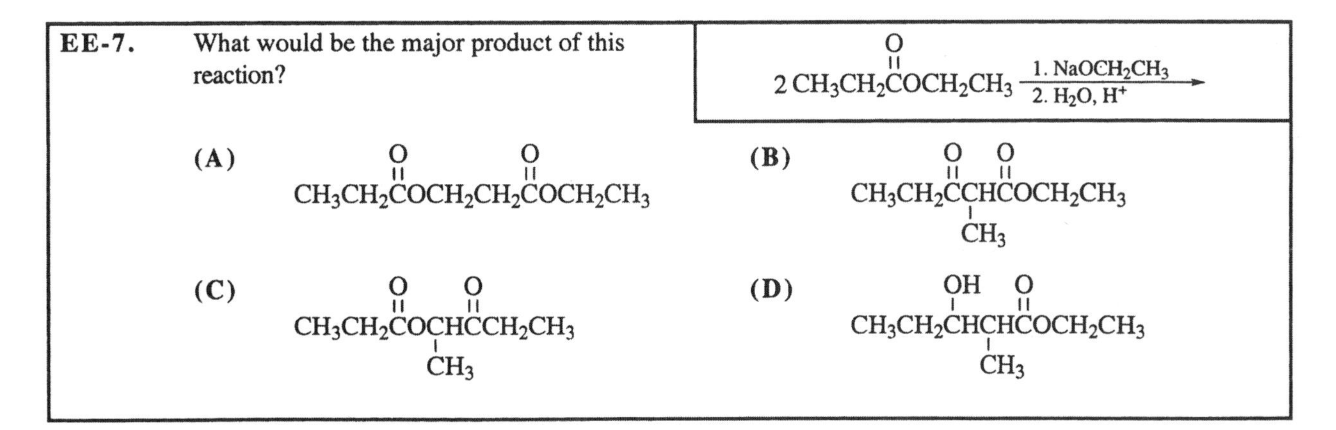

Knowledge Required: (1) Formation of enolate ions from esters that contain α–hydrogen atoms.
(2) Product of reaction between ester enolates ions and the carbonyl group of other esters (Claisen condensation).

Thinking it Through: The ester reacts with $NaOCH_2CH_3$ to produce a small amount of the enolate ion. This enolate ion is nucleophilic, just as is the enolate formed from a ketone or aldehyde, and it adds to the carbonyl group of another ester molecule to give a tetrahedral intermediate. Unlike the tetrahedral intermediates formed from the addition of nucleophiles to ketones or aldehydes, this tetrahedral intermediate has a reasonable leaving group (ethoxide ion) so the carbonyl group reforms, expelling ethoxide ion, to give the Claisen condensation product.

$CH_3CH{=}C(O^-)OCH_2CH_3$ + $CH_3CH_2C(=O)OCH_2CH_3$ → $CH_3CH(C(=O)OCH_2CH_3)$–$C(O^-)(OCH_2CH_3)CH_2CH_3$ → $CH_3CH(C(=O)OCH_2CH_3)$–$C(=O)CH_2CH_3$

The product is a β–keto ester that immediately reacts with the expelled ethoxide ion to form a very stable enolate ion, which must be neutralized with aqueous acid to obtain the β–keto ester itself. The β–keto ester shown above is identical to that in choice **(B)**, and is the correct answer. Choices **(A)** and **(C)** are eliminated because they require the formation of non-stabilized carbanions (on the β–carbon atom for **(A)** and on the ethoxy methylene carbon atom for **(C)**), which then attack the carbonyl group of another ester molecule. Neither carbanion can be generated when sodium ethoxide is the base. Choice **(D)** is illogical because it is the reduction product of choice **(B)**, but there is no reducing agent indicated.

EE-8. What is the expected product of this reaction?

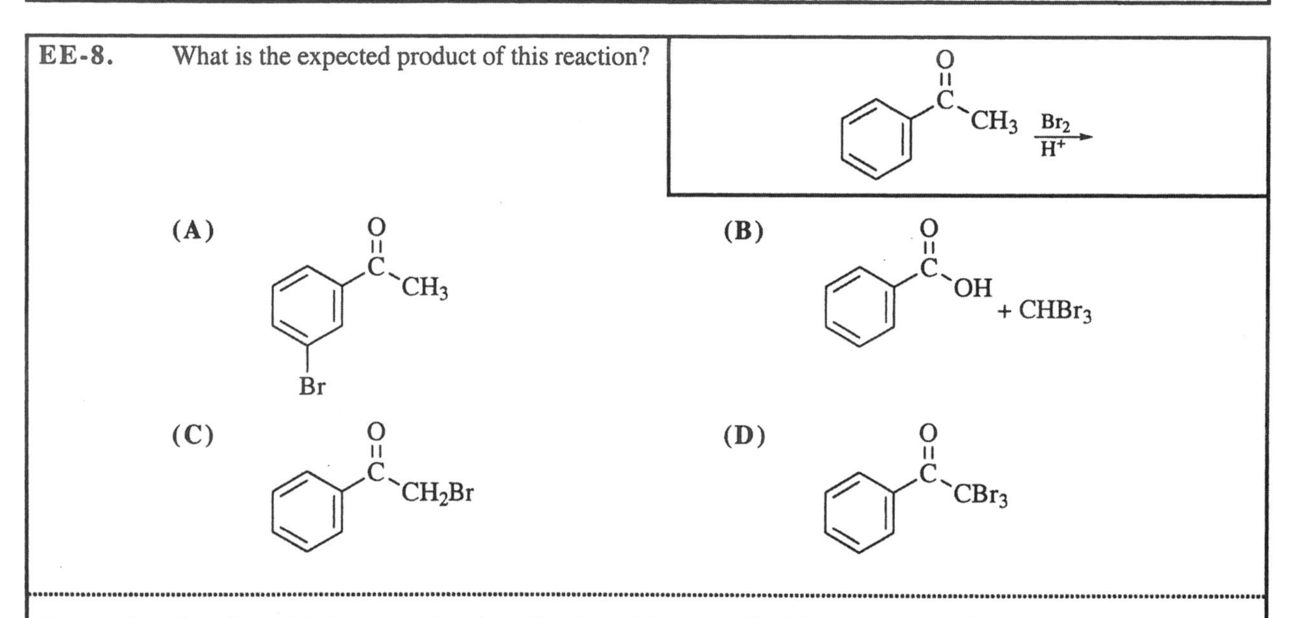

Knowledge Required: (1) Acid-catalyzed enolization of ketones. (2) Mechanism of acid-catalyzed α–halogenation of ketones.

Thinking it Through: Ketones will undergo α–bromination with bromine in the presence of either an acid or a base. Choice **(A)** would be the result of an electrophilic aromatic substitution, which requires the presence of a Lewis acid catalyst, such as $FeBr_3$, and is eliminated. Choices **(B)**, **(C)** and **(D)** all involve α–bromination. However, choices **(B)** and **(D)** are products that would form under base-catalyzed α–bromination, which proceeds through an enolate ion. The acid-catalyzed reaction proceeds through the enol tautomer, so choices **(B)** and **(D)** are eliminated. Once the ketone has been mono–α–brominated, it resists further tautomerization to the enol, and the monobrominated ketone can be prepared in high yield. Under basic conditions, each successive bromination increases the acidity of the remaining α–hydrogen atoms, and the rate at which the ketone forms the enolate ion. Consequently, it is difficult to stop the reaction at the monobromination stage. Thus, choice **(C)** is the correct answer.

Practice Questions

1. Which base should be used to quantitatively convert this *N,N*–dimethylamide into its enolate ion?

$$CH_3CH_2\overset{\overset{O}{\|}}{C}N(CH_3)_2$$

(A) $LiO\overset{\overset{O}{\|}}{C}CH_3$ (B) $LiOCH_2CH_3$

(C) $LiN[CH(CH_3)_2]_2$ (D) $LiOH$

2. Which sets of hydrogen atoms in this compound will undergo deuterium exchange with CH_3OD in the presence of CH_3ONa?

$$\underset{4}{CH_3}-\underset{3}{CH_2}-\overset{\overset{O}{\|}}{C}-\underset{2}{CH_2}-\overset{\overset{O}{\|}}{C}-\underset{1}{OCH_3}$$

(A) 2 and 3 only (B) 1, 2, and 3 only

(C) 2, 3, and 4 only (D) 1 and 4 only

3. Which compound is the product of an aldol condensation?

(A) $CH_3CH_2CH{=}\underset{CH_3}{\underset{|}{\overset{\overset{O}{\|}}{C}}}H$

(B) $CH_3CH_2\overset{\overset{OH}{|}}{C}HCH_2CH_2\overset{\overset{O}{\|}}{C}H$

(C) $CH_3\underset{\underset{OH}{|}}{C}H\overset{\overset{O}{\|}}{C}CH_2CH_3$

(D) $CH_3\overset{\overset{O}{\|}}{C}CH_2CH{=}CH_2$

4. Which set of the reagents would accomplish this conversion?

$$CH_3CH_2O\overset{\overset{O}{\|}}{C}CH_2\overset{\overset{O}{\|}}{C}OCH_2CH_3 \rightarrow CH_3CH_2O\overset{\overset{O}{\|}}{C}\underset{\underset{CH_2CH_3}{|}}{C}H\overset{\overset{O}{\|}}{C}OCH_2CH_3$$

(A) NaH followed by CH_3CH_2OH

(B) CH_3CH_2ONa followed by CH_3CH_2Br

(C) NaH followed by $CH_2{=}CH_2$

(D) CH_3CH_2OH with H^+ as a catalyst

5. Which compound forms the greatest equilibrium concentration of the enol tautomer?

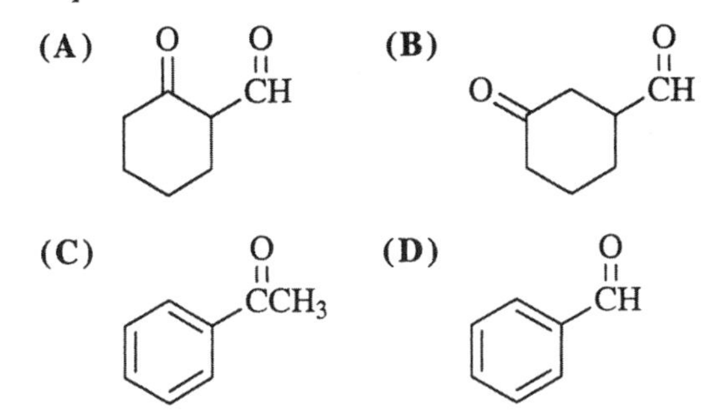

6. An aldol reaction can produce an addition product or the dehydrated addition product (condensation product). What factor favors the formation of the dehydration product?

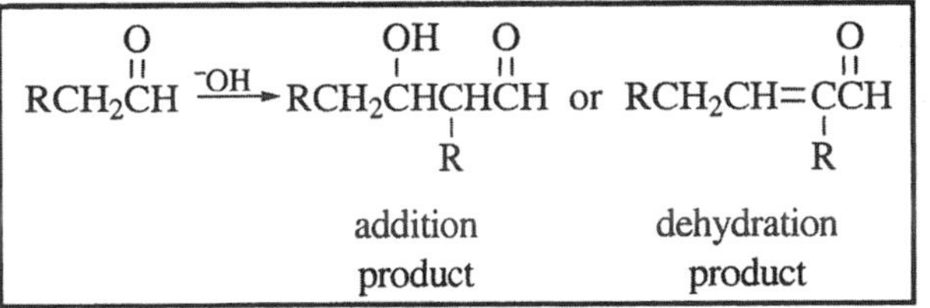

(A) excess aldehyde

(B) excess base

(C) no α–hydrogen atoms

(D) stabilization by conjugation

7. Which is *not* true of the aldol reaction?

(A) It accomplishes the formation of a new carbon–carbon bond.

(B) The key step in the mechanism is attack of the α–carbon atom of an enolate ion on a carbonyl carbon atom.

(C) Dehydration of the aldol product is often observed (an aldol condensation)

(D) The enolate is favored at equilibrium.

$$RCH_2\overset{\overset{O}{\|}}{C}H + {}^{-}OH \rightleftharpoons R\overset{-}{C}H\overset{\overset{O}{\|}}{C}H + H_2O$$

8. What would be the major product of this reaction?

$$CH_3CH_2CH_2\overset{O}{\overset{\|}{C}}H \xrightarrow[H_2O]{^-OH}$$

(A) $CH_3CH_2CH_2\overset{O}{\overset{\|}{C}}\underset{CH_2CH_3}{\underset{|}{C}}H\overset{O}{\overset{\|}{C}}H$

(B) $CH_3CH_2CH_2CH_2OH + CH_3CH_2CH_2COO_2^-$

(C) $CH_3CH_2CH_2\overset{OH}{\overset{|}{C}}H\underset{CH_2CH_3}{\underset{|}{C}}H\overset{O}{\overset{\|}{C}}H$

(D) $CH_3CH_2CH_2\overset{OH}{\overset{|}{C}}HCH_2CH_2CH_2\overset{O}{\overset{\|}{C}}H$

9. Which of these is *not* a step in the mechanism of the aldol reaction?

(A) $H\overset{O}{\overset{\|}{C}}-\overset{H}{\underset{}{C}}HR$ + ^-OH (curved arrows: ^-OH removes H; C–H electrons move to C=C; C=O electrons move to O)

(B) $H\overset{^-O}{\overset{|}{C}}=CHR$ $\quad H\overset{O}{\overset{\|}{C}}-CH_2R$ (curved arrows)

(C) $H\overset{O}{\overset{\|}{C}}-\underset{R}{\underset{|}{C}}H\overset{O^-}{\overset{|}{C}}HCH_2R$ $\quad$ H–OH (curved arrows)

(D) $RCH=\underset{O^-}{\underset{|}{C}}H \quad H\overset{O}{\overset{\|}{C}}-CH_2R$ (curved arrows)

10. What would be the product of this reaction?

$$C_6H_5\overset{O}{\overset{\|}{C}}H + CH_3CH_2CN \xrightarrow[CH_3CH_2OH]{CH_3CH_2ONa}$$

(A) $C_6H_5CH=CHCH_2CN$

(B) $C_6H_5CH=\overset{CH_3}{\overset{|}{C}}CN$

(C) $C_6H_5\overset{O}{\overset{\|}{C}}\underset{CH_3}{\underset{|}{C}}HCN$

(D) $C_6H_5\overset{OH}{\overset{|}{C}}HCH_2CH_2CN$

11. What would be the major organic product of this base-catalyzed reaction?

$$(CH_3)_3C-\overset{O}{\overset{\|}{C}}H + C_6H_5\overset{O}{\overset{\|}{C}}CH_3 \xrightarrow{^-OH}$$

(A) $C_6H_5\overset{CH_3}{\overset{|}{C}}=CH-\underset{CH_3}{\underset{|}{\overset{CH_3}{\overset{|}{C}}}}-\overset{O}{\overset{\|}{C}}H$

(B) $C_6H_5\underset{CH_3}{\underset{|}{\overset{OH}{\overset{|}{C}}}}-\overset{O}{\overset{\|}{C}}-\underset{CH_3}{\underset{|}{\overset{CH_3}{\overset{|}{C}}}}-CH_3$

(C) $C_6H_5\overset{O}{\overset{\|}{C}}CH_2-\overset{O}{\overset{\|}{C}}-\underset{CH_3}{\underset{|}{\overset{CH_3}{\overset{|}{C}}}}-CH_3$

(D) $C_6H_5\overset{O}{\overset{\|}{C}}CH_2\overset{OH}{\overset{|}{C}}H-\underset{CH_3}{\underset{|}{\overset{CH_3}{\overset{|}{C}}}}-CH_3$

12. Which of these reactants can be used to make the compound shown by an intramolecular aldol reaction?

(Structure: cyclohexenone ring with C=O and CH_3 on the ring double bond)

(A) $CH_3CH_2\overset{O}{\overset{\|}{C}}CH_2CH_2CH_2\overset{O}{\overset{\|}{C}}H$

(B) $CH_3\overset{O}{\overset{\|}{C}}CH_2CH_2CH_2\overset{O}{\overset{\|}{C}}CH_3$

(C) $CH_3CH_2\overset{O}{\overset{\|}{C}}CH_2CH_2\overset{O}{\overset{\|}{C}}CH_3$

(D) $CH_3\overset{O}{\overset{\|}{C}}-\overset{O}{\overset{\|}{C}}CH_2CH_2CH_2\overset{O}{\overset{\|}{C}}H$

13. Which compound would *not* react in the iodoform test to produce a yellow precipitate?

(A) $CH_3\overset{O}{\overset{\|}{C}}CH_2CH_2CH_3$

(B) $CH_3CH_2\overset{O}{\overset{\|}{C}}CH_2CH_3$

(C) $CH_3CH_2CH_2CH_2\overset{O}{\overset{\|}{C}}CH_3$

(D) $CH_3\overset{O}{\overset{\|}{C}}CH_2OH$

14. Which of these compounds ***cannot*** undergo an aldol reaction in the presence of dilute base?

(A) $C_6H_5CH_2\overset{O}{\overset{\Vert}{C}}H$ (B) $(CH_3)_3\overset{O}{\overset{\Vert}{C}}H$

(C) $(CH_3)_2CH\overset{O}{\overset{\Vert}{C}}H$ (D) $CH_3CH_2\overset{O}{\overset{\Vert}{C}}H$

15. Which set of reactants, in the presence of a basic catalyst, could be used to prepare this compound?

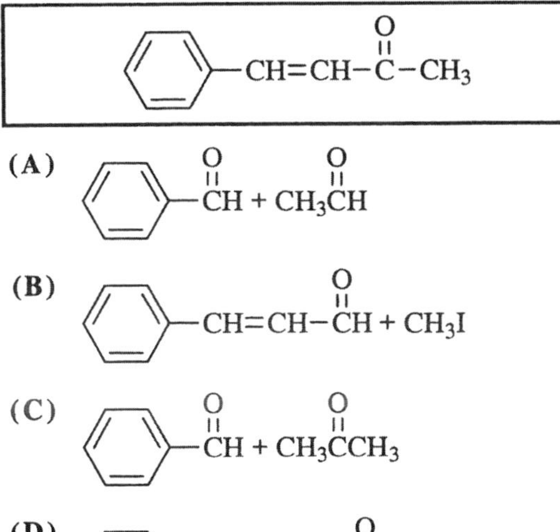

16. The first two steps in the base-catalyzed condensation of acetaldehyde would be described as

(A) attack of ^-OH on the carbonyl carbon atom, then loss of water.

(B) attack of ^-OH on the carbonyl carbon atom, then the resultant anion attacks the carbonyl carbon atom on a second molecule of acetaldehyde.

(C) ^-OH abstracts an α–hydrogen atom, then the resultant anion attacks the carbonyl carbon atom on a second molecule of acetaldehyde.

(D) ^-OH abstracts the hydrogen atom from the carbonyl carbon atom, then the resultant anion attacks the carbonyl carbon atom on a second molecule of acetaldehyde.

17. What is the product of this reaction?

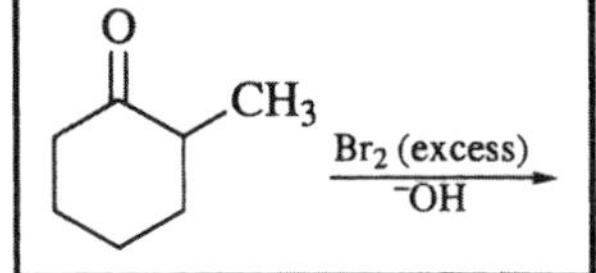

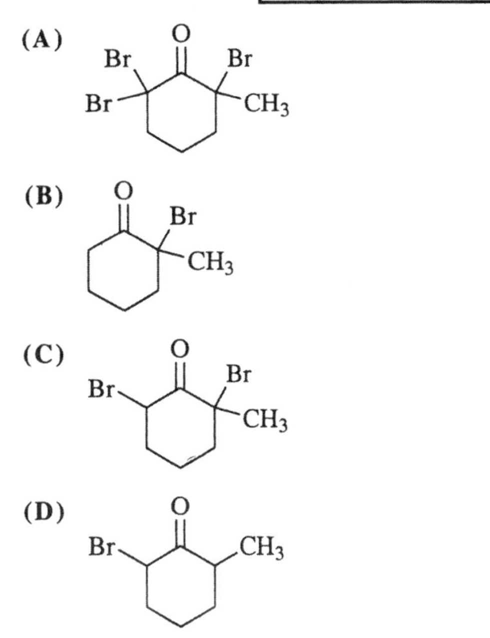

18. If this steroid tetracarboxylic acid is heated, which carboxyl group will be readily lost as carbon dioxide?

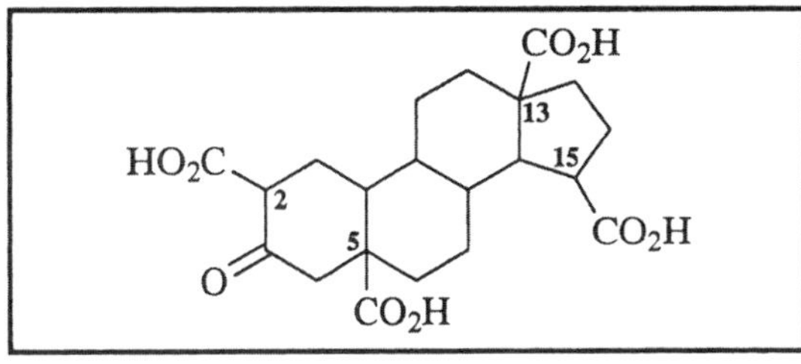

(A) CO_2H at C–15 **(B)** CO_2H at C–13

(C) CO_2H at C–5 **(D)** CO_2H at C–2

19. What product is expected from this reaction sequence?

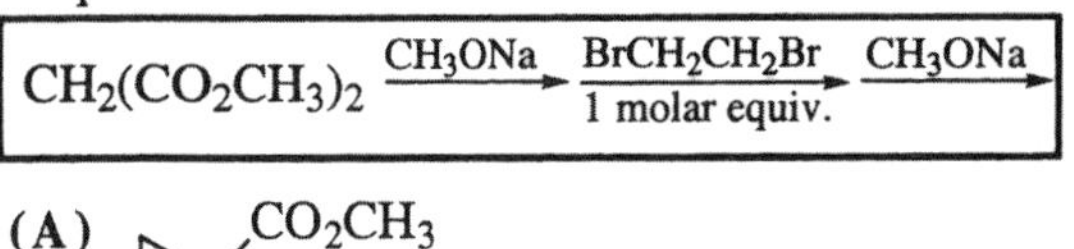

(A) CO_2CH_3 / CO_2CH_3

(B) $(CH_3O_2C)_2CHCH_2CH_2CH(CO_2CH_3)_2$

(C) $CH_3OCH_2CH_2CH(CO_2CH_3)_2$

(D) $CH_2{=}CHCH(CO_2CH_3)_2$

20. Which represents a keto–enol tautomerization?

(A) $\overset{-}{C}H_2-\overset{O}{\overset{\|}{C}}-CH_3 \longleftrightarrow CH_2=\overset{O^-}{\overset{|}{C}}-CH_3$

(B) $CH_3-\overset{O}{\overset{\|}{C}}-CH_3 \longleftrightarrow CH_3-\overset{O^-}{\overset{|}{\underset{+}{C}}}-CH_3$

(C) $CH_3-\overset{O}{\overset{\|}{C}}-CH_3 \rightleftharpoons CH_3-\overset{OH}{\overset{|}{C}}H-CH_3$

(D) $CH_3-\overset{O}{\overset{\|}{C}}-CH_3 \rightleftharpoons CH_2=\overset{OH}{\overset{|}{C}}-CH_3$

21. What would be the major product of this reaction sequence?

$$C_6H_5\overset{O}{\overset{\|}{C}}OCH_2CH_3 + CH_3CH_2\overset{O}{\overset{\|}{C}}OCH_2CH_3 \xrightarrow[\text{2. } H_2O,\ H^+]{\text{1. } NaOCH_2CH_3}$$

(A) $C_6H_5\overset{O}{\overset{\|}{C}}CH_2CH_2\overset{O}{\overset{\|}{C}}OCH_2CH_3$

(B) $C_6H_5\overset{O}{\overset{\|}{C}}\underset{CH_3}{\underset{|}{C}H}\overset{O}{\overset{\|}{C}}OCH_2CH_3$

(C) $C_6H_5\overset{O}{\overset{\|}{C}}\underset{CH_3}{\underset{|}{C}H}O\overset{O}{\overset{\|}{C}}CH_2CH_3$

(D) $C_6H_5\overset{OH}{\overset{|}{C}H}\underset{CH_3}{\underset{|}{C}H}\overset{O}{\overset{\|}{C}}OCH_2CH_3$

22. What is the starting material that will yield this compound upon treatment with CH_3CH_2ONa in CH_3CH_2OH?

[2-oxocyclopentane with $-\overset{O}{\overset{\|}{C}}OCH_2CH_3$ substituent]

(A) $CH_3CH_2O\overset{O}{\overset{\|}{C}}CH_2CH_2CH_2\overset{O}{\overset{\|}{C}}OCH_2CH_3$

(B) $CH_3CH_2O\overset{O}{\overset{\|}{C}}CH_2CH_2CH_2\overset{O}{\overset{\|}{C}}CH_2OCH_2CH_3$

(C) $CH_3CH_2O\overset{O}{\overset{\|}{C}}CH_2CH_2CH_2CH_2CH_2\overset{O}{\overset{\|}{C}}OCH_2CH_3$

(D) $CH_3CH_2O\overset{O}{\overset{\|}{C}}CH_2CH_2CH_2CH_2\overset{O}{\overset{\|}{C}}OCH_2CH_3$

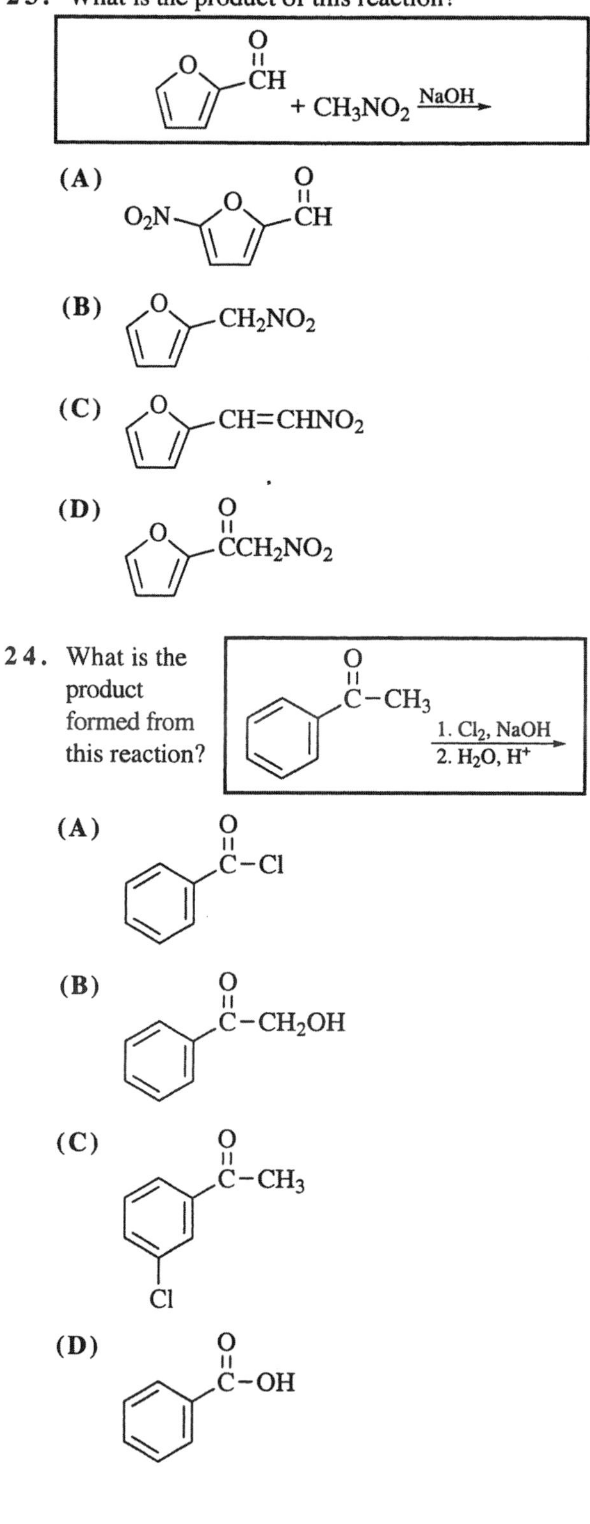

(A)

$$O_2N-\underset{\text{(furan)}}{}-\overset{\overset{\displaystyle O}{\|}}{C}H$$

(B) furan $-CH_2NO_2$

(C) furan $-CH=CHNO_2$

(D) furan $-\overset{\overset{\displaystyle O}{\|}}{C}CH_2NO_2$

24. What is the product formed from this reaction?

(A) phenyl $\overset{\overset{\displaystyle O}{\|}}{C}-Cl$

(B) phenyl $\overset{\overset{\displaystyle O}{\|}}{C}-CH_2OH$

(C) (3-chlorophenyl) $\overset{\overset{\displaystyle O}{\|}}{C}-CH_3$

(D) phenyl $\overset{\overset{\displaystyle O}{\|}}{C}-OH$

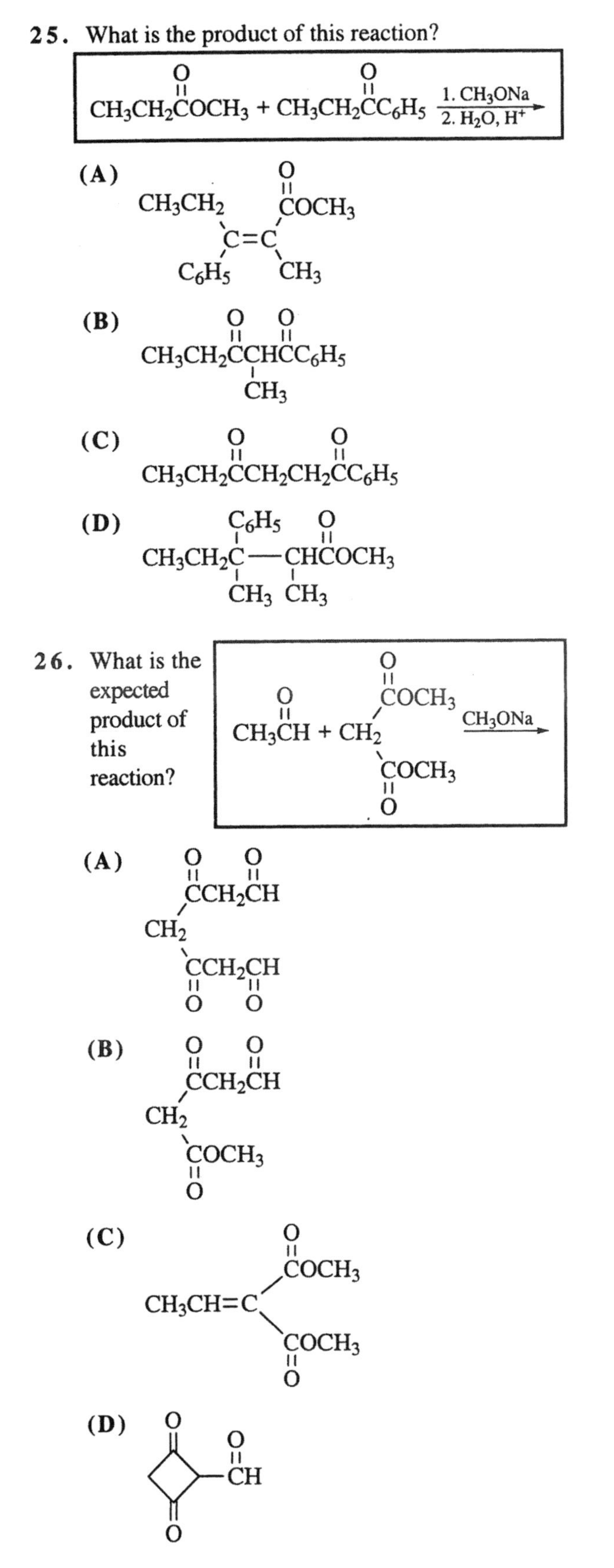

25. What is the product of this reaction?

$CH_3CH_2COCH_3$ (C=O) + $CH_3CH_2CC_6H_5$ (C=O) $\xrightarrow[2.\ H_2O,\ H^+]{1.\ CH_3ONa}$

(A) $CH_3CH_2(C_6H_5)C{=}C(CH_3)COCH_3$ (C=O)

(B) $CH_3CH_2CCHCC_6H_5$ (two C=O; CH_3 on CH)

(C) $CH_3CH_2CCH_2CH_2CC_6H_5$ (two C=O)

(D) $CH_3CH_2C(C_6H_5)(CH_3)$—$CH(CH_3)COCH_3$ (C=O)

26. What is the expected product of this reaction?

CH_3CH (C=O) + $CH_2(COCH_3)_2$ (two C=O) $\xrightarrow{CH_3ONa}$

(A) CH_2 with two CCH_2CH groups (each with two C=O)

(B) CH_2 with CCH_2CH (two C=O) and $COCH_3$ (C=O)

(C) $CH_3CH{=}C(COCH_3)_2$ (two C=O)

(D) cyclobutane-1,3-dione bearing a CH=O group

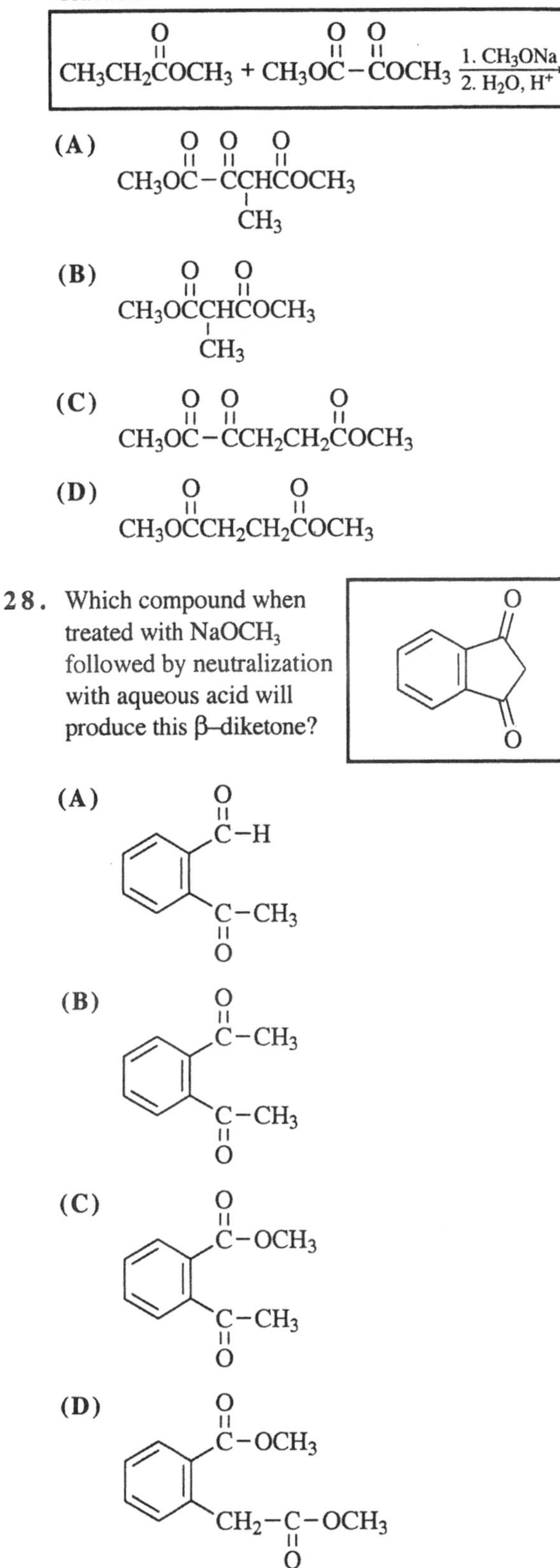

27. What compound will be produced by this reaction?

$CH_3CH_2COCH_3$ (C=O) + $CH_3OC-COCH_3$ (two C=O) $\xrightarrow[2.\ H_2O,\ H^+]{1.\ CH_3ONa}$

(A) $CH_3OC-CCHCOCH_3$ (three C=O; CH_3 on CH)

(B) $CH_3OCCHCOCH_3$ (two C=O; CH_3 on CH)

(C) $CH_3OC-CCH_2CH_2COCH_3$ (three C=O)

(D) $CH_3OCCH_2CH_2COCH_3$ (two C=O)

28. Which compound when treated with $NaOCH_3$ followed by neutralization with aqueous acid will produce this β-diketone?

(A) benzene ring with $C(=O)-H$ and $C(=O)-CH_3$

(B) benzene ring with $C(=O)-CH_3$ and $C(=O)-CH_3$

(C) benzene ring with $C(=O)-OCH_3$ and $C(=O)-CH_3$

(D) benzene ring with $C(=O)-OCH_3$ and $CH_2-C(=O)-OCH_3$

29. Aldol condensations, even with ketones, can occur under acid conditions where the active nucleophile is the enol, not the enolate ion. For the aldol condensation shown for acetone, what would be the structure of the electrophilic species that the enol attacks?

$$CH_3\overset{O}{\overset{\|}{C}}CH_3 \xrightarrow{H^+} (CH_3)_2C{=}CH{-}\overset{O}{\overset{\|}{C}}CH_3$$

(A) $CH_2{=}\overset{OH}{\overset{|}{C}}{-}CH_3$

(B) $CH_2{=}\overset{O^-}{\overset{|}{C}}{-}CH_3$

(C) $CH_3{-}\overset{O}{\overset{\|}{C}}{-}CH_3$

(D) $CH_3{-}\overset{^+O{-}H}{\overset{\|}{C}}{-}CH_3$

30. Which of these reactions could be used to prepare this β-keto ester?

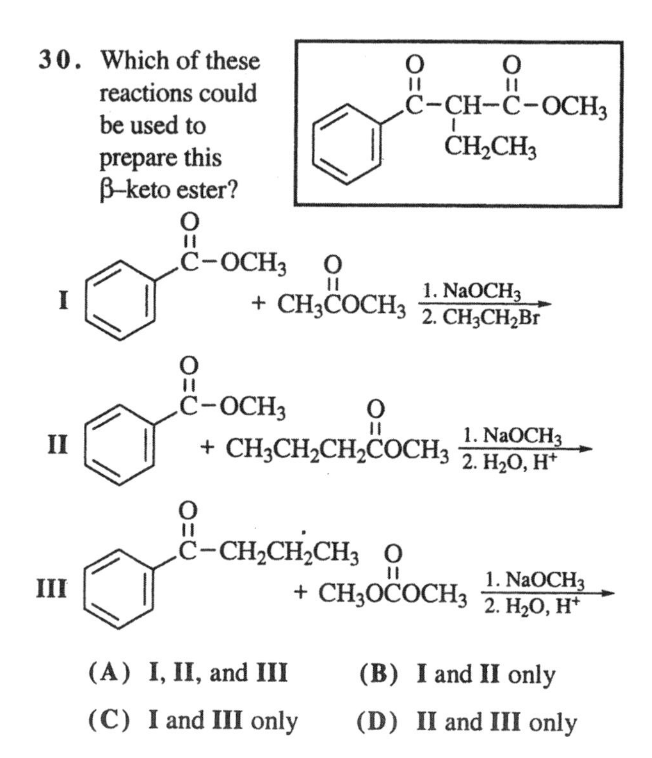

(A) I, II, and III (B) I and II only

(C) I and III only (D) II and III only

Answers to Study Questions

1. D
2. B
3. A
4. B
5. C
6. B
7. B
8. C

Answers to Practice Questions

1. C
2. A
3. A
4. B
5. A
6. D
7. D
8. C
9. D
10. B
11. D
12. B
13. B
14. B
15. C
16. C
17. A
18. D
19. A
20. D
21. B
22. D
23. C
24. D
25. B
26. C
27. A
28. C
29. D
30. A

Electrophilic and Nucleophilic Aromatic Substitution

Electrophilic substitution is the most common reaction type to occur on aromatic rings. In the fundamental reaction an electrophile E^+ is generated by any of several reagents. The electrophile accepts a pair of electrons from the aromatic ring to generate a cyclohexadienyl cation as shown in equation *1*. Rather than add a nucleophile to yield an addition product, the cyclohexadienyl cation loses a proton to regenerate the aromatic system. The result is that E^+ substitutes for a hydrogen atom, as shown in equation *2*. This is the essence of electrophilic aromatic substitution.

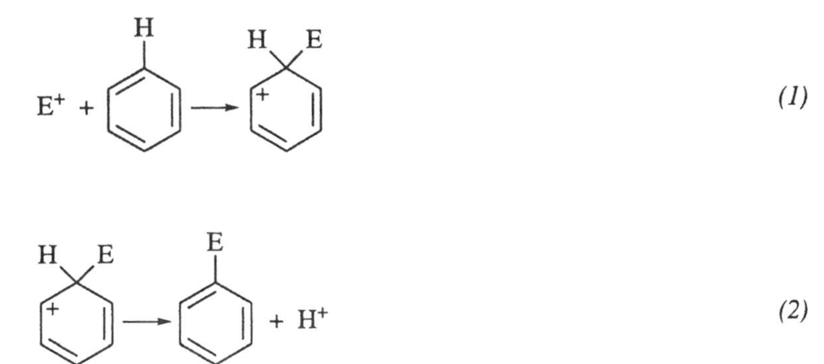

Groups attached to the aromatic ring destroy the six-fold symmetry of the ring, and the incoming electrophile has three possible locations to substitute relative to the group. Groups attached to the aromatic ring influence both the reactivity of the ring and the location of attack by the electrophile. If a group reduces the electron density of the ring by either resonance or induction, the ring will be less reactive than benzene. Groups that increase the electron density of the ring by resonance or induction will be more reactive than benzene.

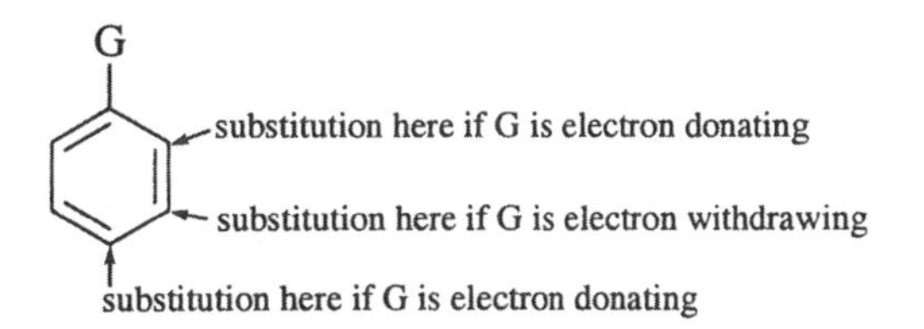

Electron-donating groups increase the electron density at the *ortho* and *para* positions more than at the *meta* position, leading electrophiles to preferentially attack those positions. For electron withdrawing groups, electron density is reduced more at the *ortho* and *para* position more than it is reduced at the *meta* position, rendering the *meta* position as the most susceptible to electrophilic attack.

Resonance has the greatest influence on the electron density of the aromatic ring. Groups that have an atom with nonbonded electrons next to the aromatic ring increase the electron density, as illustrated by the generalized resonance forms shown in equation *3*. Methoxybenzene is an example of such a compound. They are more reactive than benzene, and are *ortho/para* directors.

Groups that have a multiple bond to an atom more electronegative than carbon reduce the electron density, as illustrated by the generalized resonance forms shown in equation *4*. Benzoic acid is an example of such a compound. They are less reactive than benzene, and are *meta* directors.

(4)

Although the electrophilic aromatic substitution mechanism is the most common reaction type for aromatic systems, aromatic rings will react with nucleophiles under some conditions. Two distinct pathways for nucleophilic aromatic substitution include addition–elimination and elimination–addition.

In the addition–elimination mechanism, the aromatic ring first accepts a pair of electrons from a nucleophile to form an anionic intermediate. This is possible only when the aromatic ring is substituted with strong electron-withdrawing groups, such as nitro ($-NO_2$) groups, which stabilize the anion through resonance. Formation of the anion is followed by elimination of a suitable leaving group, as is shown in equation *5*.

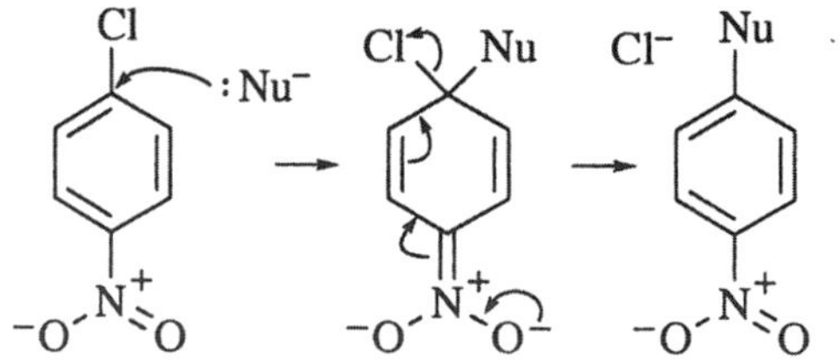

(5)

Hydride ion (H^-) is an extremely poor leaving group; therefore, the nucleophile must add at the site of a good leaving group. Often this is a halogen such as F, Cl, Br or I.

In the elimination–addition mechanism, a strong base initiates an elimination to form a highly reactive benzyne, which rapidly undergoes nucleophilic addition (equation *6*).

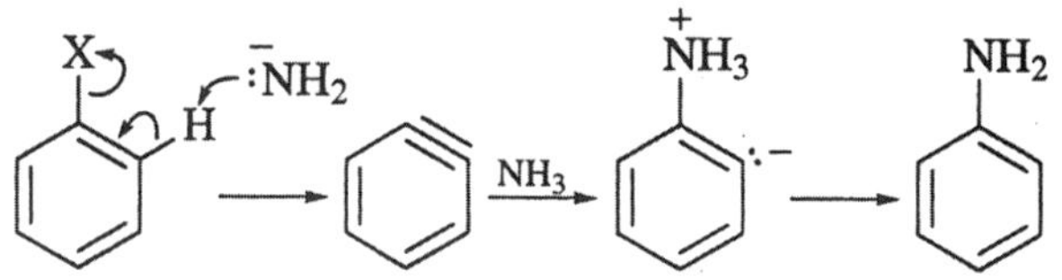

(6)

In summary, aromatic systems that undergo nucleophilic aromatic substitution have a good leaving group and one or more electron withdrawing groups *ortho* or *para* to that leaving group. As the number of electron-withdrawing groups attached to the ring increases, reactivity toward nucleophiles increases. The elimination–addition mechanism is not as common as the addition–elimination mechanism because a strongly basic nucleophile (such as amide ion, NH_2^-), a good leaving group (usually a halogen), and at least one *ortho* hydrogen atom are all three required. An important difference between these two nucleophilic substitution mechanisms is the site of attachment of the nucleophile. For the addition–elimination pathway, the nucleophile becomes attached at the site bearing the leaving group. However, for the elimination–addition pathway, the nucleophile may become attached either at the site bearing the leaving group *or* at the site bearing the *ortho* hydrogen atom.

Study Questions

AS-1. What type of reaction is this?

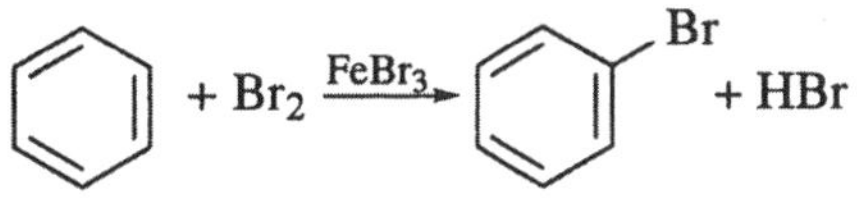

$$\text{C}_6\text{H}_6 + Br_2 \xrightarrow{FeBr_3} \text{C}_6\text{H}_5Br + HBr$$

(A) nucleophilic substitution (B) electrophilic substitution

(C) nucleophilic addition (D) electrophilic addition

Knowledge Required: (1) Definition of addition and substitution reactions. (2) Definitions of electrophiles and nucleophiles. (3) That $FeBr_3$ can accept an electron pair, acting as a Lewis acid.

Thinking it Through: In the reaction, a hydrogen atom on benzene is indicated as being replaced by a bromine atom, which makes it a substitution reaction. Choices **(C)** and **(D)** are thus eliminated. $FeBr_3$ is a Lewis acid and can coordinate with Br_2 to form a complex that can act as a source of Br^+. Br^+ does not have a complete octet of electrons, and is powerful electrophile.

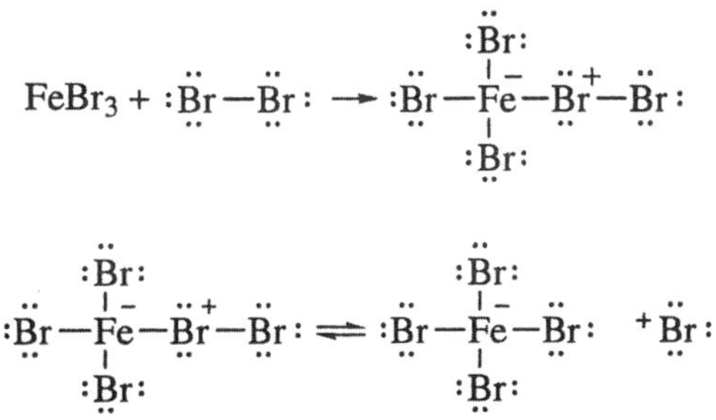

$$FeBr_3 + :\ddot{Br}-\ddot{Br}: \longrightarrow :\ddot{Br}-\overset{:\ddot{Br}:}{\underset{:\ddot{Br}:}{\overset{|-}{Fe}}}-\ddot{Br}^{+}-\ddot{Br}:$$

$$:\ddot{Br}-\overset{:\ddot{Br}:}{\underset{:\ddot{Br}:}{\overset{|-}{Fe}}}-\ddot{Br}^{+}-\ddot{Br}: \rightleftharpoons :\ddot{Br}-\overset{:\ddot{Br}:}{\underset{:\ddot{Br}:}{\overset{|-}{Fe}}}-\ddot{Br}: \quad {}^{+}\ddot{Br}:$$

Br^+ accepts a pair of electrons from the aromatic ring to form the cyclohexadienyl cation, which subsequently loses a hydrogen atom to form the substitution product.

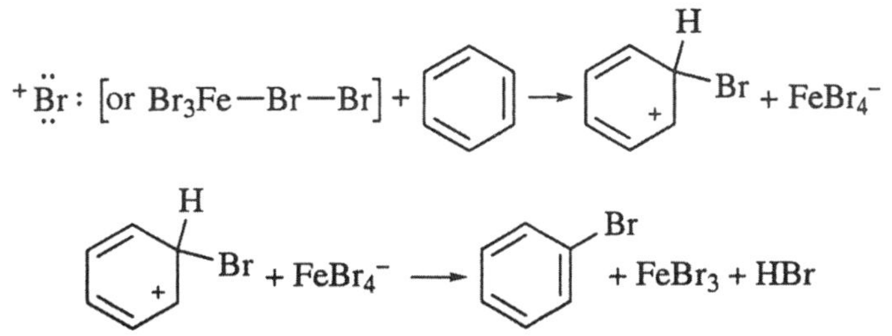

$${}^{+}\ddot{Br}: \ [\text{or } Br_3Fe-Br-Br] + \text{C}_6\text{H}_6 \longrightarrow [\text{C}_6\text{H}_6Br]^{+} \ (\text{H, Br}) + FeBr_4^-$$

$$[\text{C}_6\text{H}_6Br]^{+} \ (\text{H, Br}) + FeBr_4^- \longrightarrow \text{C}_6\text{H}_5Br + FeBr_3 + HBr$$

Choice **(B)** must be the correct answer. This is an *electrophilic* substitution reaction.

AS-2. Which substituent would be classified as an activating *ortho/para* director in an electrophilic aromatic substitution reaction?

(A) $-\overset{O}{\overset{\|}{C}}-OH$ (B) $-O-\overset{O}{\overset{\|}{C}}-CH_3$ (C) $-Cl$ (D) $-\overset{+}{N}H_3$

Knowledge Required: (1) How to draw resonance structures. (2) Inductive effect of groups. (3) The rate-limiting step in an electrophilic aromatic substitution reaction.

Thinking it Through: The rate-limiting step in electrophilic aromatic substitution reactions is the addition of the electrophile (E^+) to the aromatic ring. Any group that increases the electron density of the aromatic ring by resonance will increase the reactivity of the ring, or be an activating group. Any group that reduces the electron density of the ring by resonance or induction will act as a deactivating group. The carboxyl group withdraws electron density from the aromatic ring by pulling the electrons toward the carbonyl oxygen atom through resonance forms such as **II** below.

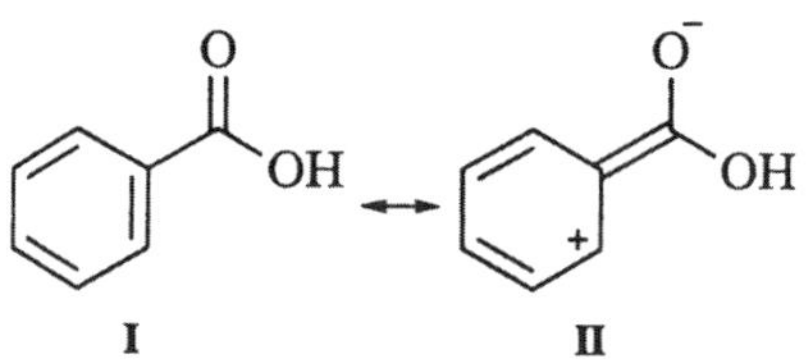

Choice **(A)** is eliminated because the carboxyl group deactivates the ring. The acetoxy group, choice **(B)** would donate electrons to the aromatic ring through resonance forms such as **IV**. It is a weakly activating group because the electrons on the oxygen atom are also delocalized toward the carbonyl group through resonance forms such as **V**. Choice **(B)** is a better answer than choice **(A)**, but we must still consider whether choices **(C)** or **(D)** are stronger activating groups.

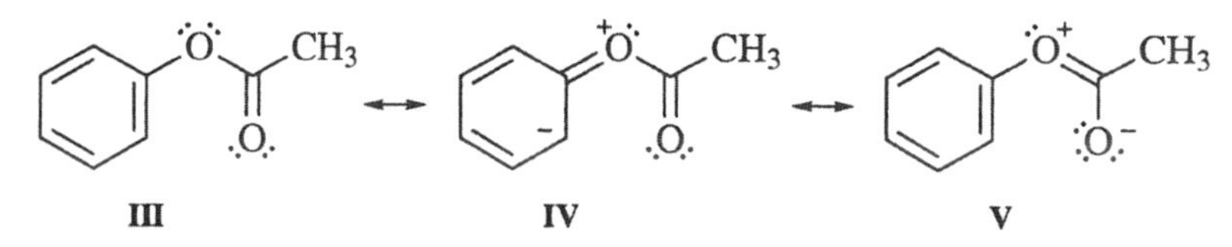

The chlorine atom, choice **(C)**, poses an interesting problem. Considering resonance, it should be an activating group because the electron pairs on the chlorine atom can be delocalized into the aromatic ring through resonance forms such as structure **VII**. Structure **VII** resembles structure **IV** for the ester, but there is an important difference. Oxygen and carbon, both second row elements, have atoms of about the same size. Chlorine, a third row element, has significantly larger atoms. The consequence is that **VII** contributes less to the structure of chlorobenzene than **IV** contributes to phenyl acetate, making choice **(C)** less likely than choice **(B)**.

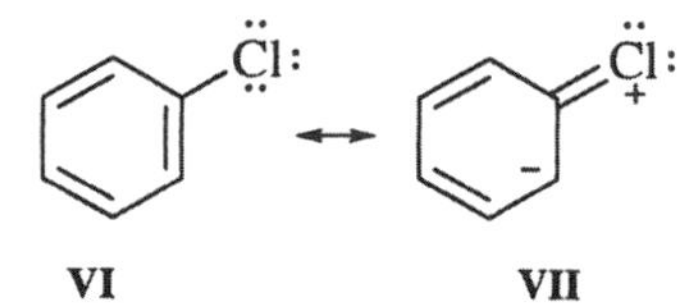

Before ruling out choice **(C)**, we must have to consider one other phenomenon—induction. Chlorine is more electronegative than hydrogen and will reduce the electron density of the aromatic ring. Induction should make chlorine a deactivating group, further decreasing the likelihood of choice **(C)**. On the other hand, oxygen is more electronegative than hydrogen. Considering induction, the acetoxy group, choice **(B)**, should also be electron withdrawing and act as a deactivating group. The question is whether resonance or induction dominates. The answer is that induction is more important when the group is a chlorine atom, and resonance is more important when considering an acetoxy group. An oxygen atom bonded directly the aromatic ring turns out to be a net electron donating group, and behaves as an activating group in electrophilic aromatic substitution reactions. A chlorine atom bonded directly to an aromatic ring is a mild deactivating group. The ammonium group $-NH_3^+$, choice **(D)**, has neither an unshared pair of electrons nor an empty orbital, so it cannot interact with the aromatic ring by resonance. Its only influence on the electron density of the ring is by induction. Since nitrogen is more electronegative than hydrogen, ammonium is a deactivating group, eliminating choice **(D)**.

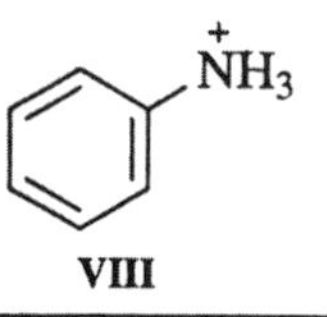

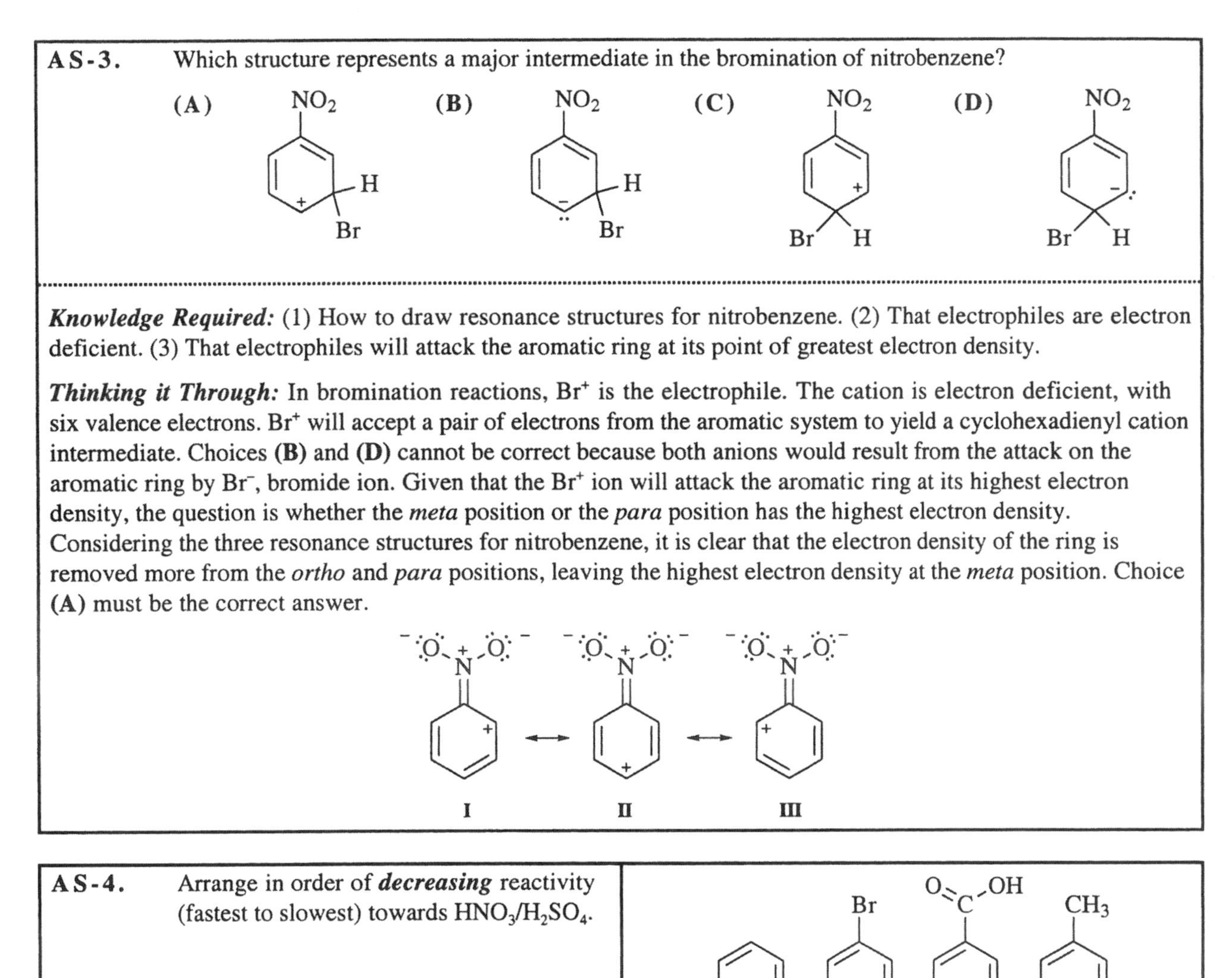

AS-3. Which structure represents a major intermediate in the bromination of nitrobenzene?

Knowledge Required: (1) How to draw resonance structures for nitrobenzene. (2) That electrophiles are electron deficient. (3) That electrophiles will attack the aromatic ring at its point of greatest electron density.

Thinking it Through: In bromination reactions, Br^+ is the electrophile. The cation is electron deficient, with six valence electrons. Br^+ will accept a pair of electrons from the aromatic system to yield a cyclohexadienyl cation intermediate. Choices **(B)** and **(D)** cannot be correct because both anions would result from the attack on the aromatic ring by Br^-, bromide ion. Given that the Br^+ ion will attack the aromatic ring at its highest electron density, the question is whether the *meta* position or the *para* position has the highest electron density. Considering the three resonance structures for nitrobenzene, it is clear that the electron density of the ring is removed more from the *ortho* and *para* positions, leaving the highest electron density at the *meta* position. Choice **(A)** must be the correct answer.

AS-4. Arrange in order of ***decreasing*** reactivity (fastest to slowest) towards HNO_3/H_2SO_4.

I II III IV

(A) III > IV > II > I

(B) I > II > IV > III

(C) IV > I > II > III

(D) IV > II> I > III

Knowledge Required: (1) How to draw resonance structures. (2) How the inductive effect of groups affects reactivity. (3) That electrophiles will attack the aromatic ring at its point of greatest electron density.

Thinking it Through: A good approach is to use benzene as the reference compound, and to compare the reactivity of the other compounds to benzene. The reactivity will depend on how the groups attached influence the electron density of the aromatic ring relative to benzene. If the group increases the electron density of the ring, the compound will be more reactive than benzene. However, if the group decreases the electron density of the ring relative to benzene, the compound will be less reactive than benzene. Considering benzoic acid (**III**) first, the carbonyl group of the benzoic acid withdraws electron density from the ring through resonance forms such as **V**. This would indicate that benzoic acid should be less reactive than benzene. Therefore, structure **I** must be more reactive than structure **III**. Only choice **(A)** is eliminated.

V

For toluene (**IV**), the $-CH_3$ group is a mild electron-donating group. This electron donating ability is usually attributed to hyperconjugation forms such as **VI**. The hyperconjugation form **VI** increases the electron density in the ring relative to benzene, and should make toluene more reactive than benzene. Therefore, structure **IV** must be more reactive than structure **I**. Choice **(B)** is eliminated. To choose between choices **(C)** and **(D)**, it is necessary to determine whether structure **II** is more or less reactive than benzene, structure **I**. Two attributes of a bromine atom attached to an aromatic ring must be considered: (1) the resonance effect, and (2) the inductive effect. The resonance effect would indicate that bromine should be an electron-donating group thanks to contributions of resonance forms such as **VII–IX**, and, thus, would be more reactive than benzene. On the other hand, the electronegativity of bromine is greater than carbon and a bromine atom should withdraw electron density from the ring by induction, which would make the ring less reactive than benzene. Resonance is less important when atoms significantly larger than carbon atoms are involved; consequently structures **VII–IX** contribute little to increasing electron density in the ring. The inductive, electron-withdrawing effect predominates with regard to reactivity. Benzene (structure **I**) is more reactive than bromobenzene (structure **II**), so choice **(D)** cannot be correct. This leaves choice **(C)** as the correct answer.

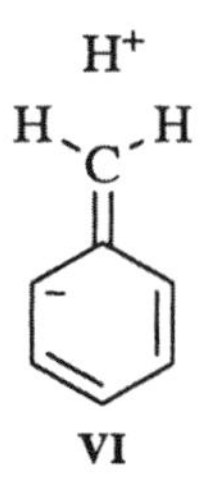

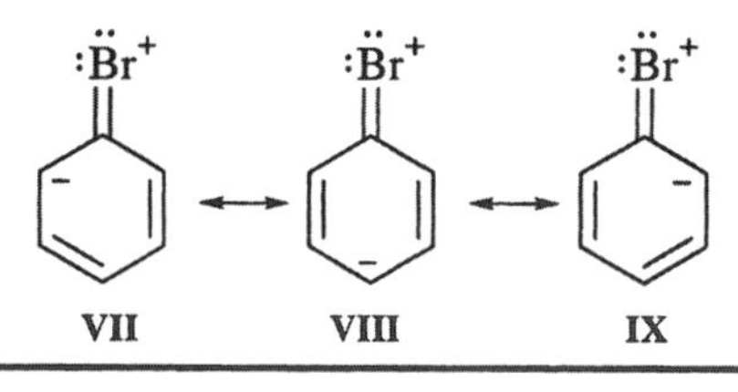

AS-5. The electrophilic aromatic substitution reaction is exothermic. Which potential energy reaction diagram best describes the three steps in this reaction?

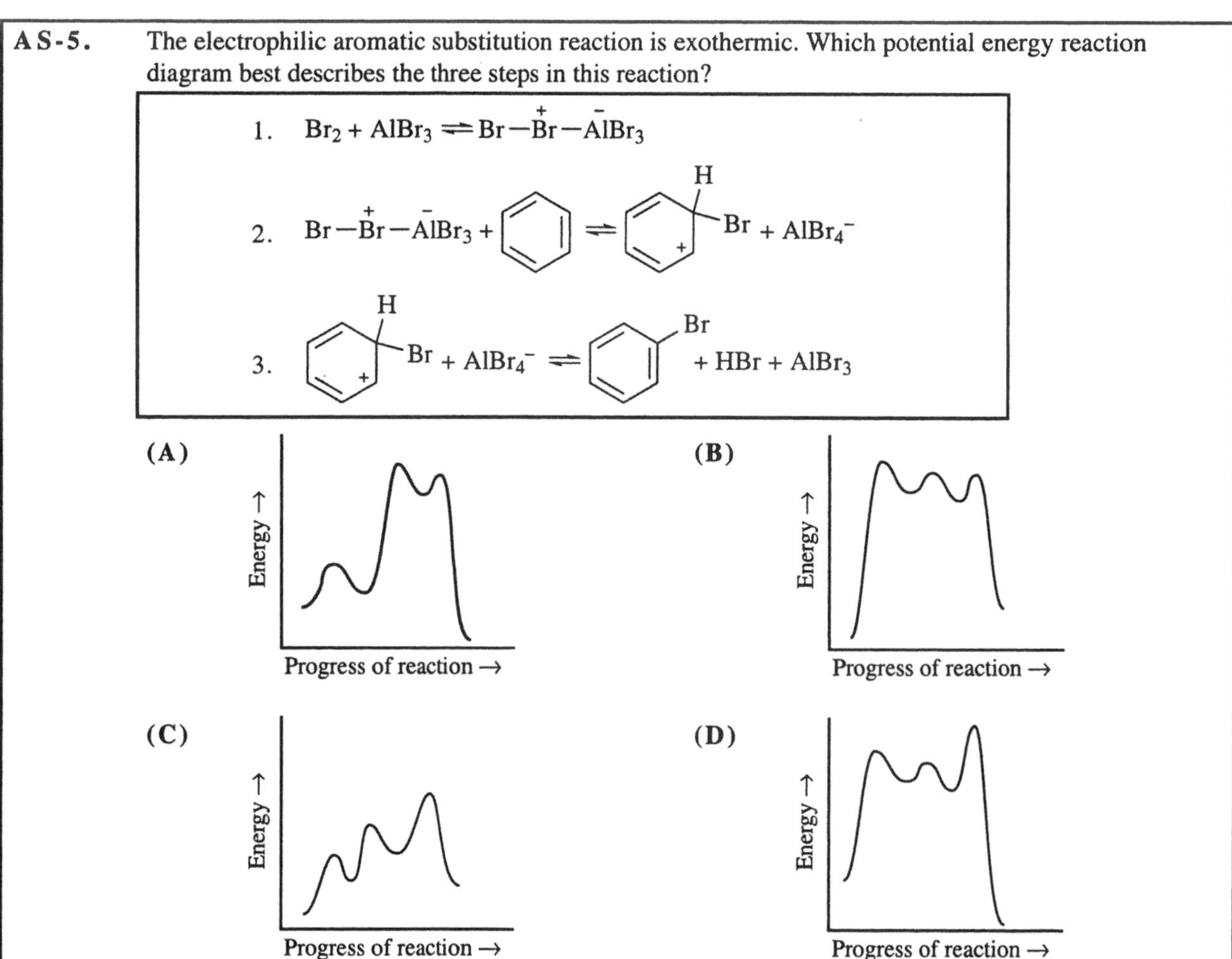

Knowledge Required: (1) Relationship between activation energy and rates of reaction. (2) That the slow step in electrophilic aromatic substitution is the addition of the electrophile to the aromatic ring. (3) The connection between exothermic reactions and potential energy of reactants and products.

Thinking it Through: The reaction is exothermic, indicating the products are more stable than the reactants. This eliminates choices **(B)** and **(C)**, because they represent endothermic reactions. In choice **(A)**, the *second* step is rate determining because this step is represented as having the highest potential energy. In choice **(D)**, the *third* step is represented as having the highest potential energy, so it would be rate determining. In most electrophilic aromatic substitution reactions the rate of the reaction depends on the rate of addition of the electrophile to the aromatic ring, which is the second step. Consequently, choice **(A)** the correct answer.

AS-6. What would be the major product of this reaction?

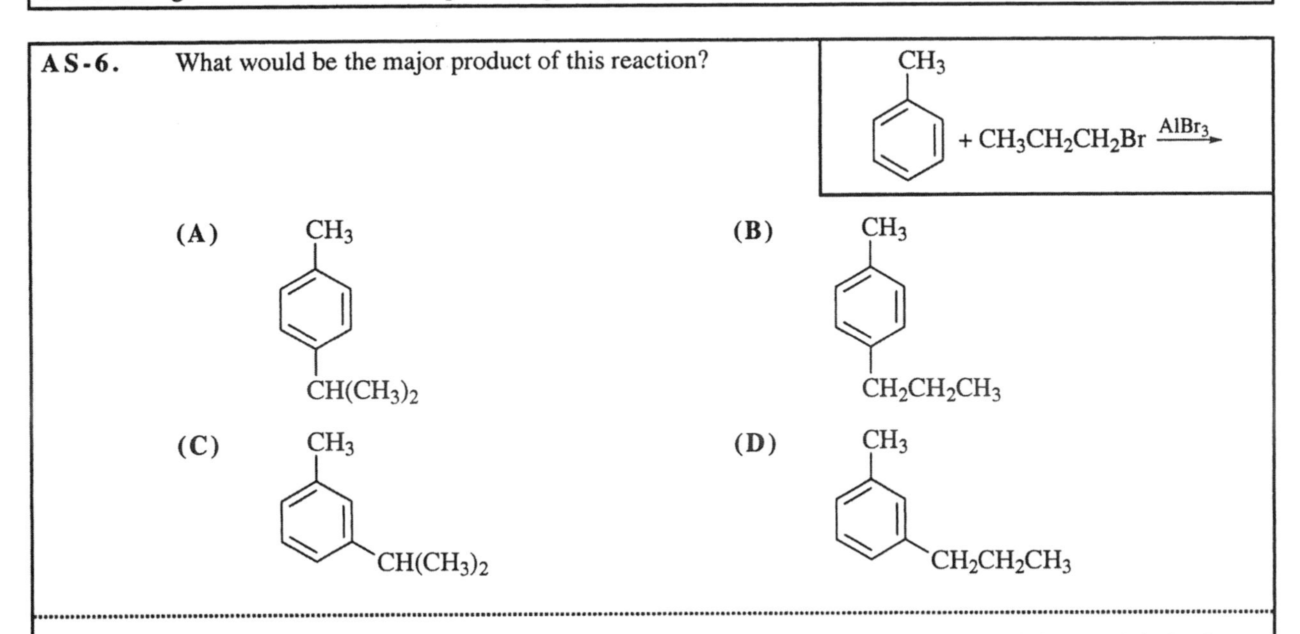

Knowledge Required: (1) Directing effect of a methyl substituent. (2) The mechanism of the Friedel–Craft reaction. (3) Carbocation rearrangement and carbocation stability.

Thinking it Through: The methyl group is a mild electron-donating group in an electrophilic aromatic substitution reaction. This is important, because the Friedel–Crafts alkylation reaction does not occur on aromatic rings less reactive than chlorobenzene. Toluene is reactive enough for the reaction to occur. As an electron-donating group, the methyl group would be *ortho/para* directing. Consequently, choices **(C)** and **(D)** are eliminated. The Friedel–Crafts alkylation is believed to involve a carbocation intermediate (or a species that resembles a carbocation). The 1-bromopropane initially forms the complex **I** with the Lewis Acid, $AlBr_3$, which then dissociates to yield the carbocation. Since secondary carbocations are more stable than primary carbocations, the secondary carbocation, structure **II**, would be the alkylating species. Choice **(A)** is the correct answer.

$$AlBr_3 + CH_3CH_2CH_2Br \rightleftharpoons \underset{\textbf{I}}{CH_3CH_2CH_2{-}\overset{+}{Br}{-}\overset{-}{Al}Br_3}$$

$$\underset{\textbf{I}}{CH_3CH_2CH_2{-}\overset{+}{Br}{-}\overset{-}{Al}Br_3} \rightleftharpoons \underset{\textbf{II}}{CH_3\overset{+}{C}HCH_3} + AlBr_4^-$$

AS-7. What is the major product of this reaction?

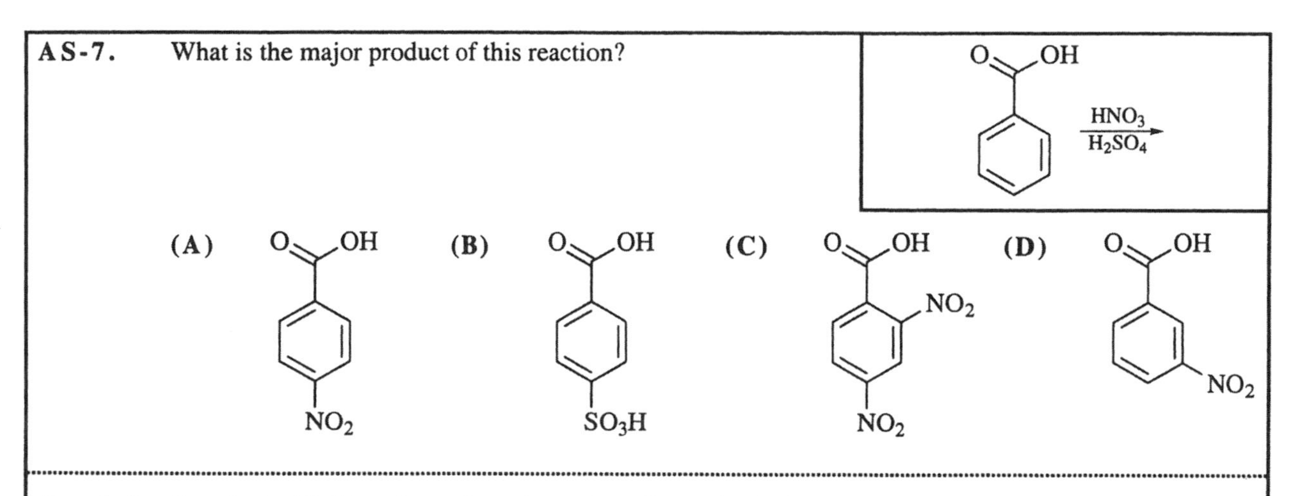

Knowledge Required: (1) Directing effect of a carboxylic acid substituent. (2) The mechanism of the nitration reaction.

Thinking it Through: In the nitration reaction, the electrophile is NO_2^+. This electrophile is generated from nitric acid and concentrated sulfuric acid. The sulfuric acid protonates the nitric acid to yield structure **I**, which then loses water to form the NO_2^+, structure **II**. The point of equilibrium for each of these two reactions is to the left. However, once water is formed, it is protonated by concentrated sulfuric acid to form H_3O^+. The point of equilibrium for the protonation reaction is far to the right, which draws all the other equilibria to the right. The mixture of HNO_3/H_2SO_4 is a good source of the electrophile NO_2^+.

$$HNO_3 + H_2SO_4 \rightleftharpoons \underset{\mathbf{I}}{O{=}N(O^-)\text{–}O^+H_2} + HSO_4^-$$

$$\underset{\mathbf{I}}{O{=}N(O^-)\text{–}O^+H_2} \rightleftharpoons \underset{\mathbf{II}}{O{=}N^+{=}O} + H_2O$$

With the electrophile being NO_2^+, choice **(B)** cannot be correct. The problem now is to determine the directing effect of the carboxyl group. Resonance structures **III–V** indicate that the carboxyl group should reduce the electron density at the *ortho* and *para* positions more than at the *meta* position. Since the highest electron density on the ring is at the *meta* position, the electrophile will attack there, making choice **(D)** the correct answer.

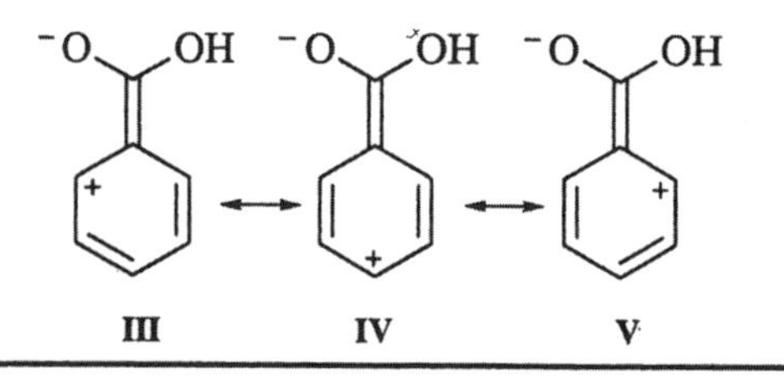

AS-8. Which would be the major product of this reaction?

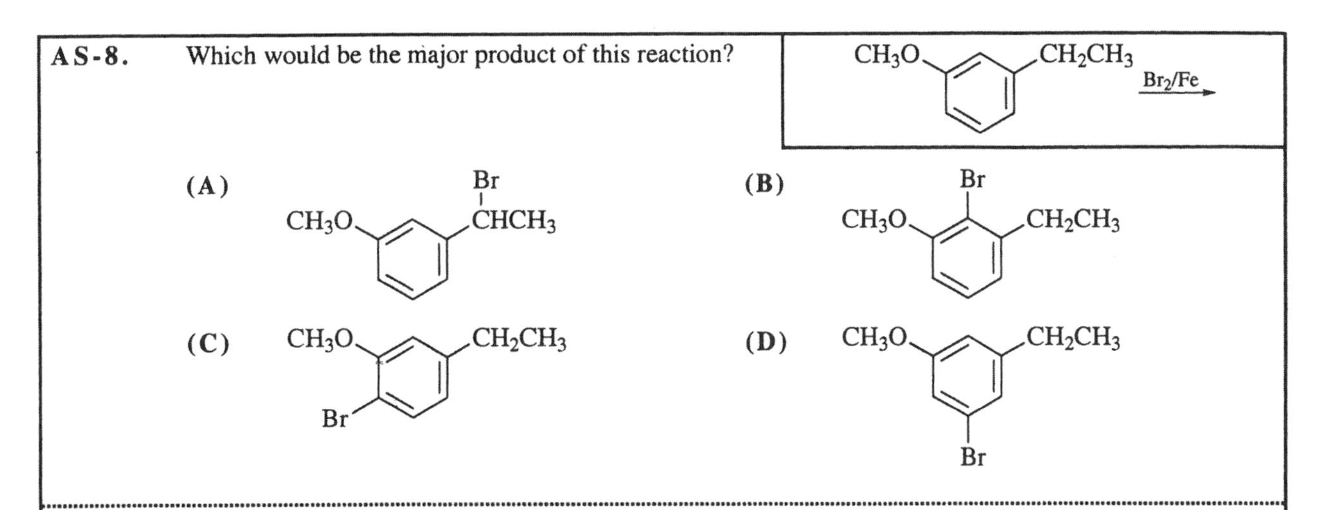

Knowledge Required: (1) Directing effect of alkyl and methoxy substituents. (2) The relative activating ability of the two groups.

Thinking it Through: The reaction conditions (Br_2/Fe) are those for the generation of Br^+ in an electrophilic aromatic substitution reaction. Choice **(A)** can be immediately eliminated because it would be the expected product for the free radical substitution reaction on the starting material. For choice **(A)** to form, Br_2 and a free radical initiator such as AIBN or UV light would be necessary. When considering electrophilic substitution on an aromatic ring having multiple substituents, the directing effect is dominated by the group that activates the ring the most. The ethyl group is a mild activating group. On the other hand, the methoxy group can activate the ring by resonance, and behaves as a strong activating group. Resonance structures **I–III** show how the electron density would be increased at the positions *ortho* and *para* to the methoxy group.

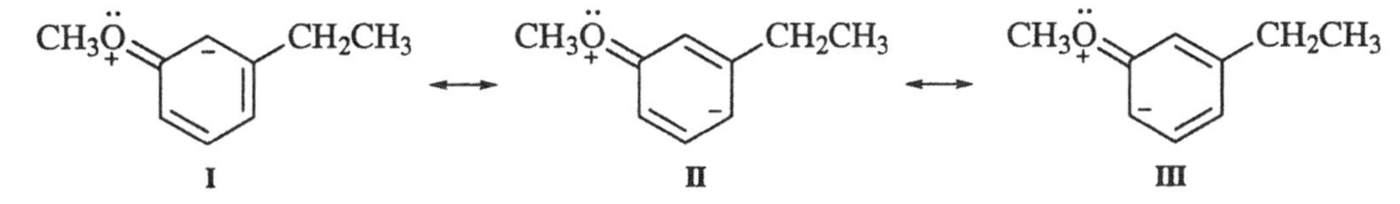

Three possible products could be imagined each coming from the electrophile, Br^+, adding to the positions represented by **I–III**. Choice **(D)** is eliminated because the Br^+ would have to attack *meta* to the methoxy group. Both choices **(B)** and **(C)** with the bromine attached *ortho* to the methoxy group, are possible. However, putting the large bromine atom between two substituent groups in the structure in choice **(B)** is less likely than the less strained position of the structure in choice **(C)**. Choice **(C)** is therefore the best answer. Another likely product, structure **IV**, was not given as a choice.

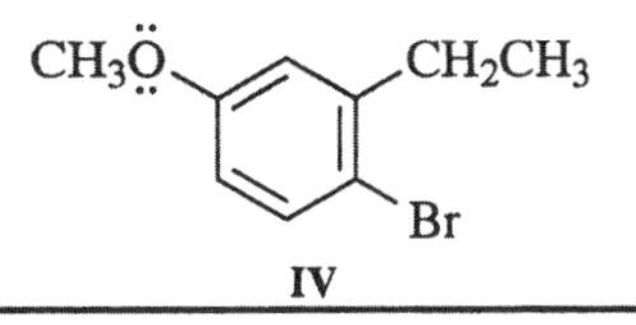

IV

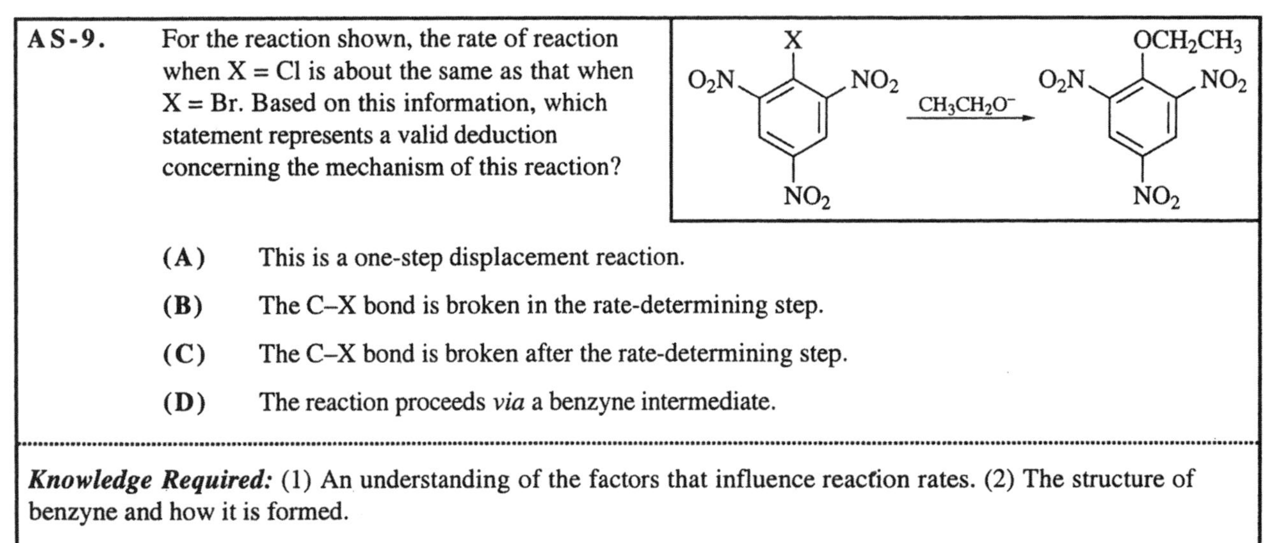

AS-9. For the reaction shown, the rate of reaction when X = Cl is about the same as that when X = Br. Based on this information, which statement represents a valid deduction concerning the mechanism of this reaction?

(A) This is a one-step displacement reaction.

(B) The C–X bond is broken in the rate-determining step.

(C) The C–X bond is broken after the rate-determining step.

(D) The reaction proceeds *via* a benzyne intermediate.

Knowledge Required: (1) An understanding of the factors that influence reaction rates. (2) The structure of benzyne and how it is formed.

Thinking it Through: This reaction is classified as a nucleophilic aromatic substitution reaction. A nucleophile, $CH_3CH_2O^-$, displaces Cl^- or Br^-. Usually, aromatic systems react with electrophiles rather than nucleophiles, because the aromatic π electron system has high electron density. In this case, two factors facilitate nucleophilic substitution: (1) Three electron-withdrawing nitro groups attached to the ring decrease its electron density, and (2) Cl and Br are good leaving groups. They are also in a favorable position to leave. (*Ortho* or *para* positions relative to the nitro groups have the lowest electron density.) Choice **(D)** can be eliminated immediately because there must be at least one hydrogen atom *ortho* to the leaving group for benzyne to form.

The rate of reaction depends on the activation energy of a reaction. The energy of the transition state will be influenced by the bonds that are being broken and formed in the transition state. If a bond is not involved in the highest energy transition state, the making or breaking of that bond will not influence the rate of the reaction. The problem states that the rate of the reaction is the same for both X = Cl and X = Br, indicating that the C–X bond is not involved in the rate-determining step. Consequently, choices **(A)** and **(B)** are eliminated. This leaves choice **(C)** as the only possible correct answer. The nucleophile must add first in the rate-determining step, and this slow step is followed by a rapid loss of X.

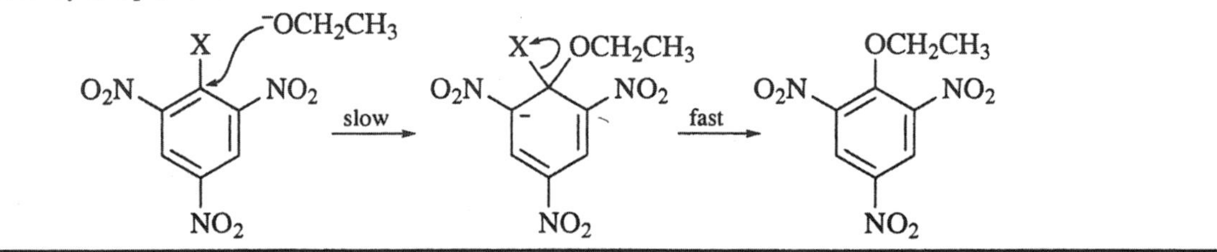

AS-10. What would be the major product of this reaction?

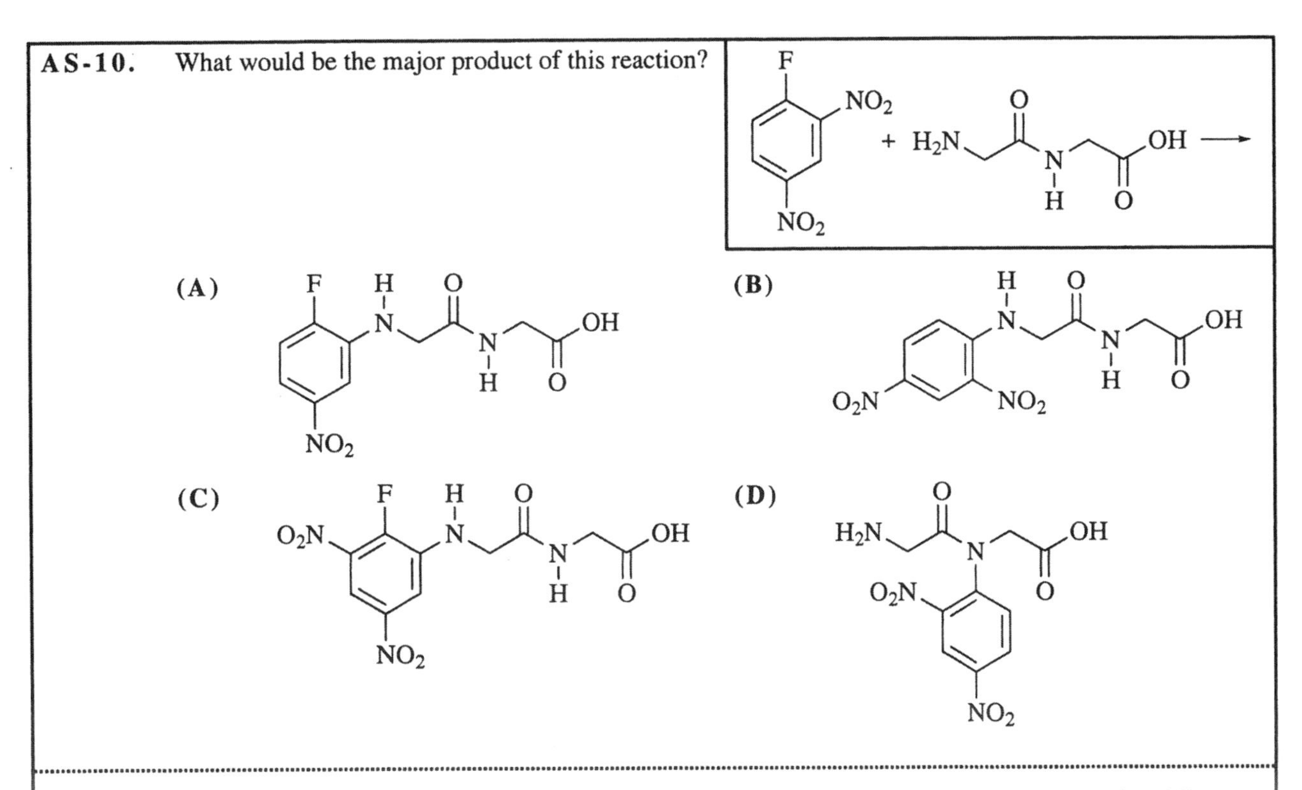

Knowledge Required: (1) That 2,4-dinitrofluorobenzene, Sanger's Reagent, readily undergoes nucleophilic aromatic substitution, and is used for end-group analysis in peptide chemistry. (2) Relative base strengths of amine nitrogen atoms and amide nitrogen atoms.

Thinking it Through: Sanger's Reagent, 2,4-dinitrofluorobenzene, is a classic reagent for nucleophilic aromatic substitution. The two nitro groups withdraw electron density from the aromatic ring to make the ring reactive toward nucleophiles, and the fluorine atom withdraws electron density toward it through induction, which accentuates the reactivity to nucleophiles at that position. Choice **(C)** can be eliminated because this product could only be formed by the loss of a hydride, H^-. Choice **(A)** can also be eliminated because the leaving group would have to be nitrite, NO_2^-, and the $-NH_2$ nucleophile would not attack the aromatic ring at the position where the $-NO_2$ is attached. The most likely point of attack is at the carbon atom carrying the fluorine atom. Choices **(B)** and **(D)** both would form as the result of nucleophilic attack at the carbon atom attached to fluorine. The problem is to decide which nitrogen atom is most nucleophilic. The electron pair on the amine nitrogen atom is localized on the nitrogen atom and is very available to act as a base or nucleophile. The electron pair on the amide nitrogen atom is delocalized toward the carbonyl group through resonance such as in structure **I**. This delocalization reduces the base strength and nucleophilicity of this nitrogen atom. The terminal amine will be the one to react with the 2,4-dinitrofluorobenzene, leading to choice **(B)** as the correct answer.

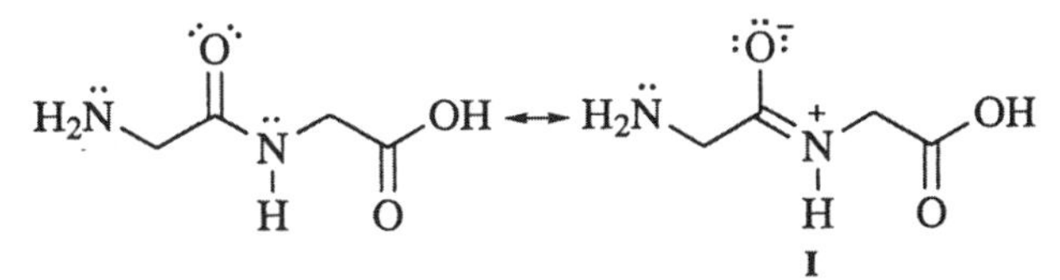

When a protein is treated with Sanger's Reagent, all the terminal amino groups react with the reagent. The protein can then be hydrolyzed to the amino acids and those that contain the dinitrophenyl group can be identified. This information allows one to know which amino acids are at the end of the protein chain.

AS-11. In addition to the product indicated, what other compound is expected to be formed in the reaction?

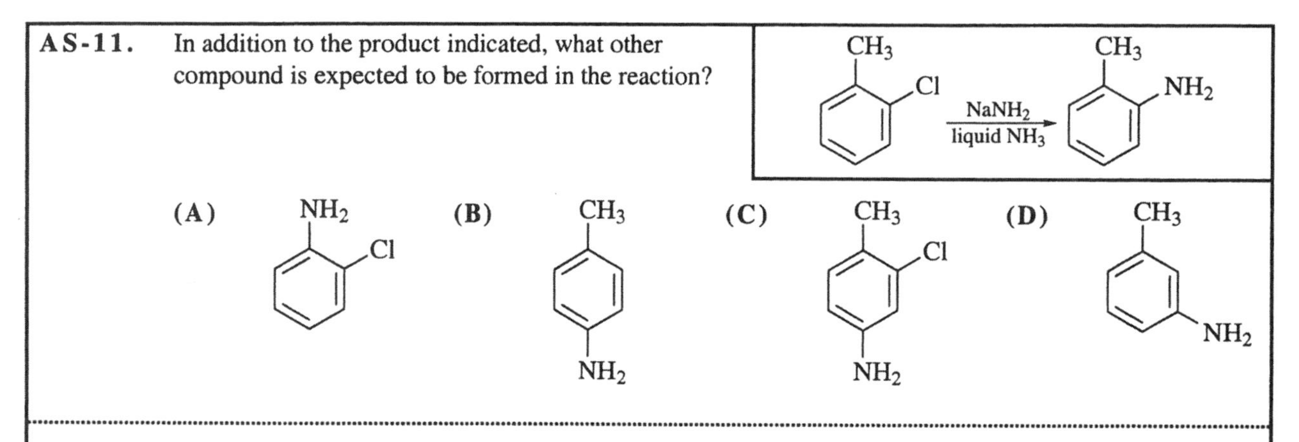

Knowledge Required: (1) Conditions necessary for the formation of benzynes. (2) Mechanism for reactions that proceed through a benzyne intermediate.

Thinking it Through: Aryl halides that are not activated by electron-withdrawing groups towards nucleophilic substitution *via* an addition–elimination mechanism will undergo nucleophilic substitution *via* an elimination–addition mechanism when treated with strong bases. The elimination of HX yields a highly reactive benzyne intermediate that rapidly undergoes nucleophilic addition. The nucleophile can attack either carbon atom of the benzyne triple bond. Thus, substituted benzynes can give rise to two addition products. Choices **(A)** and **(C)** are eliminated because they still contain the chlorine atom. Choice **(B)** is also eliminated because it could not possibly be formed from the benzyne (I) produced in the reaction. This leaves choice **(D)** as the only possible answer. The choice **(D)** product would be formed by NH_3 attacking the carbon atom of the triple bond that is *meta* to the $-CH_3$ group.

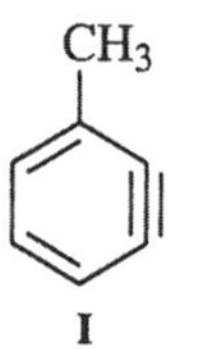

Practice Questions

1. Which substituents would deactivate benzene toward electrophilic aromatic substitution reactions?

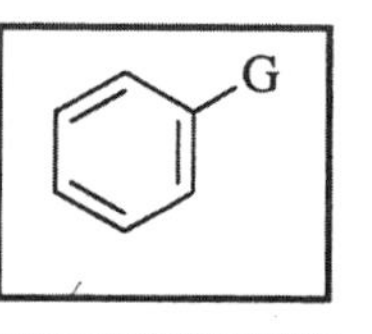

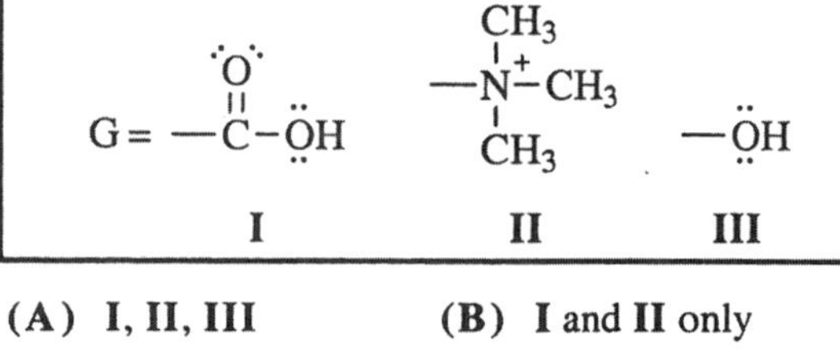

(A) I, II, III **(B)** I and II only

(C) II only **(D)** I and III only

2. Which structure represents a major intermediate in the nitration of toluene?

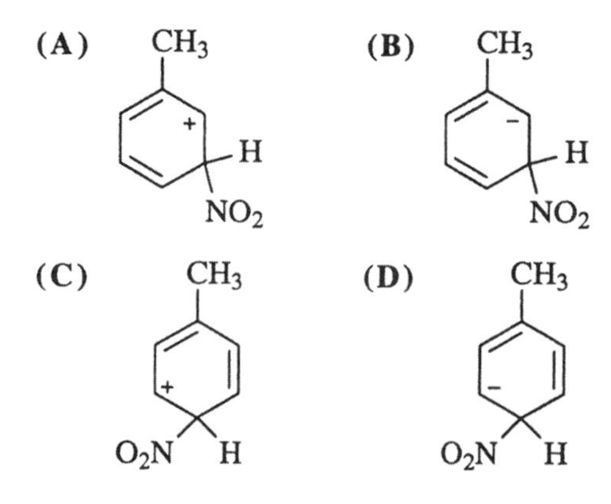

3. What would be the major product of this reaction?

Benzene $\xrightarrow[AlCl_3]{CH_3CH_2CH_2CH_2Cl}$

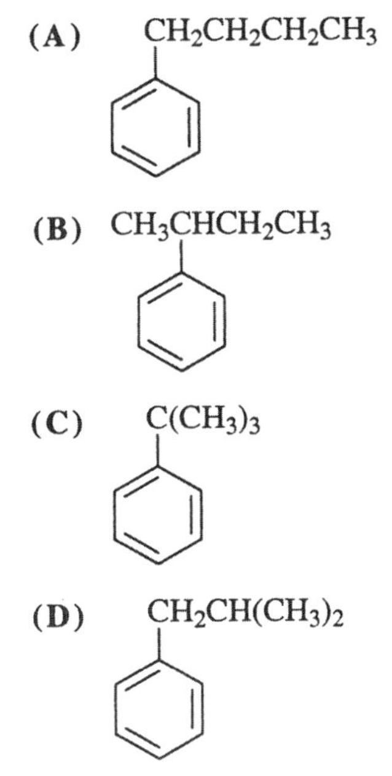

4. Which set of reagents would most likely bring about this transformation?

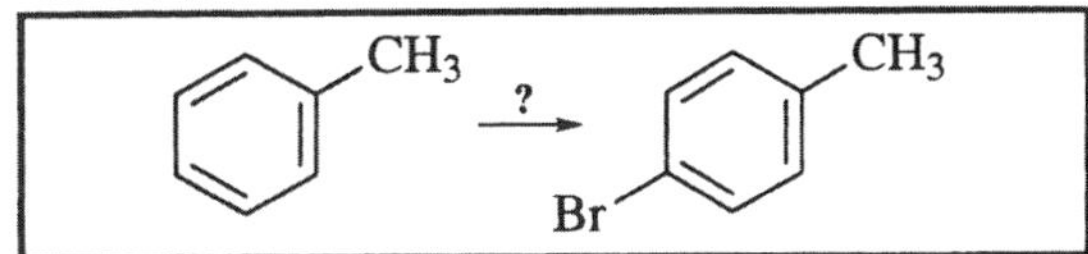

(A) Br_2 with $FeBr_3$

(B) Br_2 in CCl_4

(C) Br_2 with UV light

(D) NaBr with H_2SO_4

5. Which sequence of reagents would produce propylbenzene from benzene?

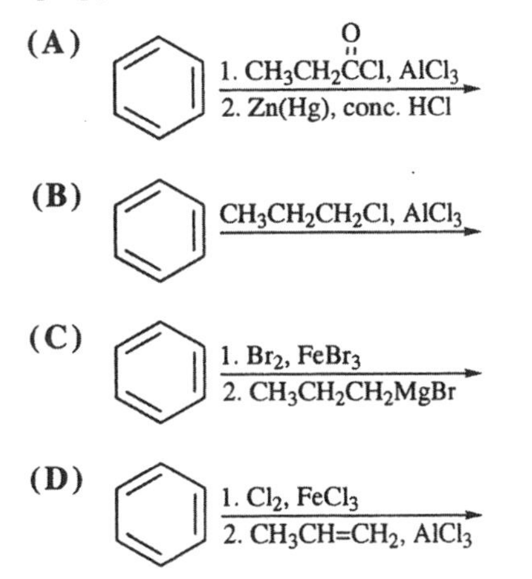

6. Which position will be attacked most rapidly by the nitronium ion (NO_2^+) when the compound undergoes nitration with HNO_3/H_2SO_4?

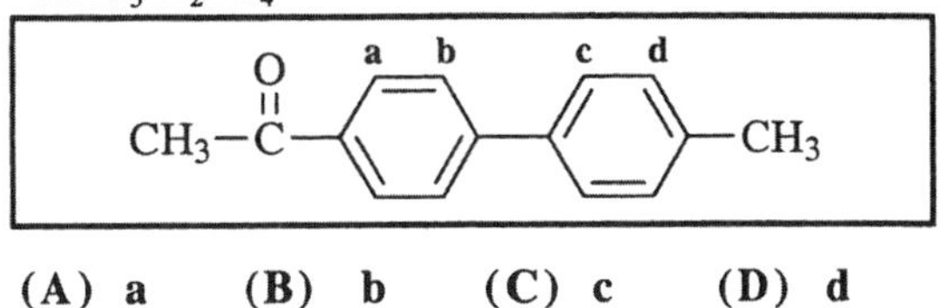

(A) a (B) b (C) c (D) d

7. Which description is most applicable to the mechanism of this reaction?

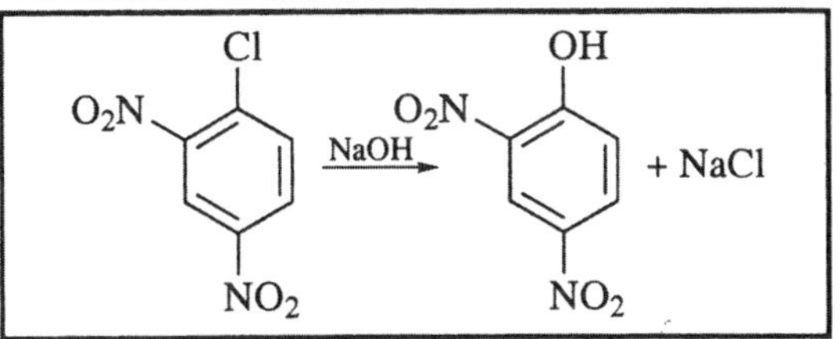

(A) It takes place in one step with no ionic intermediates.

(B) It takes place in two steps; elimination, then addition.

(C) It takes place in two steps; addition, then elimination.

(D) It takes place in three steps and involves a carbocation intermediate.

8. For the nitration of bromobenzene with HNO_3/H_2SO_4, which intermediate is used to explain why halogens are *ortho/para* directors?

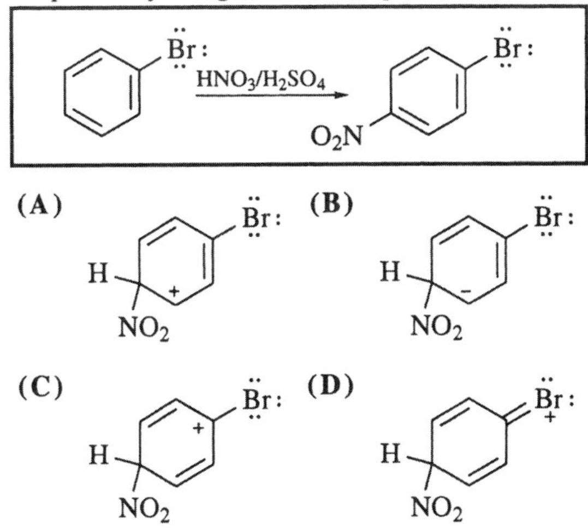

9. Which statement best describes why anisole brominates faster than benzene?

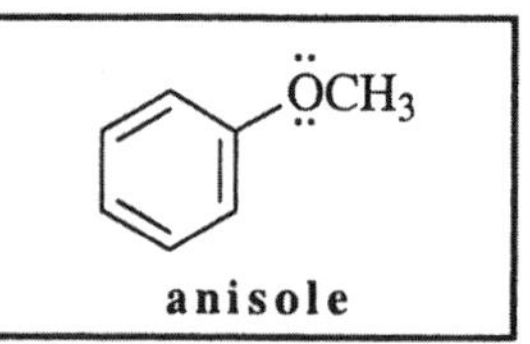

(A) The inductive effect of the methoxy group stabilizes of the cationic intermediate.

(B) The inductive effect of the methoxy group stabilizes the anionic intermediate.

(C) The resonance effect of the methoxy group stabilizes the cationic intermediate.

(D) The resonance effect of the methoxy group stabilizes the anionic intermediate.

10. Which aryl chloride reacts the fastest with NaOH?

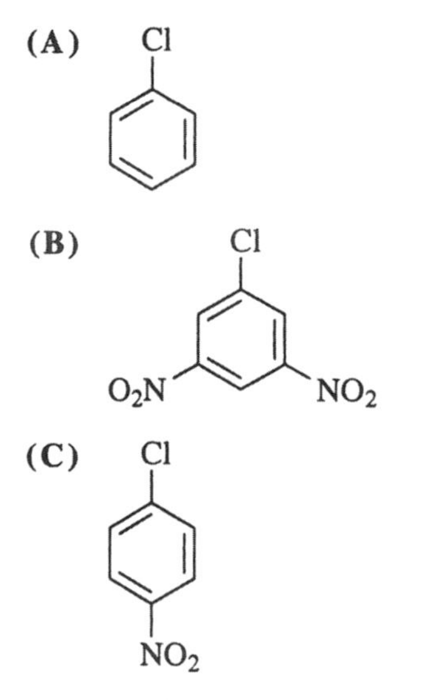

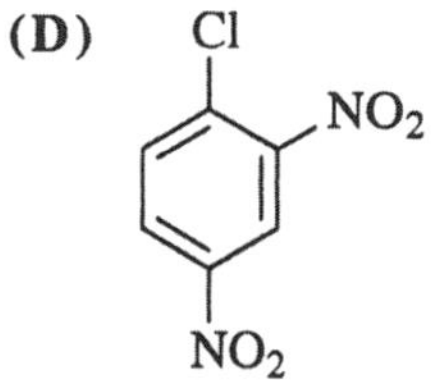

11. What is the expected product of this reaction?

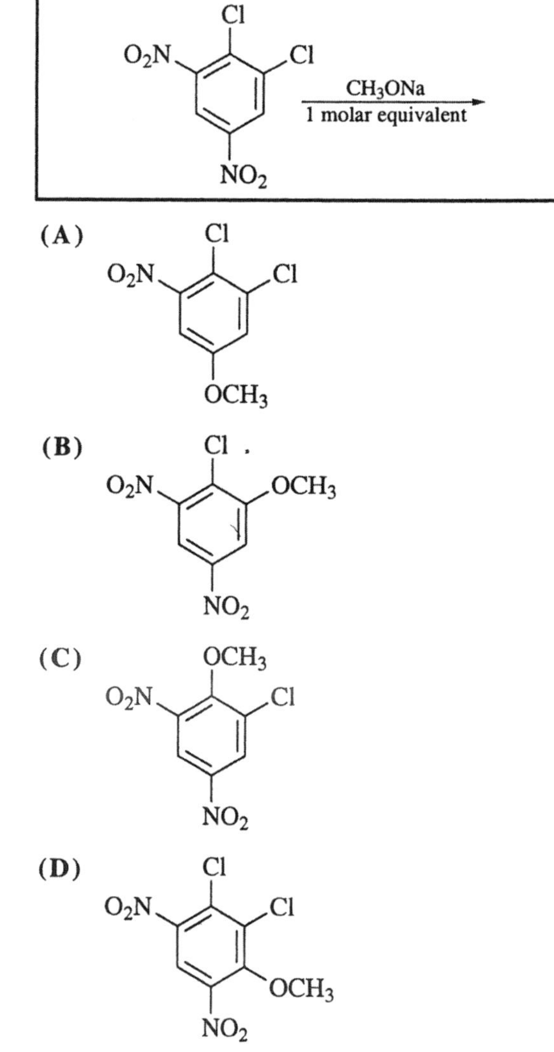

12. Which intermediate is involved in this reaction?

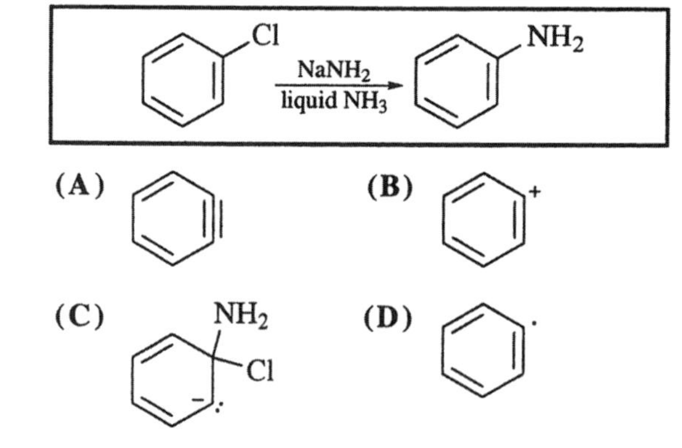

13. Which sequence of reactions is expected to give the best yield of 3-nitrobenzoic acid?

3-nitrobenzoic acid

(A) C_6H_5–NO_2 $\xrightarrow[AlCl_3]{CH_3Cl}$ $\xrightarrow[heat]{K_2Cr_2O_7,\ H_3O^+}$

(B) C_6H_5–CH_3 $\xrightarrow[heat]{K_2Cr_2O_7,\ H_3O^+}$ $\xrightarrow[heat]{HNO_3,\ H_2SO_4}$

(C) C_6H_5–CH_3 $\xrightarrow[heat]{HNO_3,\ H_2SO_4}$ $\xrightarrow[heat]{K_2Cr_2O_7,\ H_3O^+}$

(D) C_6H_5–CH_3 $\xrightarrow[heat]{K_2Cr_2O_7,\ H_3O^+}$ $\xrightarrow[°C]{NaNO_2,\ HCl}$

14. What is the major product of this reaction?

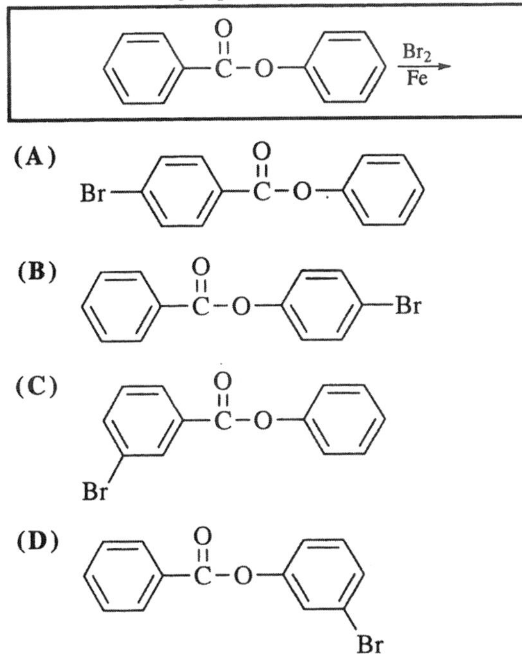

15. Which of these reactions would produce *tert*-butylbenzene in high yield?

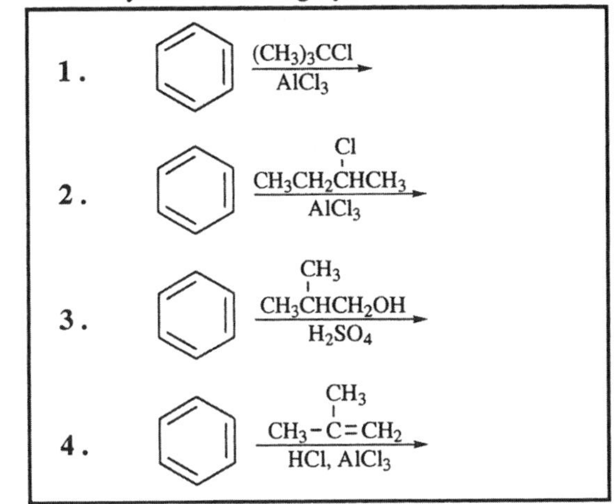

(A) 1, 2, 3, and 4 (B) 1, 2, and 3 only

(C) 1 and 2 only (D) 1, 3, and 4 only

Answers to Study Questions

1. B
2. B
3. A
4. C
5. A
6. A
7. D
8. C
9. C
10. B
11. D

Answers to Practice Questions

1. B
2. C
3. B
4. A
5. A
6. D
7. C
8. D
9. C
10. D
11. C
12. A
13. B
14. B
15. D

Free-Radical Substitutions and Additions

Free radicals (often just called "radicals") are species such as Cl• or CH_3• that have an unpaired electron. Radicals are electron-deficient because they do not have a full octet of electrons, and they are usually highly reactive. Radicals are neutral, meaning that they have no formal charge. Reactions involving ions (S_N1, E2, etc.) require a solvent to dissolve the ionic species, but because radicals are electrically neutral, a solvent is not necessary. Many radical reactions, such as the reaction of CCl_2F_2 with ozone, take place in the gas phase.

There are two general types of free-radical reactions in organic chemistry: substitution and addition. The halogenation of a hydrocarbon with chlorine is an example of a free-radical substitution reaction (equations *1a–1b*).

$$Cl\bullet + CH_4 \rightarrow CH_3\bullet + HCl \qquad (1a)$$

$$CH_3\bullet + Cl_2 \rightarrow CH_3Cl + Cl\bullet \qquad (1b)$$

Two free-radical addition reactions often discussed in undergraduate organic chemistry are the addition of HBr to alkenes in the presence of peroxides, and the polymerization of alkenes or dienes. These reactions can be carried out in solution or in the gas phase, and the reactions can be initiated by heat, peroxides, or ultraviolet (UV) light.

All radical reactions are chain reactions. That is, the reaction proceeds in a series of discrete steps. Regardless of whether the reaction is substitution or addition, radical reactions proceed through three general steps. The first step, **initiation**, generates radicals. **Propagation**, the second step, involves a radical reacting with a stable molecule to produce a new radical. The new radical reacts in turn with another stable molecule to produce yet another radical, and so on. The last step, **termination**, destroys radicals and stops the reaction. In a typical radical reaction, each radical takes part in hundreds or thousands of propagation steps before it is destroyed in a termination step.

The energy required for the generation of organic radicals reflects their relative stability. The most stable radicals require the least energy to form, and they are generated most rapidly. The stability of radicals generally increases with increasing substitution on the carbon atom that carries the unpaired electron. Allylic and benzylic radicals are easy to generate because they are stabilized through resonance. Resonance forms of an allylic radical are illustrated in equation 2.

$$\bullet CH_2{-}CH{=}CH_2 \longleftrightarrow CH_2{=}CH{-}CH_2\bullet \qquad (2)$$

The general trend in stability for organic radicals is:

allylic/benzylic > 3° alkyl > 2° alkyl > 1° alkyl > methyl > vinylic/aryl

A reaction usually involves radicals if any of these reactants or conditions are present: AIBN (a radical initiator), peroxides (RO–OR), UV light (often indicated by the symbol *hν*) or high temperature (300–500 °C).

Study Questions

FR-1. In which classification is this reaction properly placed?

$$CH_4 + Cl_2 \xrightarrow{400\,°C} CH_3Cl + HCl$$

(A) electrophilic substitution **(B)** electrophilic addition

(C) elimination **(D)** free-radical substitution

Knowledge Required: Identification of reaction types.

Thinking it Through: Examination of the reaction equation shows it to be a substitution reaction, and the reaction takes place at high temperature. Methane and molecular chlorine are gases under these conditions. Recognition that the reaction is a substitution reaction allows choices **(B)** and **(C)** to be discarded immediately. Electrophilic substitution, choice **(A)**, involves the formation of ions in the course of the reaction. Ions are difficult to form in the gas phase, and electrophilic substitutions almost all require a solvent. Free radicals, on the other hand, do not require a solvent because they are uncharged, and are typically formed in the gas phase. The use of high temperatures is an additional clue that a free-radical mechanism is likely. Choice **(D)** is the best answer.

FR-2. Which of the indicated sites would most readily undergo hydrogen atom abstraction to generate a radical?

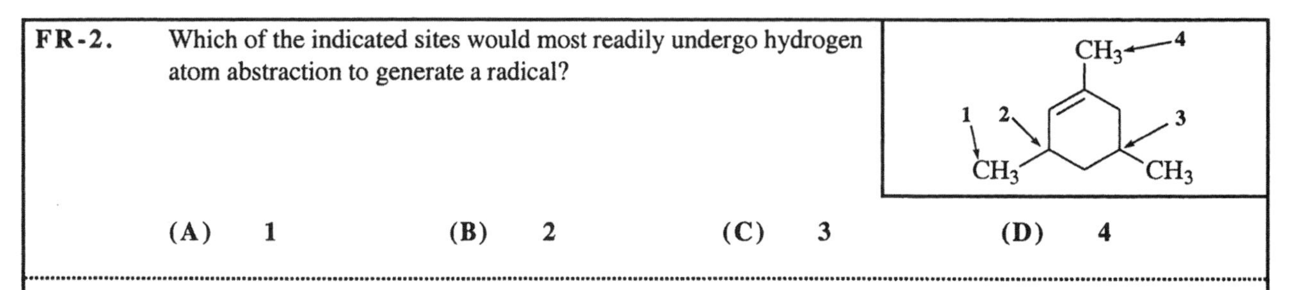

(A) 1 (B) 2 (C) 3 (D) 4

Knowledge Required: Relative stability of different types of radicals.

Thinking it Through: The hydrogen atom that is most easily abstracted will be determined by which of the possible radical intermediates is most stable. It is necessary to recall that generation of an allylic radical is favored over generation of any alkyl (methyl, 1°, 2°, 3°) radical, and that substitution stabilizes alkyl radicals. Methyl radicals are the least stable, and tertiary alkyls are the most stable.

Considering the molecule in this problem, removal of a hydrogen atom from carbon atom **1** will produce a primary radical. Removal of a hydrogen atom from carbon atom **2** will generate a tertiary radical that is also allylic. Removal of a hydrogen atom from carbon atom **3** will generate a tertiary radical. Finally, removal of a hydrogen atom from carbon atom **4** will give rise to a primary radical which is also allylic.

Given that allylic radicals are more stable than any alkyl radicals (which also means that homolytic breaking of the C—H bond to produce them requires less energy), the correct answer must be either choice **(B)** or **(D)**. Choices **(A)** and **(C)** cannot be correct because these give rise to ordinary alkyl radicals. The decision between choice **(B)** and **(D)** is made based on the extent to which the carbon atom carrying the radical is substituted. Since removal of the hydrogen atom at **2** forms a tertiary allylic radical, while removal of a hydrogen atom at **4** forms a primary allylic radical, choice **(B)** is the best answer. It is easier to remove a tertiary allylic hydrogen atom than it is to remove a primary allylic hydrogen atom.

FR-3. Which radical is the most stable?

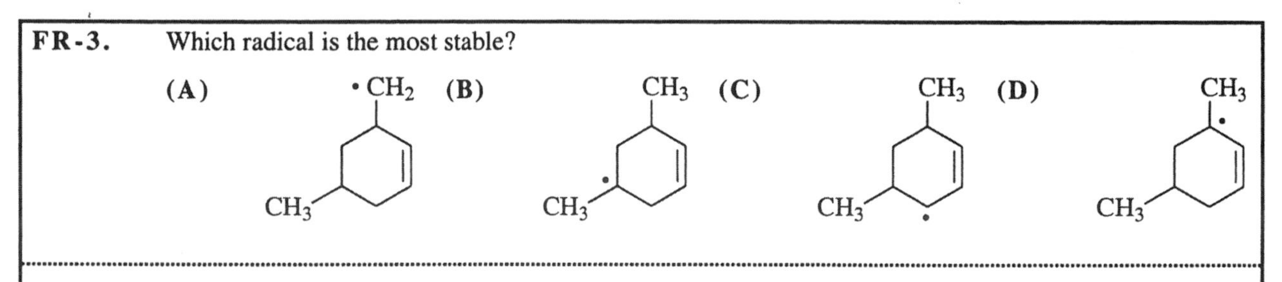

Knowledge Required: Relative stability of various radical structures.

Thinking it Through: This problem is best approached by first classifying each of the radicals in terms of substitution (1°, 2°, 3°, etc.) and whether the radical is alkyl or allylic. Choice **(A)** is a primary alkyl radical. Choice **(B)** is a tertiary alkyl radical. Choice **(C)** is a secondary allylic radical. Choice **(D)** is a tertiary allylic radical. Recall that allylic radicals are more stable than alkyl radicals, so choices **(A)** and **(B)** can be eliminated because they are alkyl radicals. The remaining choices are a secondary allylic radical(choice **(C)**) and a tertiary allylic radical (choice **(D)**). The stability of allylic radicals follows the same trend as alkyl radicals (3° allylic is more stable than 2° allylic, and so on). Choice **(D)** is both tertiary and allylic, and is the best answer.

FR-4 Each of these reactions occurs during the free-radical chlorination of methane. Which step corresponds to the first propagation step of the reaction?

(A) $Cl{-}Cl \rightarrow 2Cl\cdot$

(B) $Cl\cdot + Cl\cdot \rightarrow Cl{-}Cl$

(C) $Cl\cdot + CH_4 \rightarrow CH_3\cdot + HCl$

(D) $CH_3\cdot + Cl_2 \rightarrow CH_3Cl + Cl\cdot$

Knowledge Required: Mechanism of free-radical halogenation reactions.

Thinking it Through: Recalling that free-radical reactions take place in three general steps—initiation (radicals generated); propagation (radicals reacting with neutral molecules to generate new radicals); and termination (radicals destroyed)—the choices can be analyzed to determine which equation represents the first radical generated reacting with a neutral molecule. Choice **(A)** is incorrect because this equation is for an initiation step. Radicals are being formed from neutral molecules. Choice **(B)** can be eliminated because it is the equation for a termination step. Radicals are recombining to form a neutral molecule. Only choices **(C)** and **(D)** involve the reaction of a radical and a neutral molecule to produce a new radical, which is the requirement for a propagation step. The question asks for the *first* propagation step, which requires that the first radical formed appear on the reactant side of the equation. Since Cl• is the first radical formed, choice **(C)** must be the correct answer. The products of the first propagation reaction are $CH_3\cdot + HCl$, and the methyl radical reacts with Cl_2 in the second propagation step.

FR-5 What is the expected stereochemistry of the organic product from this reaction?

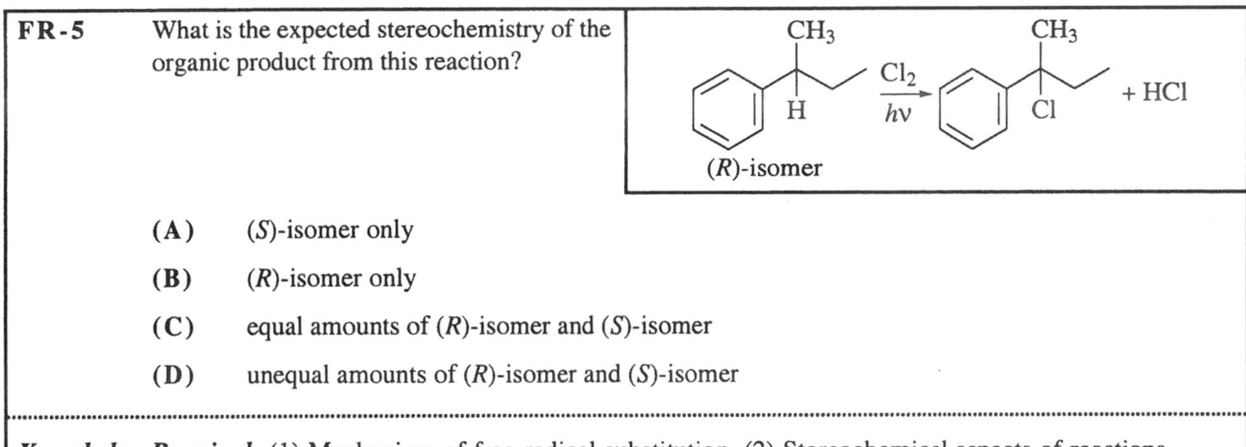

(A) (*S*)-isomer only

(B) (*R*)-isomer only

(C) equal amounts of (*R*)-isomer and (*S*)-isomer

(D) unequal amounts of (*R*)-isomer and (*S*)-isomer

Knowledge Required: (1) Mechanism of free-radical substitution. (2) Stereochemical aspects of reactions involving radicals.

Thinking it Through: This reaction is a free-radical substitution in which a benzylic hydrogen atom is replaced by a chlorine atom. The starting material is a single enantiomer, and the initiation step involves generation of chlorine radicals. A benzylic hydrogen atom is abstracted from the starting material, leading to the radical shown below. Typical radicals—including benzylic radicals—are planar because the radical carbon atom is sp^2-hybridized. The radical has a plane of symmetry, which makes it achiral.

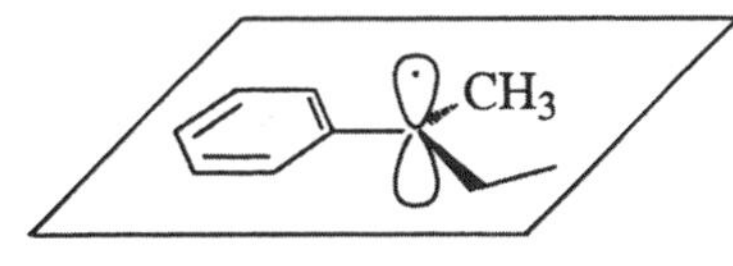

When this achiral radical reacts with Cl_2, the chlorine can attack from either face of the radical with equal probability. Attack on one face leads to one enantiomer and attack on the other face leads to the other enantiomer, consequently equal amounts of the (*R*)- and the (*S*)-products will be formed. Racemization occurs because the reaction proceeds through an achiral radical. Choices **(A)** and **(B)** can be eliminated because they imply complete inversion and retention, respectively, neither of which is consistent with formation of an achiral intermediate. Choice **(D)** is also incorrect because this response implies that only partial racemization takes place. Choice **(C)** is the correct answer.

FR-6 What is the expected product of this reaction?

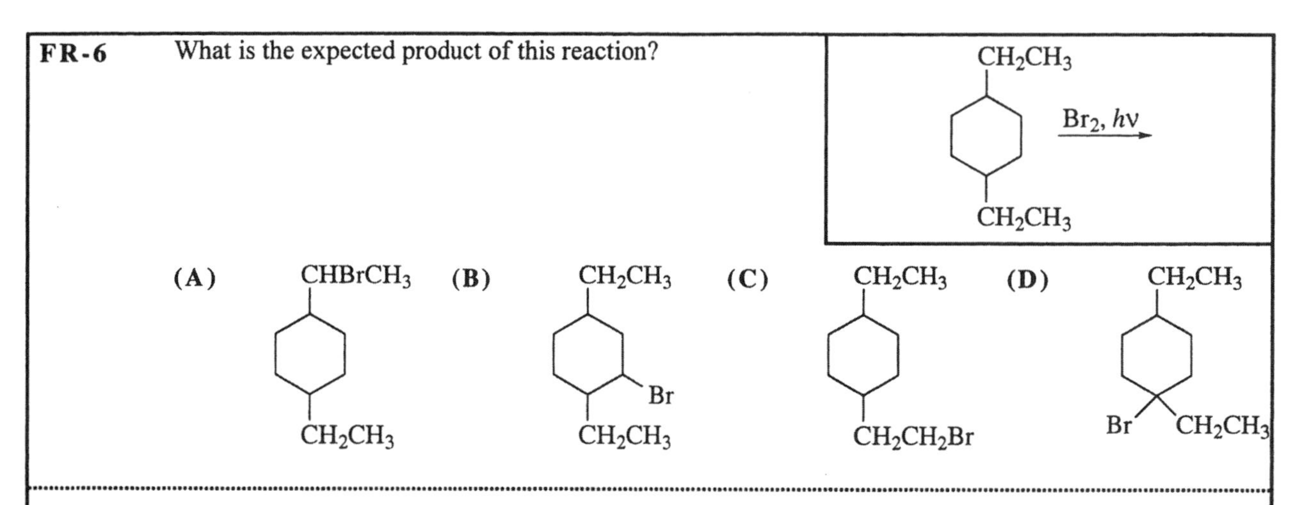

Knowledge Required: Selectivity of bromine in free-radical halogenation reactions.

Thinking it Through: A good way to approach this problem is to determine which type of hydrogen atom is being replaced in each choice. In products **(A)** and **(B)**, a secondary alkyl hydrogen atom is replaced. The product in choice **(C)** has a primary alkyl hydrogen atom being replaced, while the product in choice **(D)** is formed by the replacement of a tertiary alkyl hydrogen atom.

In free-radical bromination reactions, the reaction

$$\text{R–H} + \text{Br}\cdot \longrightarrow \text{R}\cdot + \text{HBr}$$

is usually endothermic. The Br• tends to be quite selective in its reactions, usually abstracting the hydrogen atom that has the weakest C–H bond. Given the choices of a primary, secondary, or tertiary hydrogen atom to be abstracted, the tertiary alkyl hydrogen atom is the most easily abstracted. Thus, choice **(D)** is the correct answer. The other choices indicate replacement of less easily abstracted hydrogen atoms.

FR-7 Which reagent would be used to carry out this transformation?

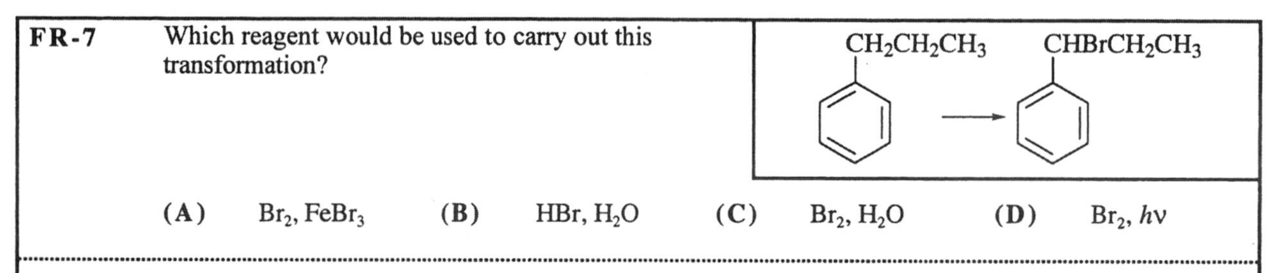

Knowledge Required: Reagents used for free-radical halogenation of alkylbenzenes.

Thinking it Through: In the reaction, propylbenzene is brominated at the benzylic position, which requires free-radical conditions. Choice **(A)** is incorrect because these reagents are used to brominate aromatic rings by electrophilic substitution. The use of these reagents with propylbenzene would result in the formation of *ortho* and *para* brominated propylbenzene, neither of which is the desired product. The reagents in choice **(B)** will not react with the starting material, so this choice cannot be correct. Bromine and water are used to form halohydrins from alkenes. The starting material is not an alkene, and no reaction would occur, eliminating choice **(C)**. Bromine and light (choice **(D)**) can be used to brominate alkanes by a free-radical mechanism. The free-radical bromination of alkylbenzenes leads to substitution at the benzylic position, which is what is indicated in this case. Therefore, choice **(D)** is the correct answer.

FR-8 If radical Y is extremely regioselective in the reaction, which reaction coordinate diagram would be most appropriate?

R—H + Y• ⟶ R• + H—Y

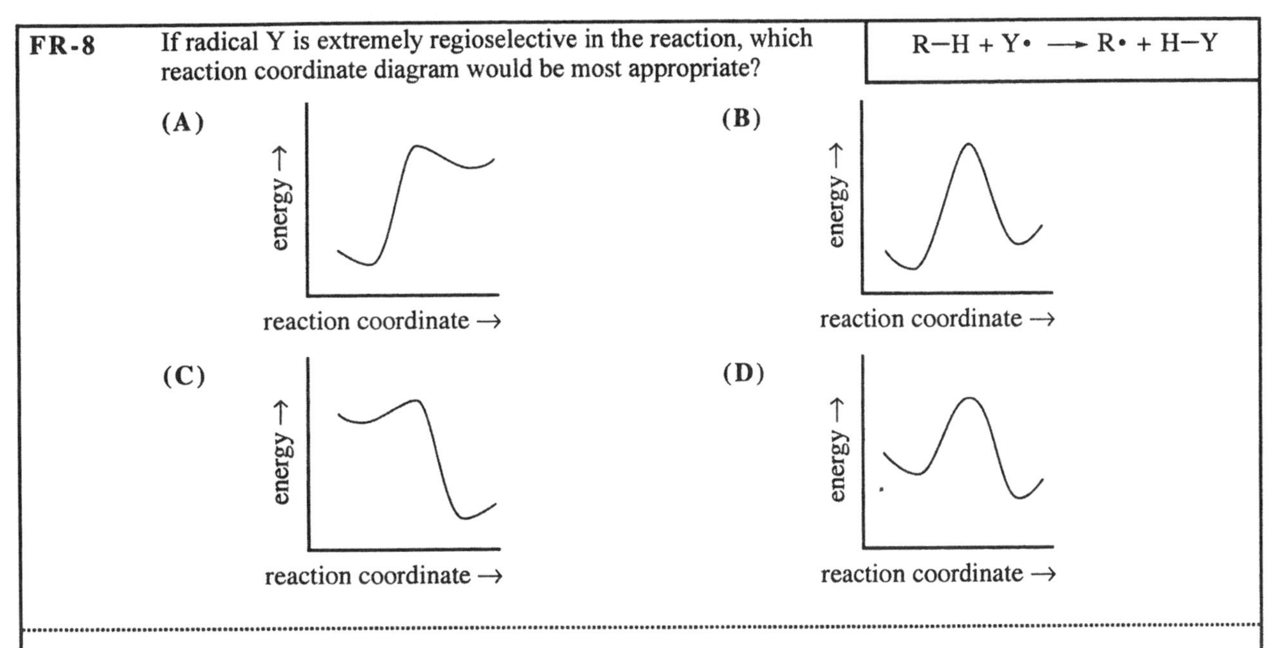

Knowledge Required: (1) Interpretation of energy diagrams. (2) The Hammond postulate and the reactivity-selectivity principle.

Thinking it Through: The hydrogen-abstraction reaction may be exothermic or endothermic. For a highly exothermic reaction with a low E_{act} (the diagram in choice **(C)**, for example), the transition state energy lies close to the energy of the reactants. Conversely, for a highly endothermic reaction with a high E_{act} (as in the diagram in choice **(A)**), the transition state energy lies close to the energy of the products. These statements are often called the Hammond postulate. The product of the present reaction is a carbon radical. The transition state energy in the choice **(C)** diagram resembles the energy of the reactants, and has very little alkyl radical character. This means that alkyl radical stability has practically no effect on the transition-state energy diagram, and that reactions that correspond to the choice **(C)** diagram should exhibit very little regioselectivity. The transition state energy in the choice **(A)** diagram resembles the product energy, and has a significant amount of alkyl radical character. Thus, alkyl radical stability should greatly affect the transition-state energy, and reactions that correspond to diagram **(A)** should exhibit high regioselectivity. Thus, choice **(A)** appears to be the best answer. Because diagrams **(B)** and **(D)** are neither highly exothermic nor endothermic, their transition state energies should have alkyl radical character that is intermediate between that for the diagrams of choices **(A)** and **(C)**. Thus, reactions corresponding to the diagrams of choices **(B)** and **(D)** would be more regioselective than those that correspond to the diagram of choice **(C)**, but less regioselective than those that corresponding to the diagram of choice **(A)**. A generalization of the above is the "reactivity-selectivity principle," which states that highly reactive species are nonselective in their reactions and relatively non-reactive species are very selective in their reactions. This holds because the reactions of highly reactive species are exothermic, whereas the reactions of relatively non-reactive species are endothermic.

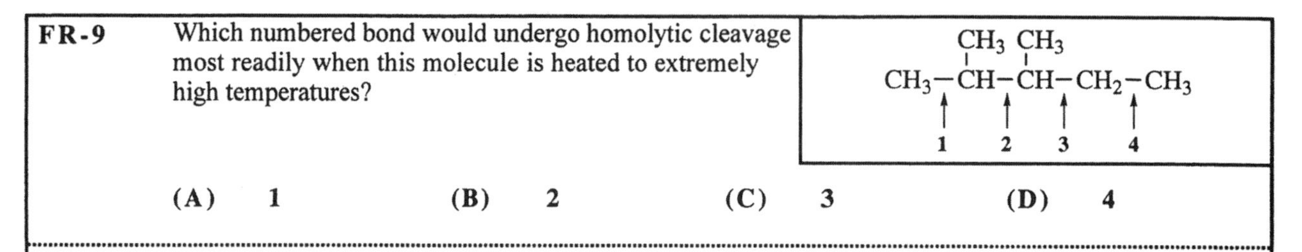

FR-9 Which numbered bond would undergo homolytic cleavage most readily when this molecule is heated to extremely high temperatures?

(A) 1 (B) 2 (C) 3 (D) 4

Knowledge Required: Relative stabilities of various types of alkyl radicals.

Thinking it Through: The heating of molecules to extremely high temperatures is called pyrolysis. A typical pyrolysis reaction is the cleavage of σ bonds to give radicals. As one might expect, cleavage typically occurs first at the weakest bond(s) in a molecule. In this example, there are several bonds that could be broken. The best way to choose the most likely bond to be broken is to compare the stabilities of the radicals that would form from each cleavage reaction. In choice **(A)**, a methyl radical and a secondary radical would form. Cleavage of the bond in choice **(B)** would lead to the formation of two secondary radicals. Choice **(C)** would give a secondary radical and a primary radical. The cleavage of the bond in choice **(D)** would lead to the formation of a primary radical and a methyl radical.

Recalling that the trend in the stability of alkyl radicals, from most to least stable, is tertiary, secondary, primary, and methyl. When considering the radicals that could be formed by pyrolysis of the above molecule, forming secondary radicals is preferable to forming primary or methyl radicals, and choice **(B)** forms two secondary radicals. It is the best answer because each of the other choices forms at least one primary or methyl radical. None of the choices form a tertiary radical.

FR-10 Which is the reactive intermediate when 1-octene is treated with HBr in the presence of peroxides?

(A) $C_6H_{13}-\dot{C}H-CH_3$

(B) $C_6H_{13}-\dot{C}H-CH_2Br$

(C) $C_6H_{13}-CH_2-\dot{C}H_2$

(D) $C_6H_{13}-CH(Br)-\dot{C}H_2$

Knowledge Required: Mechanism of reaction of alkenes with HBr plus peroxides.

Thinking it Through: The reactions of alkenes with HBr in the presence of peroxides are free-radical addition reactions that proceed through intermediate carbon radicals. What must be determined here are (1) whether the carbon radical is formed by the attack of a hydrogen radical or a bromine radical on the π bond, and (2) whether the attack occurs at C–1 or C–2 of the alkene. Regardless of which free radical initially attacks the π bond, the site undergoing the attack will be the one that produces the more stable carbon radical. Attack at C–1 would produce a secondary radical, whereas attack at C–2 would produce a primary radical. Thus, attack will occur at C–1 and choices **(C)** and **(D)** are eliminated. To decide between choices **(A)** and **(B)**, we need to know whether the attack is initiated by a hydrogen radical or a bromine radical. In the mechanism as summarized below, the peroxide first fragments into two oxygen radicals, which then abstract a hydrogen atom from the HBr to generate a bromine radical. Thus, a bromine radical attacks the π bond, not a hydrogen radical. Therefore choice **(B)** is the correct answer.

$$RO-OR \rightarrow RO\bullet + \bullet OR$$

$$RO\bullet + H-Br \rightarrow RO-H + Br\bullet$$

Propagation:

$$C_6H_{13}-CH=CH_2 + Br\bullet \rightarrow C_6H_{13}-\dot{C}H-CH_2-Br$$

$$C_6H_{13}-\dot{C}H-CH_2-Br + H-Br \rightarrow C_6H_{13}-CH_2-CH_2-Br + Br\bullet$$

FR-11 Free-radical polymerization of 2-chloro-1,3-butadiene would produce what polymer?

$$CH_2{=}CCl{-}CH{=}CH_2$$
$$\begin{matrix} 1 & 2 & 3 & 4 \end{matrix}$$

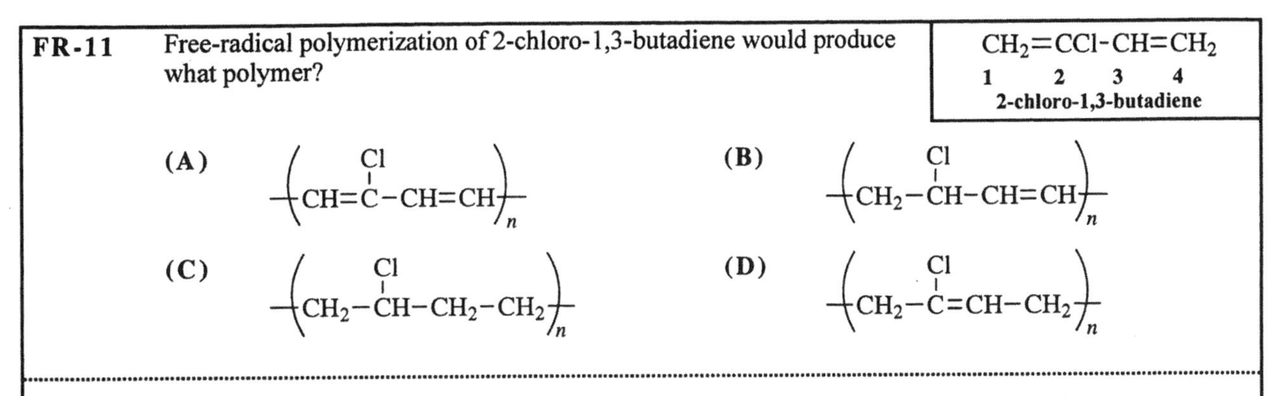

(A) $\left(\!\!\begin{array}{c} Cl \\ | \\ CH{=}C{-}CH{=}CH \end{array}\!\!\right)_n$

(B) $\left(\!\!\begin{array}{c} Cl \\ | \\ CH_2{-}CH{-}CH{=}CH \end{array}\!\!\right)_n$

(C) $\left(\!\!\begin{array}{c} Cl \\ | \\ CH_2{-}CH{-}CH_2{-}CH_2 \end{array}\!\!\right)_n$

(D) $\left(\!\!\begin{array}{c} Cl \\ | \\ CH_2{-}C{=}CH{-}CH_2 \end{array}\!\!\right)_n$

Knowledge Required: (1) Mechanism of addition polymerization. (2) Product of 1,3-diene polymerization.

Thinking it Through: Polymers are composed of repeating units derived from monomer subunits. In free-radical polymerization, a radical reacts with an alkene or diene to produce a new, larger radical. This new radical then reacts with another molecule of alkene or diene to produce a still larger radical, and so on. In the polymerization of 1,3-dienes, bonds are formed between **C–1** of one monomer unit to **C–4** of another monomer, and the repeating unit of the polymer has the same four-carbon-atom chain as the starting material. The two double bonds become single bonds in this process, and a double bond forms between C–**2** and C–**3**.

In the monomer, **C–1** carries two hydrogen atoms, **C–2** carries one chlorine atom, **C–3** carries one hydrogen atom, and **C–4** carries two hydrogen atoms. That pattern must be repeated in the polymer. The only product that matches this is choice **(D)**, the correct answer. Choice **(A)** is eliminated because the repeating unit of the polymer has two fewer hydrogen atoms than the monomer. Choice **(B)** shows an extra hydrogen atom on C–**2** and only one hydrogen atom on C–**4**. Choice **(C)** is eliminated because the repeating unit has two extra hydrogen atoms.

FR-12 Which of these polymerization reactions takes place by means of a free-radical process?

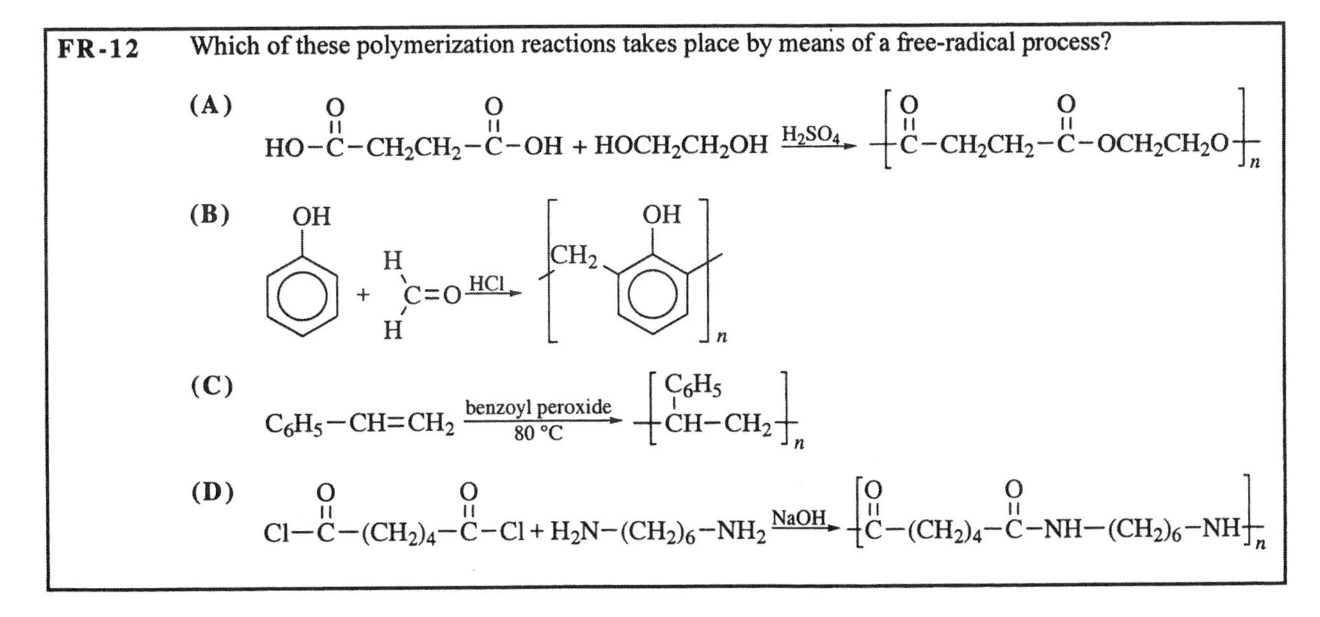

(A) $$HO{-}\overset{O}{\overset{||}{C}}{-}CH_2CH_2{-}\overset{O}{\overset{||}{C}}{-}OH + HOCH_2CH_2OH \xrightarrow{H_2SO_4} \left[\overset{O}{\overset{||}{C}}{-}CH_2CH_2{-}\overset{O}{\overset{||}{C}}{-}OCH_2CH_2O\right]_n$$

(B) (phenol) $+ \overset{H}{\underset{H}{C}}{=}O \xrightarrow{HCl} \left[CH_2\text{(phenol)}\right]_n$

(C) $$C_6H_5{-}CH{=}CH_2 \xrightarrow[80\ °C]{benzoyl\ peroxide} \left[\overset{C_6H_5}{\overset{|}{CH}}{-}CH_2\right]_n$$

(D) $$Cl{-}\overset{O}{\overset{||}{C}}{-}(CH_2)_4{-}\overset{O}{\overset{||}{C}}{-}Cl + H_2N{-}(CH_2)_6{-}NH_2 \xrightarrow{NaOH} \left[\overset{O}{\overset{||}{C}}{-}(CH_2)_4{-}\overset{O}{\overset{||}{C}}{-}NH{-}(CH_2)_6{-}NH\right]_n$$

Knowledge Required: (1) Recognition of various types of polymerization reactions. (2) Difference between addition polymers and condensation polymers.

Thinking it Through: It is important to remember that there are two main types of polymerization reactions: addition polymerization (sometimes called chain-reaction polymerization); and condensation polymerization (sometimes called step-reaction polymerization). Addition polymers give a polymer with a repeating unit that has the same chemical formula as the monomer, and the reactions typically proceed by way of a highly reactive species, such as a free radical or carbocation. Condensation polymers are formed by the reaction of two functional groups to link monomer units, accompanied by the formation of a small molecule such as water or HCl. The reaction proceeds in discrete steps to give a polymer, and the reacting molecules must each have more than one reactive site. Condensation polymers *do not* involve free radical intermediates. Free-radical polymerization reactions produce only addition polymers.

Each of the four reactions represented here is a polymerization reaction, but only one involves radicals. Since condensation polymerization never involves radicals, each choice that is a condensation polymerization can be eliminated. The polymer in choice **(A)** is eliminated because it is a condensation polymer, a polyester, that forms from repeated reactions between alcohol and carboxylic acid groups. Two molecules of water are lost per repeating unit. Choice **(B)** is eliminated because it shows the formation of another condensation polymer, a phenol-formaldehyde resin (Bakelite is an example). Polymerization has occurred with the loss of one molecule of water per repeating unit. The reaction given in choice **(C)** is the formation of polystyrene. The repeating unit in polystyrene has the same molecular formula as the styrene monomer. This is an addition polymer, and is the correct answer. This answer is supported by the presence of peroxides, which are often found in free-radical reactions. Choice **(D)** is incorrect because it represents a condensation polymerization that produces nylon. Two molecules of HCl are eliminated for every repeating unit formed.

Practice Questions

1. Which radical is the *least* stable?

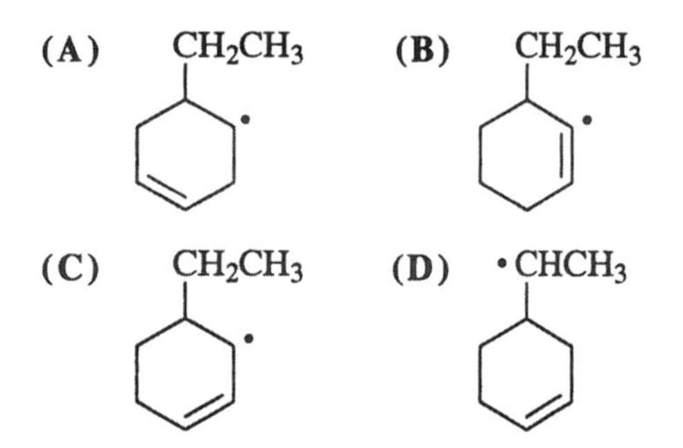

2. Which species represents an intermediate formed during the reaction shown?

$$CH_3CH_2CH_3 \xrightarrow[hv]{Cl_2} CH_3CHClCH_3 + HCl$$

(A) $CH_3\dot{C}HCH_3$ (B) $CH_3\overset{+}{C}HCH_3$

(C) $CH_3\overset{-}{C}HCH_3$ (D) $CH_3CH{=}CH_2$

3. Which is a propagation step in the free-radical chlorination of ethane?

(A) $CH_3\dot{C}H_2 + CH_3CH_3 \rightarrow CH_3CH_3 + CH_3\dot{C}H_2$

(B) $CH_3\dot{C}H_2 + CH_3\dot{C}H_2 \rightarrow CH_3CH_2CH_2CH_3$

(C) $CH_3\dot{C}H_2 + \dot{C}l \rightarrow CH_3CH_2Cl$

(D) $CH_3\dot{C}H_2 + Cl_2 \rightarrow CH_3CH_2Cl + \dot{C}l$

4. What is a key intermediate in the following reaction?

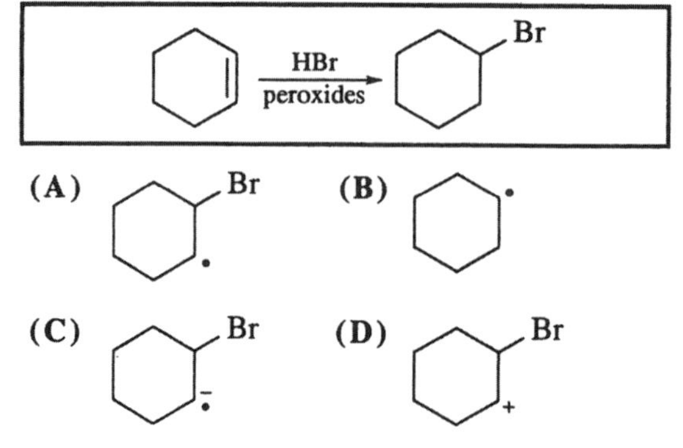

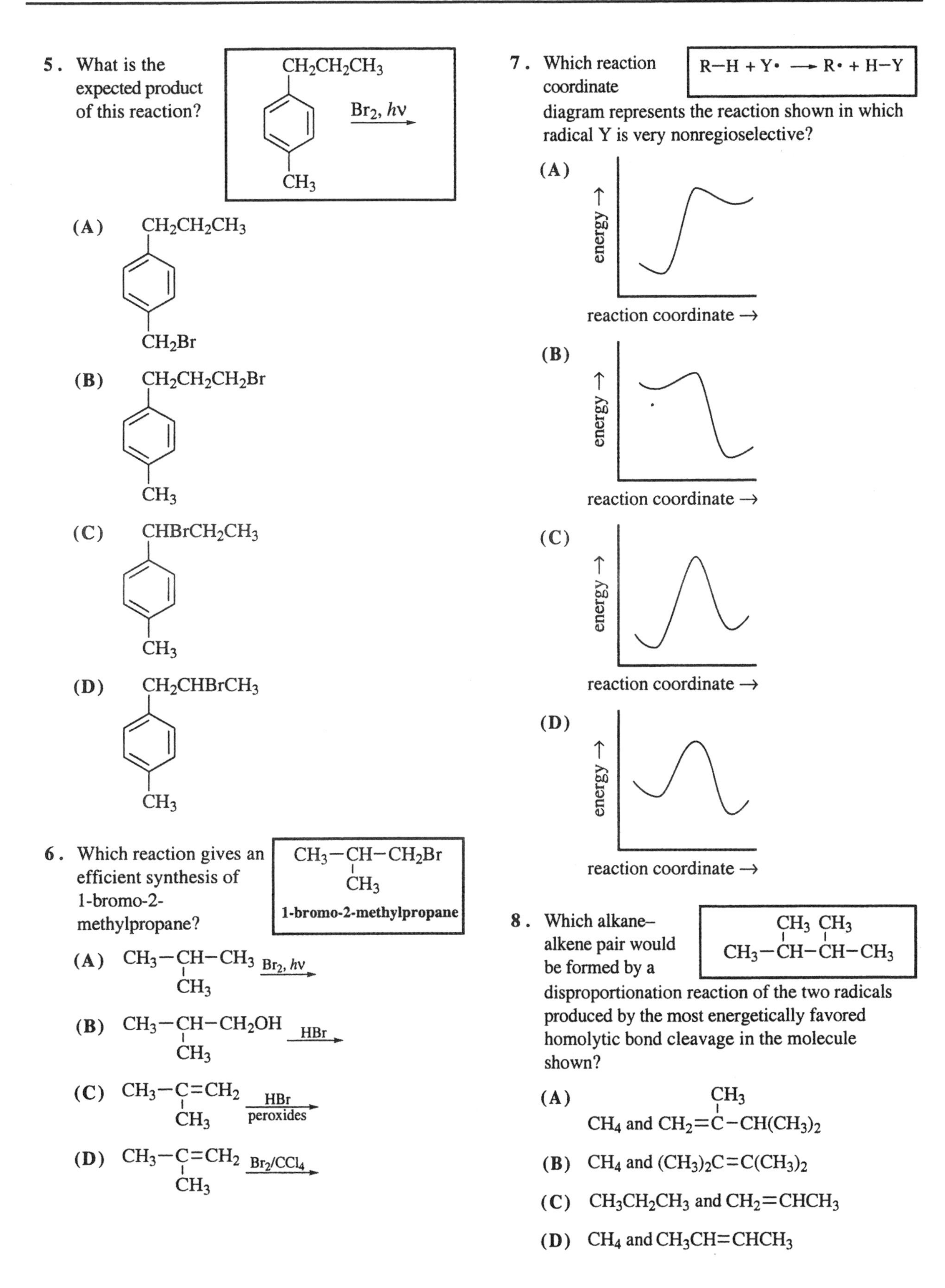

5. What is the expected product of this reaction?

(Reactant: benzene ring with $CH_2CH_2CH_3$ at the top and CH_3 at the bottom; reagent Br_2, $h\nu$)

(A) $CH_2CH_2CH_3$ (ring) CH_2Br

(B) $CH_2CH_2CH_2Br$ (ring) CH_3

(C) $CHBrCH_2CH_3$ (ring) CH_3

(D) $CH_2CHBrCH_3$ (ring) CH_3

6. Which reaction gives an efficient synthesis of 1-bromo-2-methylpropane?

$CH_3-CH(CH_3)-CH_2Br$

1-bromo-2-methylpropane

(A) $CH_3-CH(CH_3)-CH_3 \xrightarrow{Br_2,\ h\nu}$

(B) $CH_3-CH(CH_3)-CH_2OH \xrightarrow{HBr}$

(C) $CH_3-C(CH_3)=CH_2 \xrightarrow[\text{peroxides}]{HBr}$

(D) $CH_3-C(CH_3)=CH_2 \xrightarrow{Br_2/CCl_4}$

7. Which reaction coordinate diagram represents the reaction shown in which radical Y is very nonregioselective?

$R-H + Y\cdot \longrightarrow R\cdot + H-Y$

(A)

(B)

(C)

(D)

8. Which alkane–alkene pair would be formed by a disproportionation reaction of the two radicals produced by the most energetically favored homolytic bond cleavage in the molecule shown?

$CH_3-CH(CH_3)-CH(CH_3)-CH_3$

(A) CH_4 and $CH_2=C(CH_3)-CH(CH_3)_2$

(B) CH_4 and $(CH_3)_2C=C(CH_3)_2$

(C) $CH_3CH_2CH_3$ and $CH_2=CHCH_3$

(D) CH_4 and $CH_3CH=CHCH_3$

9. The copolymerization of vinyl chloride and vinylidine chloride by free radical catalysis produces a polymer used in food wrap. Which structural unit would appear in this polymer?

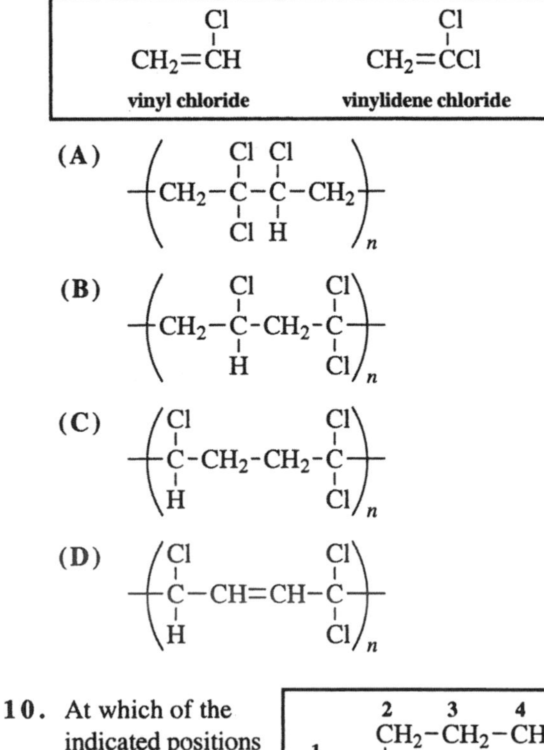

10. At which of the indicated positions would the bromine radical react at the *fastest* rate?

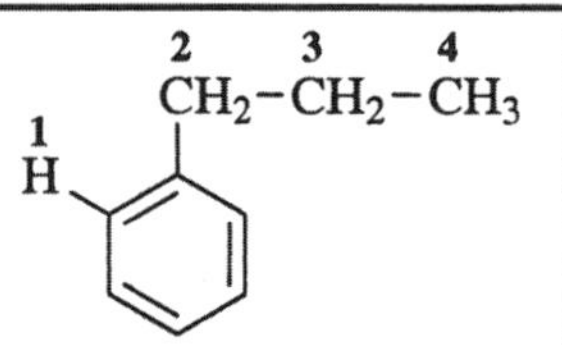

(A) 1 **(B)** 2 **(C)** 3 **(D)** 4

11. Which is the correct representation of the polymer formed by free-radical polymerization of this monomer?

$CH_2{=}CH-CH_3$

(A) $\left(CH_2-\underset{CH_3}{CH}-\underset{CH_3}{CH}-CH_2\right)_n$

(B) $\left(CH{=}\underset{CH_3}{C}-CH{=}\underset{CH_3}{C}\right)_n$

(C) $\left(\underset{CH_3}{CH}-CH_2-CH_2-\underset{CH_3}{CH}\right)_n$

(D) $\left(CH_2-\underset{CH_3}{CH}-CH_2-\underset{CH_3}{CH}\right)_n$

12. What product is formed in this reaction?

$CH_2{=}CHCN \xrightarrow[80\ °C]{(C_6H_5CO)_2O_2}$

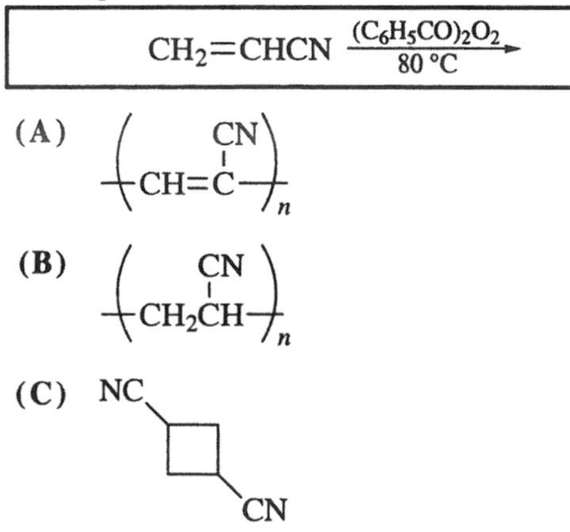

(D) $NC-CH{=}CH-CH{=}CH-CN$

Answers to Study Questions

1. D	5. C	9. B
2. B	6. D	10. B
3. D	7. D	11. D
4. C	8. A	12. C

Answers to Practice Questions

1. B	5. C	9. B
2. A	6. C	10. B
3. D	7. B	11. D
4. A	8. C	12. B

Oxidations and Reductions

Many reactions in organic chemistry are classified as oxidation–reduction reactions. Oxidation can be defined as increasing the oxygen content and/or decreasing the hydrogen content of a compound. Reduction is simply the converse of oxidation (decreasing the oxygen content and/or increasing the hydrogen content of a compound). If a reaction occurs in which the numbers of hydrogen *and* oxygen atoms change in a manner corresponding to the net addition or net elimination of water, it is neither an oxidation nor a reduction reaction. The dehydration of an alcohol is one such example. Hydration of an alkene is another.

Converting an alcohol such as ethanol (CH_2CH_2OH) to ethanal (CH_3CHO) *is* oxidation because the hydrogen atom count has been decreased. Converting ethanal to ethanoic acid (CH_3COOH) is also oxidation, because the oxygen atom count has been increased. The reverse of each of these steps is reduction. Other examples of reduction are hydrogenation of a double or triple bond.

Many different reagents for oxidation and reduction reactions are used in organic chemistry. This variety is potentially confusing, but many of these reagents behave in similar ways. Some reagents react only in specific circumstances, which makes it easier to recall when they are used.

The most widely used oxidizing agents are chromium(VI) reagents (CrO_3, $K_2Cr_2O_7$, and K_2CrO_4) and a manganese(VII) reagent ($KMnO_4$). The chromium(VI) reagents are normally used under acidic conditions, where all are converted into the same active oxidant, chromic acid (H_2CrO_4). Potassium permanganate, on the other hand, is normally used under basic conditions. Chromic acid and hot, basic permanganate are potent oxidizers and usually do not stop until *all* of the hydrogen atoms attached to the carbon atoms undergoing oxidation have been removed. However, it is possible to stop at the aldehyde stage in the oxidation of primary alcohols by the use of anhydrous chromium(VI) reagents. The anhydrous chromium(VI) reagents most commonly used are: chromium trioxide–pyridine complex (CrO_3—2 pyridine), pyridinium chlorochromate (PCC) and pyridinium dichromate (PDC). All three are readily soluble in organic solvents, and the latter two are conveniently available as stable solids. Regardless of which one is used, they all perform the same function.

Hot chromic acid and hot, basic permanganate react with alkenes, alkynes and alkylbenzenes to give oxidative cleavage products. With alkylbenzenes, the oxidation occurs at the benzylic carbon atom, but only when it has one or more benzylic hydrogen atoms, to produce benzoic acids. Other oxidative cleavage reactions are the cleavage of vicinal diols by periodic acid (HIO_4), the cleavage of alkenes by ozone, and the cleavage of methyl ketones by alkaline solutions of the halogens (haloform reaction). Cold, alkaline permanganate, osmium tetroxide and peroxy acids react with alkenes and alkynes to give oxidative addition products.

Reductions of carbonyl compounds are usually performed with complex metal hydrides, reductions of alkenes are performed with catalytic hydrogenation, and reductions of alkynes are performed with either catalytic hydrogenation or alkali metals in ammonia (dissolving metal reductions). Selective reduction of one of these functions in the presence of the others is often possible with the proper choice of reducing agent. Several special reactions are available for the reduction of C=O groups of aldehydes and ketones all the way to CH_2 groups. These are the Clemmensen reduction, the Wolff–Kischner reduction and the desulfuration of thiol acetals and ketals by Raney nickel.

Study Questions

OR-1. Which reagent would be best for carrying out this transformation?

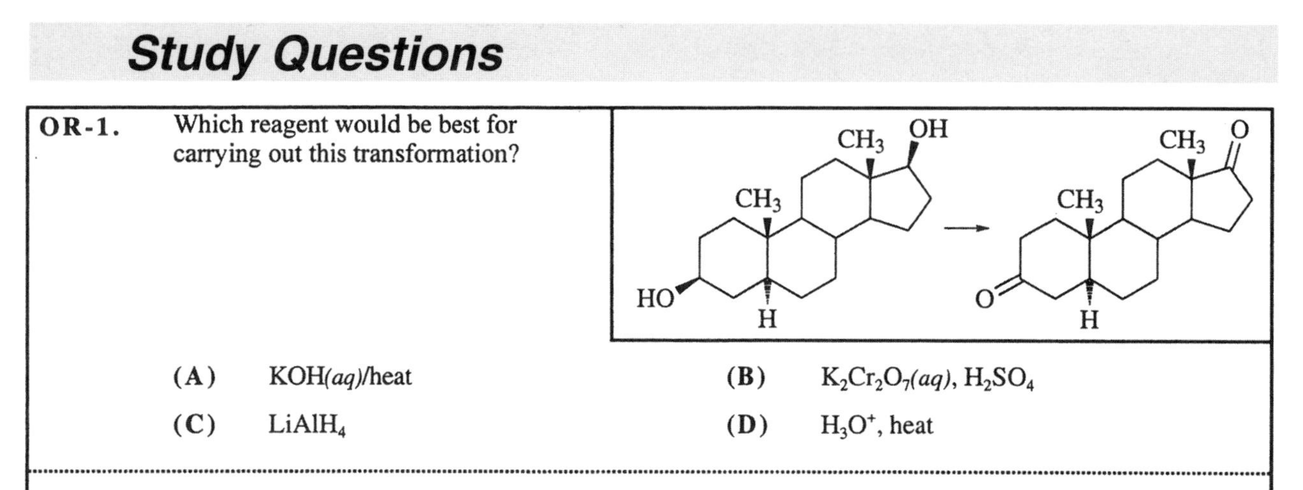

(**A**) KOH(*aq*)/heat (**B**) $K_2Cr_2O_7(aq)$, H_2SO_4

(**C**) $LiAlH_4$ (**D**) H_3O^+, heat

Knowledge Required: Reagents used to oxidize alcohols.

Thinking it Through: To solve this problem, examine the transformation that is taking place. Here, two secondary alcohol groups are being converted into ketone groups. Next, examine the reagents to be considered. Aqueous KOH and heat is not an oxidizing agent, so choice **(A)** is incorrect. Choice **(B)** gives the reagents that make up the oxidizing agent chromic acid. This reagent converts primary alcohols to carboxylic acids and secondary alcohols to ketones. This is the correct answer. Choice **(C)** is incorrect because lithium aluminum hydride ($LiAlH_4$) is a reducing agent, not an oxidizing agent. Choice **(D)** is incorrect because hot, aqueous acid is not an oxidizing agent.

OR-2. Which reagent would be best for carrying out this transformation?

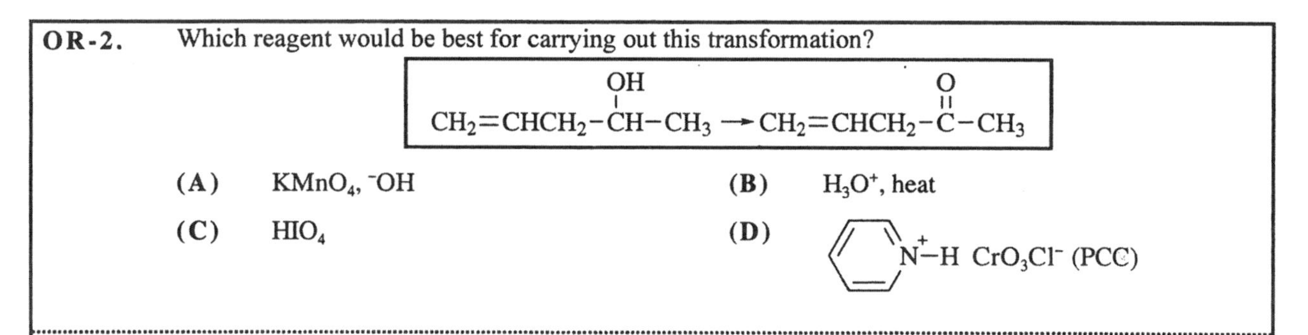

$CH_2{=}CHCH_2{-}CH(OH){-}CH_3 \rightarrow CH_2{=}CHCH_2{-}C(=O){-}CH_3$

(**A**) $KMnO_4$, ^-OH (**B**) H_3O^+, heat

(**C**) HIO_4 (**D**) pyridinium $N^+{-}H$ CrO_3Cl^- (PCC)

Knowledge Required: (1) Reactivity of alkenes with oxidizing agents. (2) Reactivity of alcohols with oxidizing agents.

Thinking it Through: The starting material contains a secondary alcohol and an alkene. The correct answer should be the one that specifically affects the alcohol group and does not react with the alkene group at all. To solve this problem, look at each reagent and recall which functional groups it affects. Choice **(A)**, alkaline $KMnO_4$, reacts with secondary alcohols to give ketones, but it also reacts rapidly with double bonds. Therefore, choice **(A)** can be eliminated. Choice **(B)**, aqueous acid and heat, can also be eliminated because it is not an oxidizing agent, although it can initiate other reactions with alkenes and alcohols. Choice **(C)**, periodic acid (HIO_4), reacts with alcohols only when they are 1,2-diols or alpha to a carbonyl group and results in oxidative cleavage of the carbon–carbon bond. Thus, choice **(C)** can be rejected. Choice **(D)** is pyridinium chlorochromate (PCC), which converts primary alcohols to aldehydes, and secondary alcohols to ketones, but it does not affect carbon–carbon double (or triple) bonds. This is the correct answer.

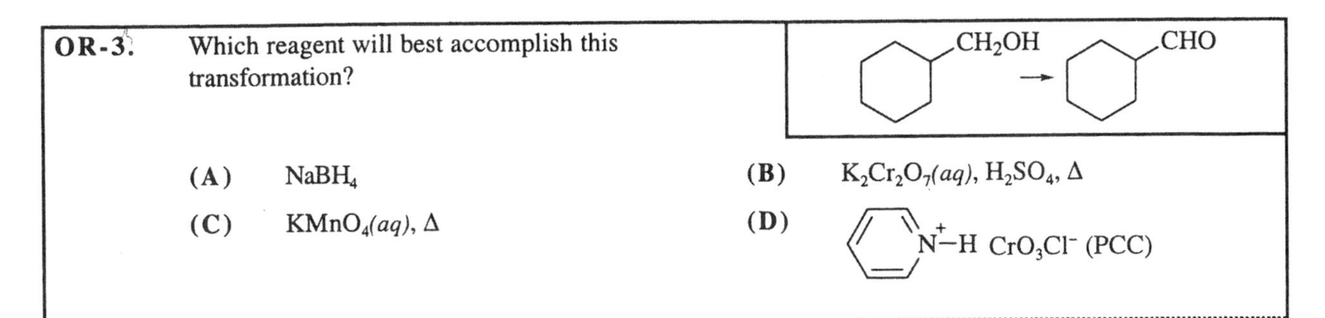

OR-3. Which reagent will best accomplish this transformation?

(A) $NaBH_4$

(B) $K_2Cr_2O_7(aq)$, H_2SO_4, Δ

(C) $KMnO_4(aq)$, Δ

(D) $\overset{+}{N}$–H CrO_3Cl^- (PCC)

Knowledge Required: Oxidation of primary alcohols to aldehydes.

Thinking it Through: In this reaction, a primary alcohol is being oxidized to an aldehyde. It is necessary to know how each of the reagents will affect primary alcohols. The reagent in choice **(A)**, $NaBH_4$, is a reducing agent. Therefore, this cannot be the correct answer. Choice **(B)**, $K_2Cr_2O_7 + H_2SO_4$, oxidizes primary alcohols to carboxylic acids. (An aldehyde intermediate is formed, but it cannot be isolated because it immediately undergoes further oxidation to give the carboxylic acid.) Thus, choice **(B)** is not corrẻct. Choice **(C)**, hot $KMnO_4$, also oxidizes primary alcohols to carboxylic acids (via an aldehyde intermediate), so this is not the correct answer either. The remaining choice, **(D)**, is the correct answer. Note that PCC is one of the few reagents that will oxidize primary alcohols to aldehydes. This is a very useful reaction to know.

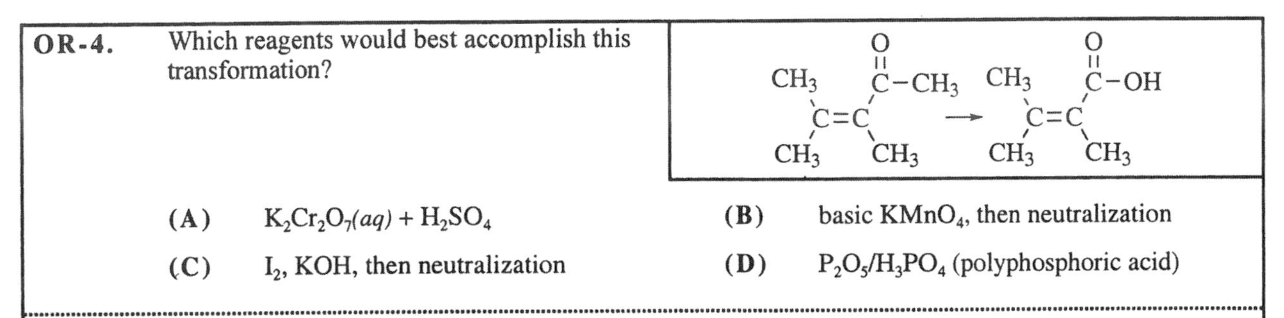

OR-4. Which reagents would best accomplish this transformation?

(A) $K_2Cr_2O_7(aq) + H_2SO_4$

(B) basic $KMnO_4$, then neutralization

(C) I_2, KOH, then neutralization

(D) P_2O_5/H_3PO_4 (polyphosphoric acid)

Knowledge Required: How to oxidize ketones to carboxylic acids.

Thinking it Through: This transformation may seem unusual because a ketone is being converted to a carboxylic acid and the product has one fewer carbon atoms than the starting material.

Choice **(A)** is incorrect because chromic acid does not oxidize ketones, but can oxidize carbon–carbon double bonds. Choice **(B)** can be rejected because permanganate reacts rapidly with carbon–carbon double bonds to give either vicinal diols or oxidative cleavage products. Polyphosphoric acid, choice **(D)**, is not an oxidizing agent. It is often used in dehydration reactions. Choice **(C)** is the correct answer because these reagents are used for the haloform reaction, which converts methyl ketones into carboxylic acids. Treatment of a methyl ketone with alkaline I_2, Br_2, or bleach (NaOCl), followed by neutralization, gives the carboxylic acid plus CHI_3, $CHBr_3$ or $CHCl_3$ (depending on which halogen is present in the reagent).

OR-5. When the compound shown is subjected to vigorous oxidization with $KMnO_4$, followed by neutralization, what is the product?

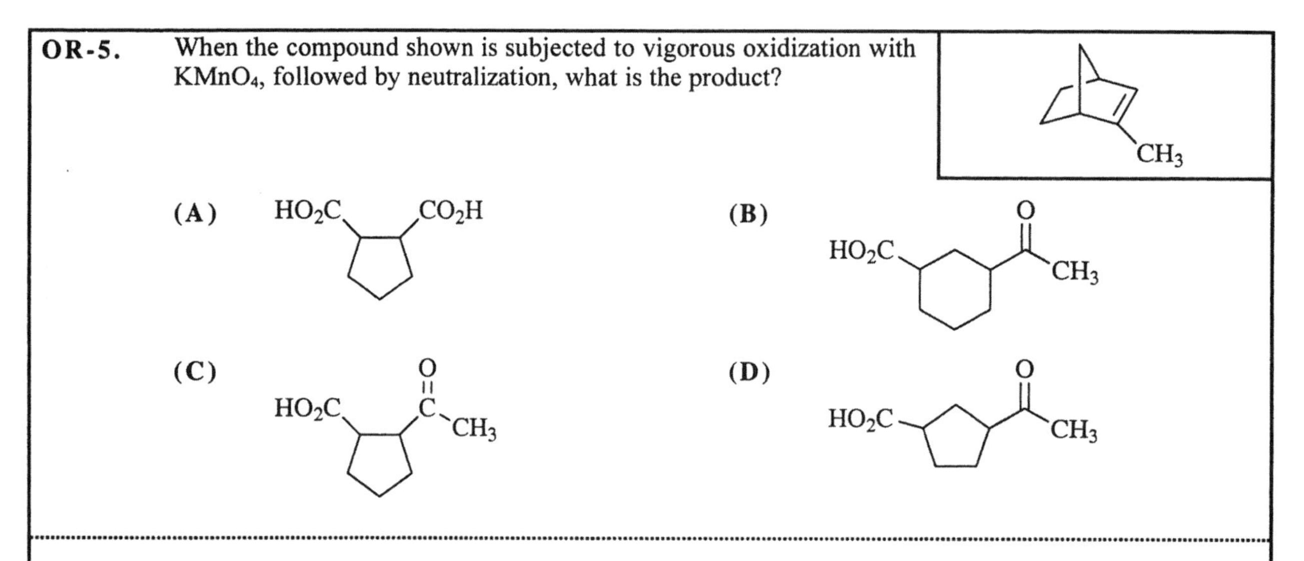

Knowledge Required: Reactivity of $KMnO_4$ with alkenes.

Thinking it Through: The oxidation of alkenes with $KMnO_4$ is well known. The double bond is cleaved, giving carbonyl compounds. For terminal alkenes, the $=CH_2$ group is oxidized to CO_2. For monosubstituted carbon atoms (=CHR), the carbon atom is oxidized to a carboxylic acid (R–COOH). Disubstituted carbon atoms (=CRR´) are oxidized to ketones (R–CO–R´). Since the double bond is cleaved, two separate molecules are formed if the double bond is part of a chain. If the double bond is part of a ring, the ring is broken.

The starting material in this exercise is a bicyclic alkene. $KMnO_4$will react with the alkene to cleave one ring. The five-membered ring on the left will remain intact. The carbon atoms that make up the right-hand ring and the methyl group will be substituents on the remaining five-membered ring. It is possible to reject choice **(B)** because a six-membered ring cannot be the result of this reaction.

To determine the nature and arrangement of the groups on the ring, examine the substitution pattern of the double bond. The vinyl carbon atom bearing a hydrogen atom will form a carboxylic acid. The vinyl carbon atom with the methyl group will form a ketone. Choice **(A)** can be eliminated because this choice has two carboxylic acid groups present. To distinguish between the remaining two choices, it is necessary to determine the relative position of the two groups. Because the substituted carbon atoms (the bridgehead carbon atoms in the starting material) have a CH_2 group between them, the substituted carbon atoms in the product will also have a CH_2 group between them. This required 1,3-relationship between the substituents clearly makes choice **(D)** the correct answer.

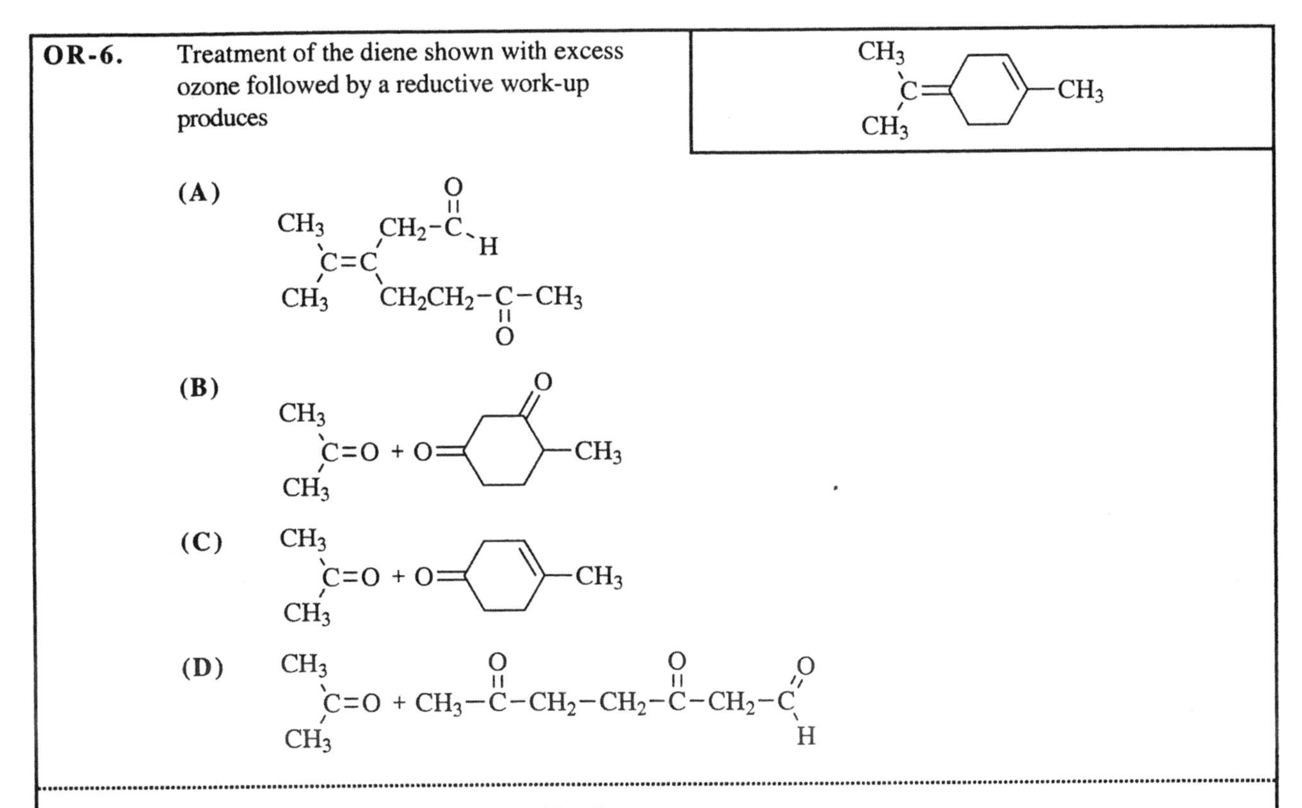

OR-6.
Treatment of the diene shown with excess ozone followed by a reductive work-up produces
(A)
(B)
(C)
(D)

Knowledge Required: How ozone reacts with alkenes.

Thinking it Through: After reductive work-up, cleavage of alkenes with ozone results in the formation of aldehydes and ketones. The vinyl carbon atom in terminal alkenes (=CH_2) is oxidized to formaldehyde, H_2C=O. A vinyl carbon atom with one substituent (=CHR) is converted to an aldehyde (R–CHO). A disubstituted vinyl carbon atom (=CRR′) is oxidized to a ketone (R–CO–R′). Cleavage of an acyclic alkene results in formation of two molecules. Cleavage of a cyclic alkene gives an acyclic product. A convenient way to remember how ozone cleaves double bonds is to imagine stretching the C=C bond and mentally replace the "C=C" with "C=O O=C". Drawing out the starting material, erasing the double bond, then drawing in two carbonyl groups works well.

The starting terpene has two double bonds, one as part of a ring and one that is exocyclic (sticking out of a ring). If the double bonds are redrawn as carbonyl groups, these structures result:

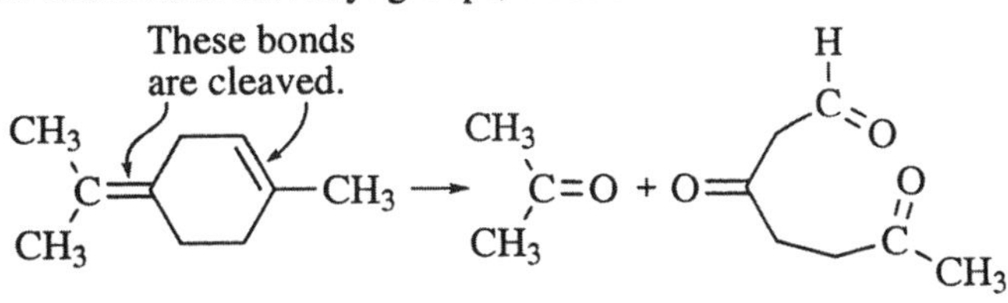

Two products are formed because the exocyclic double bond is cleaved. Further, the ring is broken into a chain. Notice that no alkene groups remain in the products. The substituents on the carbon atoms remain unchanged. Choice **(A)** can be eliminated because it does not give two products as expected and because the product contains an alkene group. Excess ozone would have reacted with the alkene group. This compound is what would result from incomplete ozonolysis, but the question stated that there was excess ozone, so the alkene group would have been cleaved. Choice **(B)** has two products, one of which is the required acetone, but the other compound formed would not be formed from ozonlysis of a cyclohexene system. The ring would be broken. Therefore **(B)** is not the correct answer. Choice **(C)** is also incorrect because one of the products contains a double bond. The correct answer is **(D)**, and inspection of the above figure shows that the projected products match those in **(D)**.

OR-7. Which diol is cleaved by periodic acid, HIO_4?

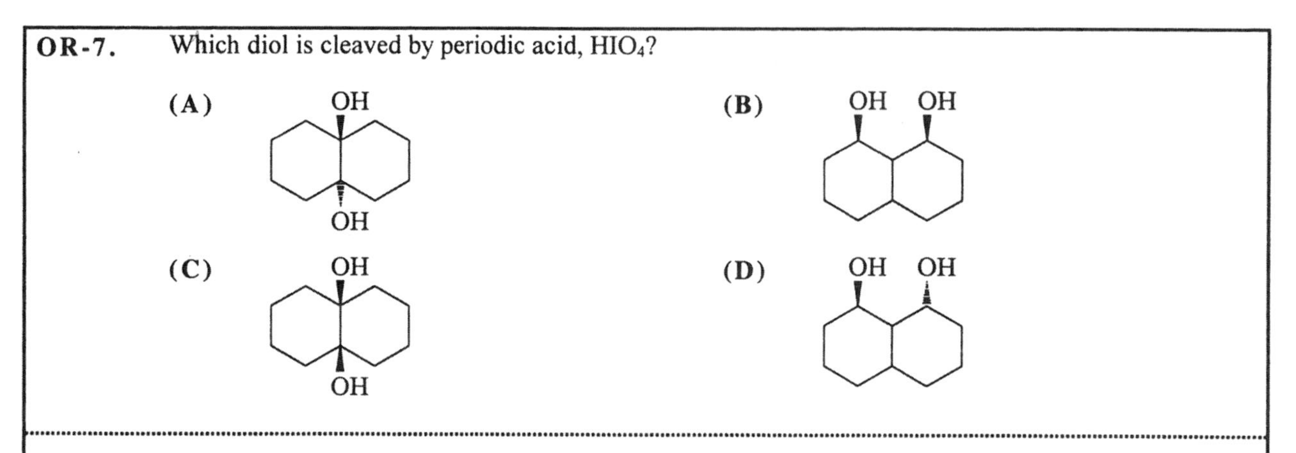

Knowledge Required: Reactivity of diols with periodic acid.

Thinking it Through: Periodic acid cleaves the carbon–carbon bond of alpha-hydroxy aldehydes, and ketones, alpha dicarbonyl compounds and 1,2-diols in which the OH groups can achieve a cisoid conformation. (The mechanism of cleavage involves a five-member intermediate and this five-member intermediate cannot be made with a *trans*-1,2-diol.)

Choice **(A)** is a 1,2-diol, but the hydroxyl groups are *trans*. Therefore, this cannot be the correct answer. Choice **(B)** contains a *cis*-diol, but can be eliminated because it is a 1,3-diol. The only diols cleaved by periodic acid are 1,2-diols. Choice **(C)** has a 1,2-diol that has *cis* hydroxyl groups, so periodic acid will cleave this compound. Choice **(D)** cannot be correct because it is a 1,3-diol.

OR-8. Which set of reagents would be best for this reaction?

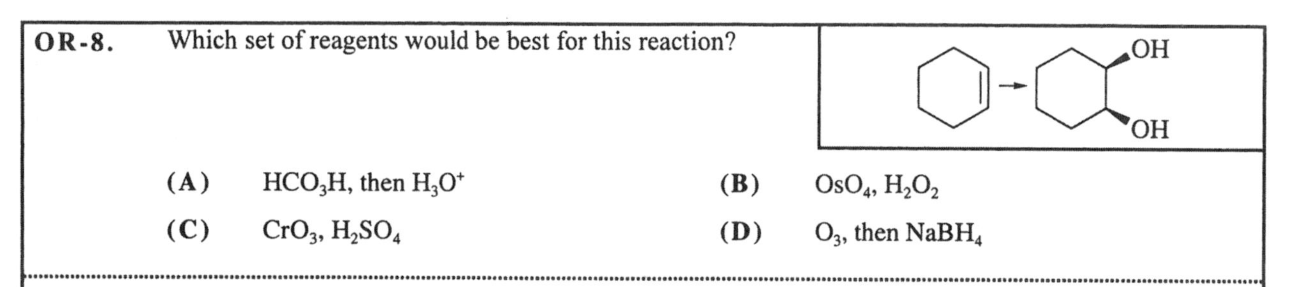

(A) HCO_3H, then H_3O^+ **(B)** OsO_4, H_2O_2

(C) CrO_3, H_2SO_4 **(D)** O_3, then $NaBH_4$

Knowledge Required: Reagents for conversion of alkenes to 1,2-diols and the stereochemistry of their reactions.

Thinking it through: The desired product is a *cis*-1,2-diol, so the object of this problem is to find which reagents would produce the *cis* diol. Choice **(A)** shows reagents that are used to make *trans*-1,2-diols. (Peracid converts the alkene to an epoxide, and the epoxide undergoes an S_N2 ring-opening reaction with the aqueous acid to give the *trans*-diol.) Therefore **(A)** can be rejected. Choice **(B)** gives the desired *cis*-1,2-diol because OsO_4 forms an intermediate cyclic osmate ester with the alkene, resulting in two oxygen atoms being added to the same face of the alkene. This is the correct answer. The chromic acid in choice **(C)** will oxidize the cyclohexene, but produces a dicarboxylic acid, not a 1,2-diol, *via* oxidative cleavage of the double bond. In choice **(D)**, ozone would cleave the double bond of cyclohexane to give hexanedial, which would then be reduced to 1,6-hexanediol by $NaBH_4$.

Note: Another reagent that would accomplish the same transformation is *cold*, alkaline $KMnO_4$, although some oxidative cleavage occurs, which results in a generally lower yield than for OsO_4.

OR-9. Which reagent(s) can be used to reduce a triple bond to a double bond having *Z* (*cis*) geometry (as shown in the figure)?

$$R{-}C{\equiv}C{-}R' \rightarrow \underset{R}{\overset{H}{C}}{=}\underset{R'}{\overset{H}{C}}$$

(A) Na, NH_3 **(B)** $NaBH_4$ **(C)** H_2, Pd/$BaSO_4$ **(D)** H_2, Pt

Knowledge Required: Reagents for reduction of alkynes to alkenes and their stereochemistry.

Thinking it Through: Each of the reagents is a reducing agent, but each has different applications. The reagents given in choice **(A)** are used in dissolving–metal reductions to convert an alkyne to an alkene with *E* (*trans*) geometry. This choice can be eliminated. Choice **(B)** can be rejected because sodium borohydride reduces aldehydes and ketones, but not alkynes.

Hydrogen gas reduces alkynes to give alkenes or alkanes, depending on the nature of the catalyst. Use of an ordinary catalyst such as Ni, Pd or Pt results in reduction all the way to the alkane. Therefore, choice **(D)** is incorrect. Choice **(C)** gives a catalyst that is partially deactivated by $BaSO_4$, so reduction stops at the alkene stage. This is the correct answer.

Note: There are a few types of catalysts that are used to produce Z alkenes from alkynes. Each one has Pd with some deactivating agent or 'poison' present, usually a barium compound. Regardless of whether the catalyst is described as "Pd/$BaSO_4$," "Pd/$BaCO_3$/quinoline," "Lindlars catalyst," "Poisoned Pd" or "deactivated Pd" the result is the same: hydrogenation of an alkyne gives a Z-alkene.

OR-10. What is the product of the indicated reaction?

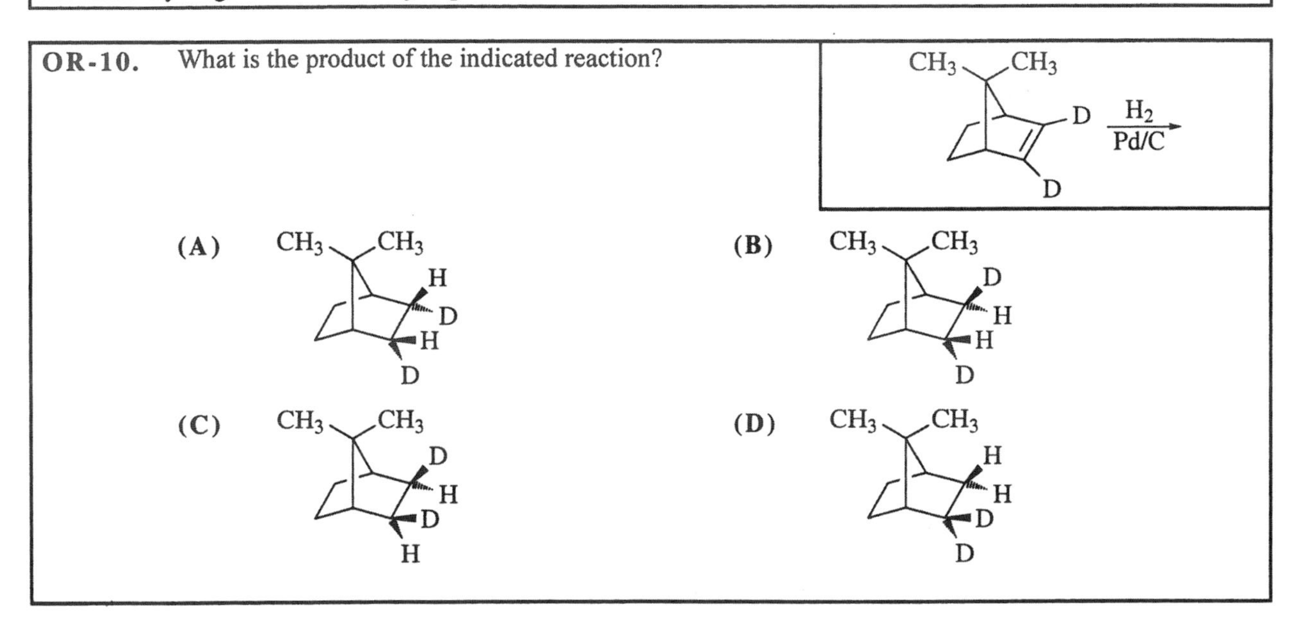

Knowledge Required: (1) Stereochemistry of hydrogenation of alkenes. (2) Steric effects and their role in reactions.

Thinking it Through: Hydrogenation of alkenes generally gives products resulting from *syn* addition of hydrogen atoms. This is due to the mechanism of hydrogenation, which is summarized below. Hydrogen is adsorbed onto the metal surface. The π bond of the alkene is attracted to the metal surface as well (**1**). While interacting with the metal, it picks up the hydrogen atoms, one after another (**2**), to give the alkane product. The alkane is not attracted to the metal surface and it moves away (**3**). The reaction gives only the product arising from *syn* addition of H_2, *cis*-1,2-dimethylcyclohexane in this case. No *trans*- product is observed. This is because in order to form *trans*-1,2-dimethylcyclohexane, the molecule would have to flip over completely between adding the first and second hydrogen atoms, which is unfavorable.

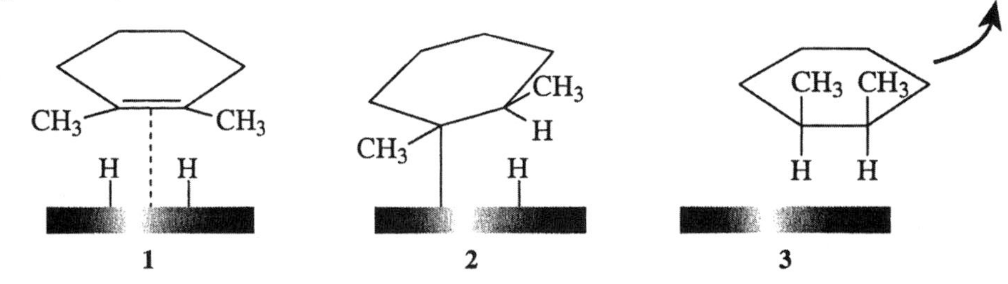

If one face of a double bond is more sterically hindered by a bridging carbon atom or other group, this face cannot easily approach the metal surface. As a result, it is unlikely that the hydrogen atoms will be attached to the more hindered face of the double bond. The hydrogen atoms will add to the face with less steric hindrance.

In this problem, two faces of the double bond in the starting material are different. The *exo* side is that nearer the bridge with the methyl groups. The *endo* side is away from the bridge with the methyl groups. If the *exo* face of the π bond approaches the metal surface, there is steric interference between the metal surface and the methyl group. This prevents the *exo* face of the π bond from getting close enough to the metal to pick up the hydrogen atoms. The less-hindered *endo* face can more closely approach the metal surface and can add the hydrogen atoms. As a result, one can expect the hydrogen atoms to add in *syn* fashion and both will end up on the *endo* face of the molecule. The diagram below illustrates the two possibilities.

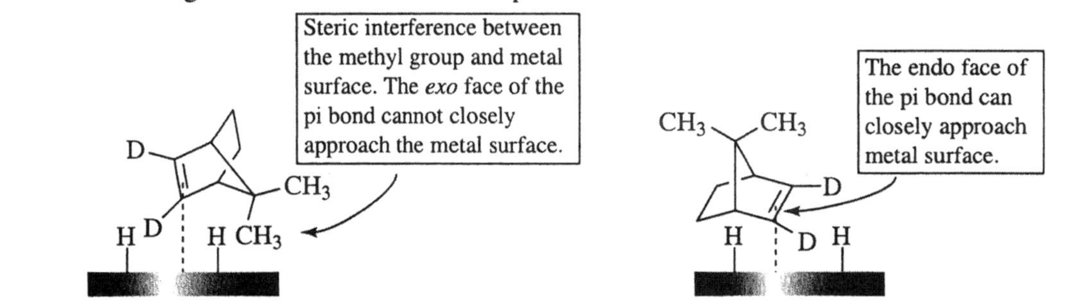

When looking at the four possible choices, one of them, (**B**), can be eliminated immediately because the hydrogen atoms are added in *anti* fashion, contrary to what is expected. In choice (**D**), not only are both hydrogen atoms adding to the same carbon atom, but, a deuterium atom is shown as having migrated from one carbon atom to another! This choice is clearly nonsense, so it can be eliminated. Of the remaining two choices, (**A**) shows the hydrogen atoms adding to the *exo* face, which is less likely because the steric interaction between the methyl group and the metal surface prevents close approach of the π bond to the metal. Choice (**C**) shows the hydrogen atoms adding to the *endo* face, which is expected because the *endo* face can more closely interact with the metal and pick up the hydrogen atoms. This is the correct answer

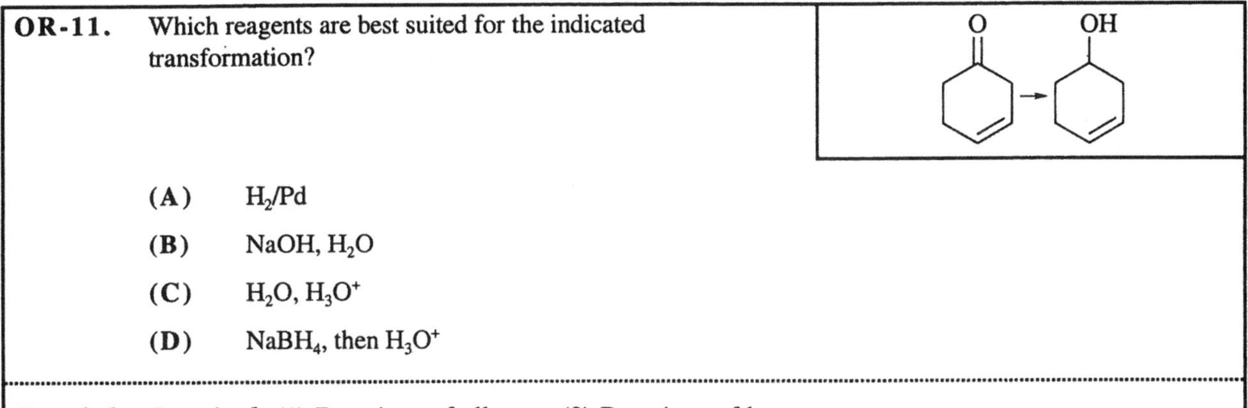

OR-11. Which reagents are best suited for the indicated transformation?

(A) H_2/Pd

(B) NaOH, H_2O

(C) H_2O, H_3O^+

(D) $NaBH_4$, then H_3O^+

Knowledge Required: (1) Reactions of alkenes. (2) Reactions of ketones.

Thinking it Through: The starting material is an unsaturated ketone and the product is an unsaturated alcohol. Each of these reagents will react with the starting material, but only one will give the desired product. This is another problem in which one must select a reagent that reacts with one of the two functional groups in the starting material. Choice (**A**), H_2 + catalyst, will reduce the ketone to an alcohol, but it will also reduce the alkene group. The resulting product would be cyclohexanol, so this cannot be the correct answer. Aqueous hydroxide, choice (**B**), does not react with alkenes, but it does react with ketones. However, it is not the correct answer, because hydroxide ion does not act as a reducing agent. Instead, hydroxide can attack the carbonyl group and form the *gem*-diol, or it can also remove an alpha-hydrogen atom to give an enolate ion. Neither of these reactions leads to the desired product, so (**B**) is not correct. Choice (**C**), aqueous acid, can also react with the starting material to give a *gem*-diol, and it can also lead to formation of the enol. (Not only that, but acid can react with double bonds to give alcohols and/or rearranged alkenes.) Aqueous acid is not a reducing agent, and it can react with the alkene, so this is not the correct answer. The correct answer is (**D**), because $NaBH_4$ reduces ketones to secondary alcohols, but does not react with alkenes.

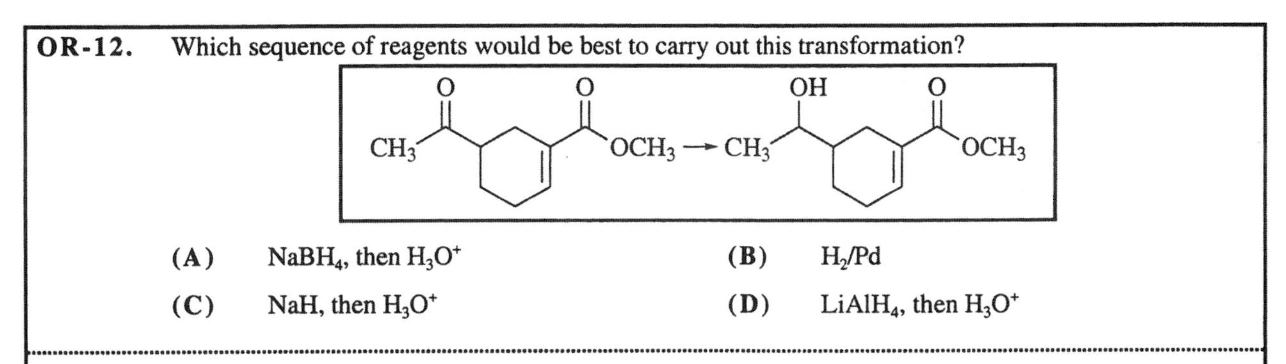

OR-12. Which sequence of reagents would be best to carry out this transformation?

(A) $NaBH_4$, then H_3O^+

(B) H_2/Pd

(C) NaH, then H_3O^+

(D) $LiAlH_4$, then H_3O^+

Knowledge Required: Reactions of ketones, alkenes and esters with various reducing agents.

Thinking it Through: The starting material contains a ketone, an alkene, and an ester, and the desired reaction affects only the ketone; the alkene and ester groups remain unchanged. Again, drawing out the reaction proposed in each choice should provide some help in selecting the correct answer.

Choice (**A**), $NaBH_4$, reduces ketones to secondary alcohols, but it does not affect esters or alkenes. This is the correct answer. Choice (**B**) can be rejected because hydrogenation will reduce the ketone to the secondary alcohol, but it will also reduce the alkene to an alkane. Choice (**C**) cannot be the correct answer because NaH will function as a base, giving any of several possible enolate ions. The aqueous acid workup will simply give back the starting material. Choice (**D**) will react with the ketone, but $LiAlH_4$ also reacts with esters to give primary alcohols. This is not the correct answer. Again, drawing out the products of the possible reactions will often reveal the correct answer.

Practice Questions

1. Which reagents are best for carrying out this reaction?

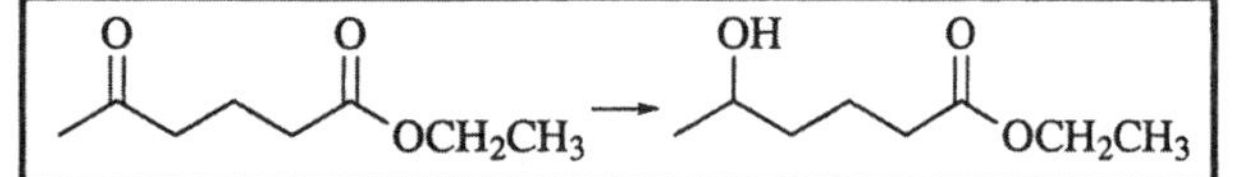

(A) $NaBH_4$, then H_3O^+

(B) Zn(Hg), conc. HCl

(C) $LiAlH_4$, ether; then aqueous workup

(D) NH_2NH_2, KOH

2. Which of these reactions will produce cyclopentanol?

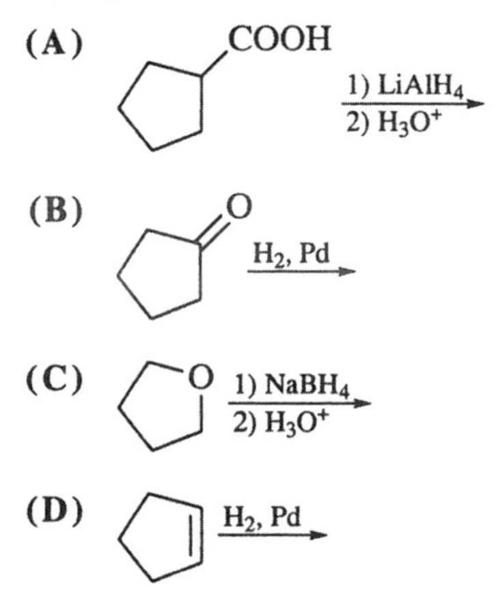

3. Which reagents would best accomplish this transformation?

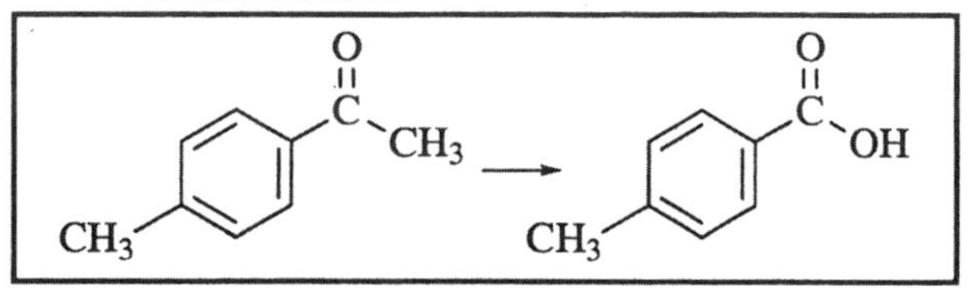

(A) $K_2Cr_2O_7 + H_2SO_4$

(B) $KMnO_4$, KOH, then neutralization

(C) I_2, KOH, then neutralization

(D) H_2O_2, KOH, then neutralization

4. Ozonolysis of which terpene would give equimolar amounts of these compounds?

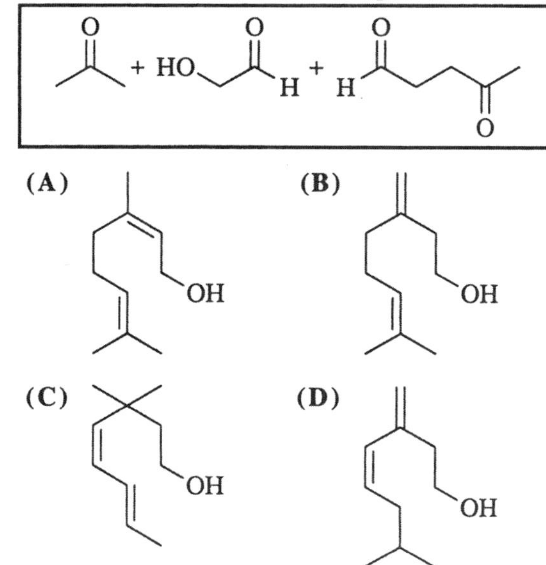

5. What is the main product of alkaline permanganate oxidation after neutralization?

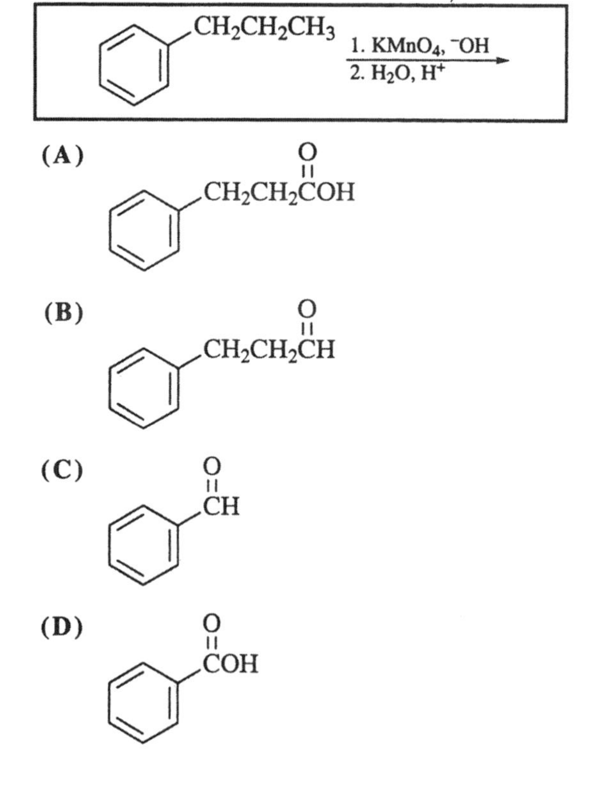

6. Reduction of a triple bond to an *E* (trans) double bond can be accomplished with which set of reagents?

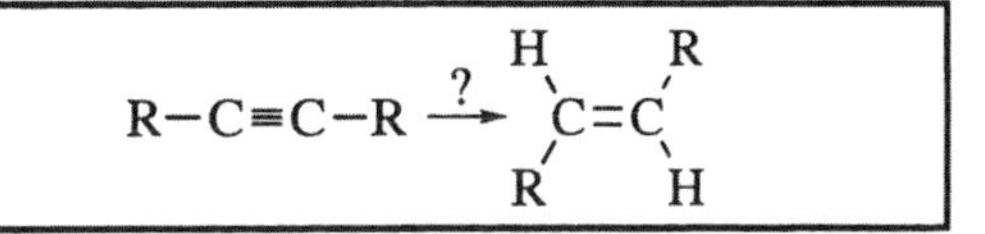

(A) Na, NH_3

(B) H_2, deactivated Pd

(C) $NaBH_4$, methanol

(D) NaH, then H_3O^+

7. Which reagents are best suited for this transformation?

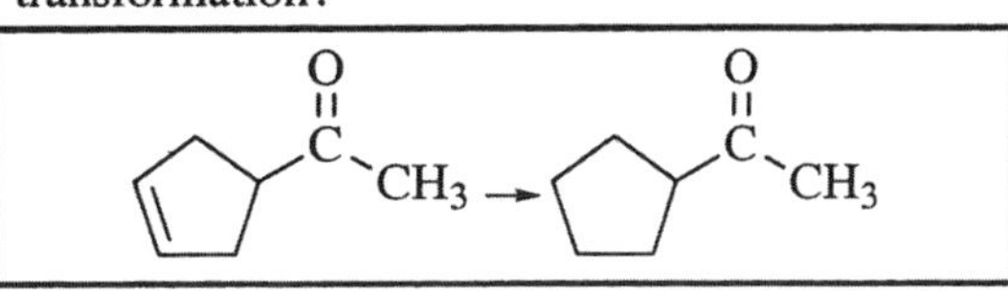

(A) $LiAlH_4$, then H_3O^+

(B) H_2O, H^+

(C) H_2, Pd/C

(D) Na, NH_3

8. Which is the expected product from this oxidation?

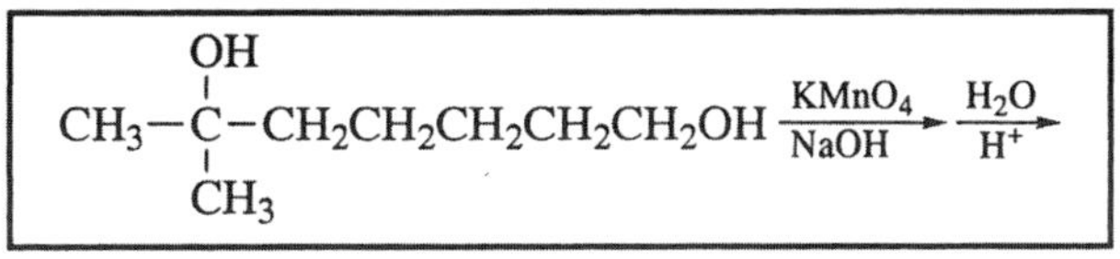

$CH_3-C(OH)(CH_3)-CH_2CH_2CH_2CH_2CH_2OH \xrightarrow[NaOH]{KMnO_4} \xrightarrow[H^+]{H_2O}$

(A) $CH_3-C(=O)-CH_2CH_2CH_2CH_2C(=O)OH$

(B) $CH_3-C(=O)-CH_2CH_2CH_2CH_2C(=O)H$

(C) $CH_3-C(OH)(CH_3)-CH_2CH_2CH_2CH_2C(=O)OH$

(D) $CH_3-C(OH)(CH_3)-CH_2CH_2CH_2CH_2C(=O)H$

9. Which of these reagents is best suited for this conversion?

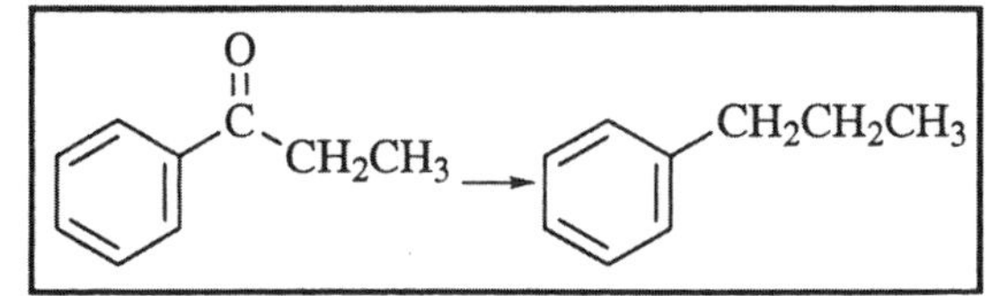

(A) $LiAlH_4$, then H_3O^+

(B) Zn(Hg), HCl

(C) H_2O, H^+

(D) Na, NH_3

10. Which compound can be most easily oxidized by $K_2Cr_2O_7 + H_2SO_4$?

(A) CH_3CH_2COOH **(B)** $(CH_3)_3COH$

(C) 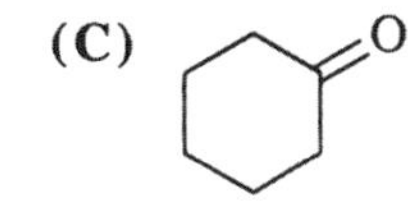

(D) 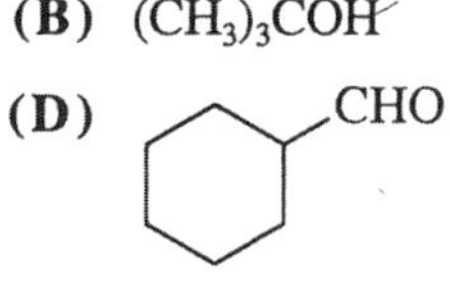

11. Which product would result from this reaction?

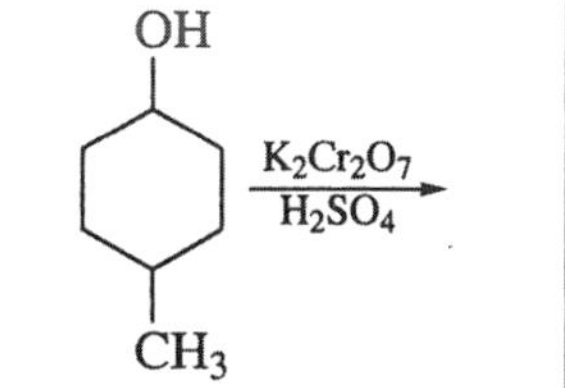

(A) 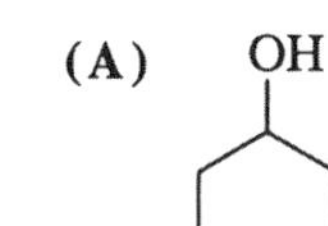

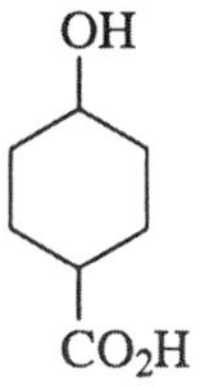

CO_2H

(B) 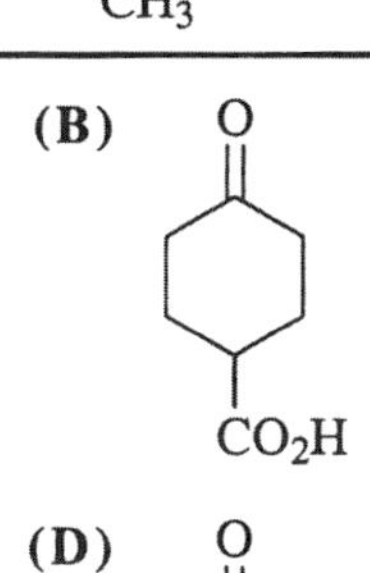

(C) 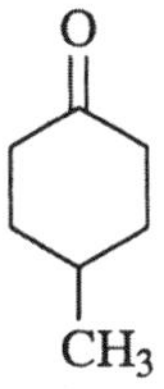 O (cyclohexanone) CH_3

(D) 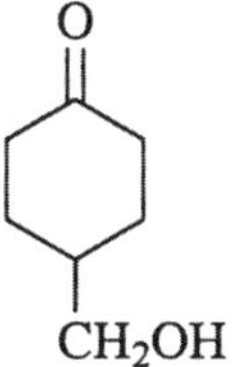

12. Which of these reagents would be best to accomplish this conversion?

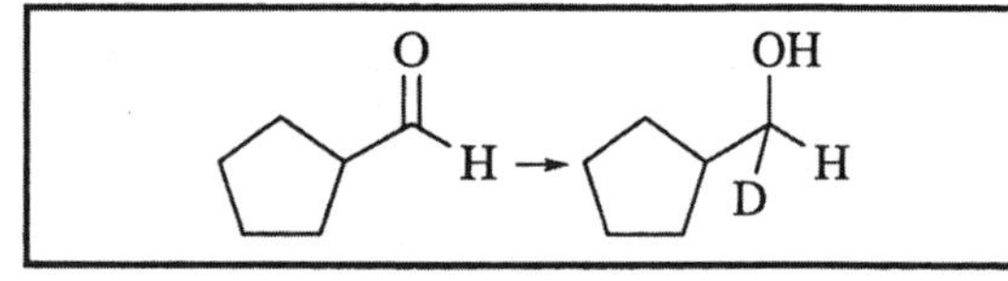

(A) D_2O, catalytic HCl

(B) D_2O_2 in CH_3COOH

(C) NaOD in CH_3CH_2OD, then aqueous work-up

(D) $NaBD_4$ in CH_3CH_2OH, then aqueous work-up

13. Which set of reagents could be used to effect this conversion?

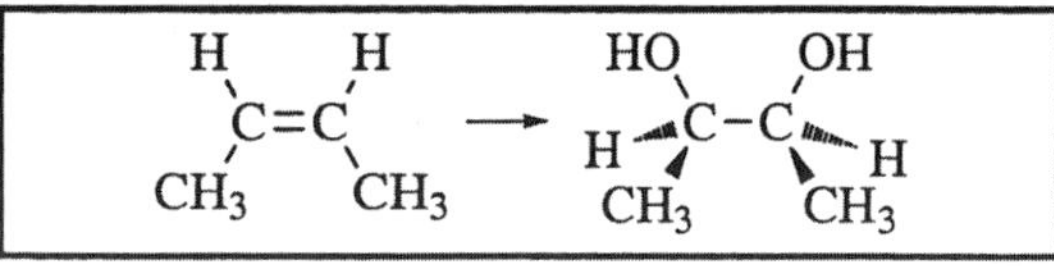

(A) BH_3, then H_2O_2 and NaOH

(B) HIO_4

(C) cold $KMnO_4$, NaOH

(D) H_2O_2, NaOH

14. What are the products when this triol is treated with excess HIO_4?

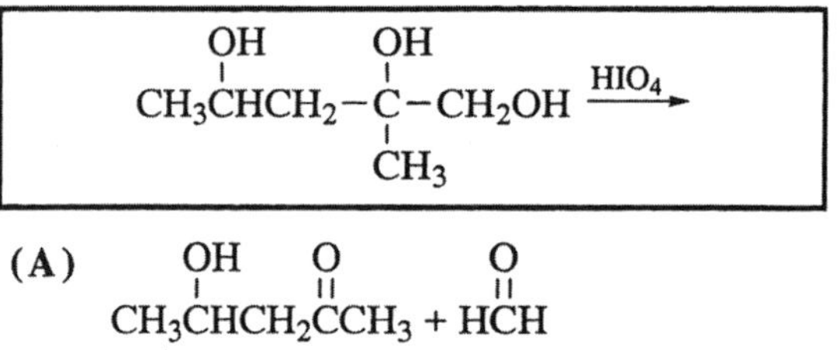

(A) $CH_3\overset{OH}{C}HCH_2\overset{O}{C}CH_3 + H\overset{O}{C}H$

(B) $CH_3\overset{OH}{C}HCH_2\overset{O}{C}CH_3 + HO\overset{O}{C}H$

(C) $CH_3\overset{O}{C}H + CH_3\overset{O}{C}CH_3 + H\overset{O}{C}H$

(D) $CH_3\overset{O}{C}H + CH_3\overset{O}{C}CH_3 + HO\overset{O}{C}H$

Answers to Study Questions

1. B	6. D	11. D
2. D	7. C	12. A
3. D	8. B	
4. C	9. C	
5. D	10. C	

Answers to Practice Questions

1. A	6. A	11. C
2. B	7. C	12. D
3. C	8. C	13. C
4. A	9. B	14. A
5. D	10. D	

Spectroscopy

Infrared (IR) and nuclear magnetic resonance (NMR) spectroscopy, and mass spectrometry, are important tools for determining the structures of unknown compounds. These techniques are also essential for confirming that the products of chemical reactions actually have the structures anticipated. Each spectral method is independently useful with simple molecules; but when used together, these methods make it possible to determine the structures of very complicated molecules. Mass spectrometry must be used when the molecular formula is not known, because neither IR nor NMR spectroscopy can provide this information.

Infrared radiation is absorbed by molecules, causing them to move to excited vibrational states. Different types of bonds, such as C=C, O–H or C=O, have different vibrational frequencies and, therefore, absorb IR radiation at different, and characteristic, frequencies (or wavelengths). Thus an IR spectrum is useful for determining whether various functional groups are present or absent in a molecule. IR spectra are usually plots of % transmission against wavelength (in micrometers, μm) or wavenumber (the reciprocal of wavelength, which is proportional to frequency, and is expressed in cm^{-1}). Wavenumber ($1/\lambda$) and frequency (c/λ) differ by a constant factor, the speed of light.

Infrared Spectroscopy

Remembering a few key absorption frequencies and the molecular vibrations responsible for them is very useful when deciphering IR spectra. Table 1 provides a concise summary of important absorption frequencies.

Table 1. Approximate Infrared Absorption Frequencies

Bond	Frequency (cm^{-1})
O–H and N–H	3400
C–H (sp^3)	3000–2850
C–H (sp^2)	3100–3000
C–H (sp)	3300
C≡C and C≡N	2100
C=O	1700
C=C, aromatic C–C	1600
aromatic C–C	1500

Nuclear Magnetic Resonance Spectroscopy

Some nuclei have magnetic moments and, in a strong magnetic field, absorb electromagnetic radiation in the radio-frequency range, giving rise to an NMR spectrum. An IR spectrum of a compound gives a picture of what functional groups are present in a molecule, but gives very little information about the number and type of hydrogen and carbon atoms in the molecule. ^1H and ^{13}C NMR spectroscopy provides this information. For ^1H NMR spectra, integration of the areas under signals in a spectrum gives the relative number of equivalent protons producing each signal, and the $n+1$ rule for spin–spin splitting of signals allows one to determine the number of protons on atoms adjacent to the proton producing the signal. In addition to being familiar with these two tools, knowing the approximate chemical shifts for six types of protons is extremely useful.

Table 2. ^1H NMR Chemical Shifts

H attached to	δ (ppm)
sp^3 C atom	1
C atom attached to π system	2
C atom attached to O atom	3–4
sp^2 C atom	5–6
aromatic C atom	7–8
aldehyde C atom	10
carboxylic acid O atom	11–12

^{13}C NMR spectra are normally recorded as proton-decoupled spectra and the signals consist of a series of single lines. Thus, the number of lines in the spectrum corresponds to the number of the different types of carbon atoms in the molecule. Several techniques are available to determine how many protons are attached to each type of carbon atom, greatly increasing the usefulness of ^{13}C NMR spectra. As with ^{1}H NMR, knowing the approximate chemical shifts for six types of carbon atoms is extremely useful.

Table 3. ^{13}C NMR Chemical Shifts

C environment	δ (ppm)
sp^3 alkyl C atom	10–30
sp^3 C atom attached to O, N, or X	50–70
sp C atom	75–90
sp^2 C atom	100–150
ester, amide, carboxyl C=O	160–180
aldehyde and ketone C=O	190–200

Mass Spectrometry

When molecules are ionized in the gas phase and the resulting ions passed through a magnetic or electric field, the ions are deflected by the field. Measuring the deflections yields a mass spectrum, in which the ions are arranged according to their mass-to-charge (*m/z*) ratio. Mass spectrometry deals with the generation and analysis of mass spectra. The usual way to ionize a molecule is to bombard it with a beam of electrons, which dislodge an electron, producing a molecular ion. The molecular ion has the same molecular mass as the molecule. If a high resolution mass spectrometer is used, the exact molecular formula for the molecule can be obtained from the mass spectrum. Molecular ions often fragment into smaller ions and neutral components (radicals or small neutral molecules). The fragment ions may themselves undergo further fragmentation. Fragmentation patterns provide clues as to the structure of the molecule. Abundances of fragment ions reflect carbocation and radical stabilities.

Study Questions

SP-1. What effect does the conjugation of a carbonyl group with a carbon–carbon double bond have on the infrared absorption due to the C=O stretch?

(A) It shifts to a lower frequency (longer wavelength).

(B) It shifts to a higher frequency (shorter wavelength).

(C) It has no effect.

(D) The absorption due to the C=O stretch disappears.

Knowledge Required: (1) Effect of resonance on the nature of double bonds. (2) Relationship between frequency, or wavelength, and energy of electromagnetic radiation. (3) Recognition that the frequency of the absorbed infrared radiation corresponds to a vibrational frequency in the molecule.

Thinking it Through: The resonance interaction between the C=C and the C=O groups introduces some single bond character into the carbonyl group as shown by the resonance structures below.

$$R_2C{=}CH{-}C({=}O){-}R \longleftrightarrow R_2\overset{+}{C}{-}CH{=}C(O^-){-}R$$

Less energy is required to stretch a carbon–oxygen single bond than a carbon–oxygen double bond. Consequently, the reduced double bond character of a conjugated carbonyl group will result in a lower stretching frequency than for a non-conjugated carbonyl group, and it will absorb lower frequency infrared radiation. Choice **(A)** is a statement of this fact and is the correct answer. Note that the stretching frequency of the carbon–carbon double bond would also be lowered.

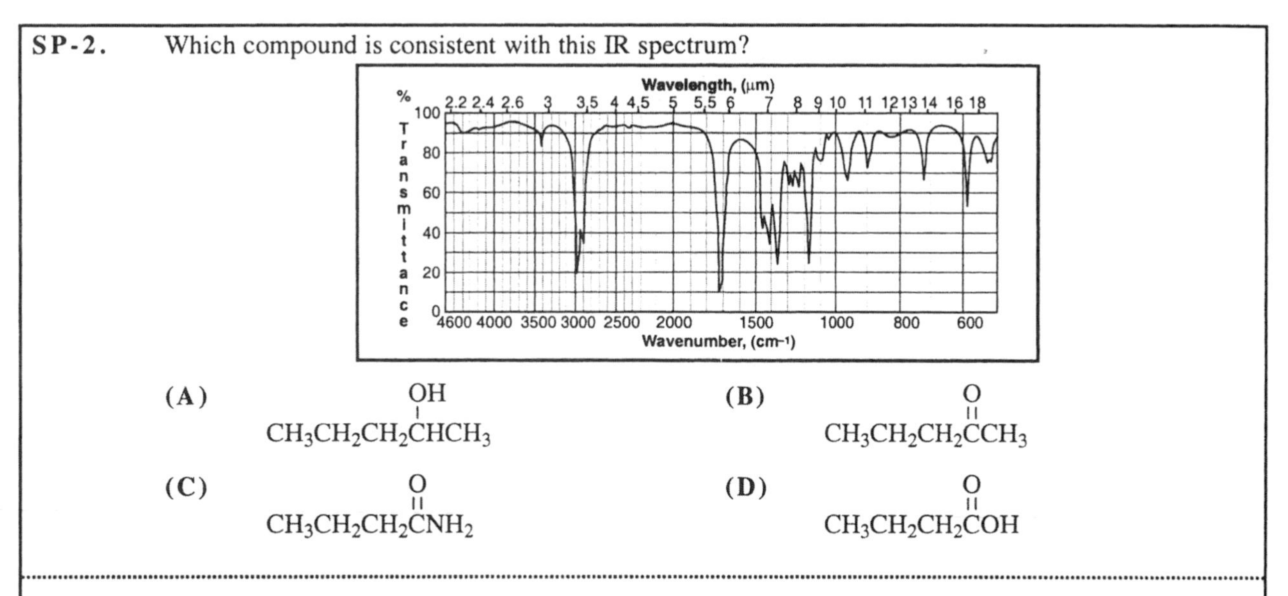

SP-2. Which compound is consistent with this IR spectrum?

(A) $CH_3CH_2CH_2CH(OH)CH_3$

(B) $CH_3CH_2CH_2C(=O)CH_3$

(C) $CH_3CH_2CH_2C(=O)NH_2$

(D) $CH_3CH_2CH_2C(=O)OH$

Knowledge Required: (1) The approximate absorption frequencies for OH, NH_2 and carbonyl groups. (2) Recognition that the absence of characteristic IR absorption bands for various functional groups allows molecules containing those groups to be ruled out.

Thinking it Through: The spectrum shows a strong absorption band at approximately 1720 cm^{-1}. Thus, the molecule contains a carbonyl group of some sort, and choice **(A)** may be ruled out. Choices **(B)**, **(C)** and **(D)** all contain a carbonyl group, but they may be distinguished because choices **(C)** and **(D)** also contain other functional groups that have other distinctive absorption bands. The strong and broad band covering the 2500–3300 cm^{-1} region that is typical for the OH of a carboxylic acid is clearly absent. Therefore, choice **(D)** cannot be correct. The NH_2 group of the amide in choice **(C)** would show two bands of medium strength near 3400 cm^{-1}, and in addition, the carbonyl band for amides normally comes below 1700 cm^{-1} (near 1650 cm^{-1}). The spectrum is clearly not that of choice **(C)**. Choice **(B)**, a ketone, must be the correct answer.

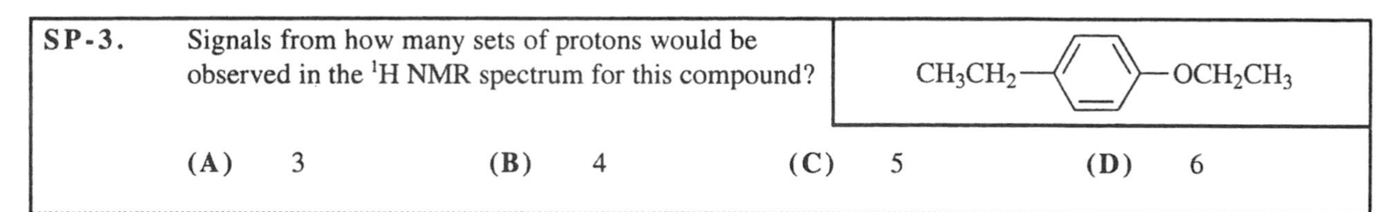

SP-3. Signals from how many sets of protons would be observed in the ^1H NMR spectrum for this compound?

$CH_3CH_2-C_6H_4-OCH_2CH_3$

(A) 3 (B) 4 (C) 5 (D) 6

Knowledge Required: (1) That chemically equivalent protons are generally magnetically equivalent. (2) That symmetry causes sets of protons to be chemically equivalent.

Thinking it Through: To give rise to different NMR signals, sets of protons must be magnetically (chemically) nonequivalent. The simplest test for chemical nonequivalence is to substitute some other atom for a proton in each set. If the substitution results in the same compound, the sets are chemically equivalent. Otherwise they are chemically nonequivalent. The two CH_3 groups and the two CH_2 groups in the two ethyl groups are chemically nonequivalent, and each one gives rise to a distinct signal (for a total of four signals). The symmetry in the molecule makes the protons that are *ortho* to the ethoxy group chemically equivalent, but nonequivalent to the likewise chemically equivalent protons *ortho* to the ethyl group. Thus, the protons on the benzene ring give rise to two distinct signals, and these two signals plus the four from the two ethyl groups gives six distinct signals for the NMR spectrum. Therefore, choice **(D)** is the correct answer. Choice **(C)** is often selected because of the incorrect assumption that the two CH_3 group will be equivalent because they are both attached to a carbon atom. Likewise, choice **(B)** would be selected if one erroneously assumed that both ethyl groups were equivalent.

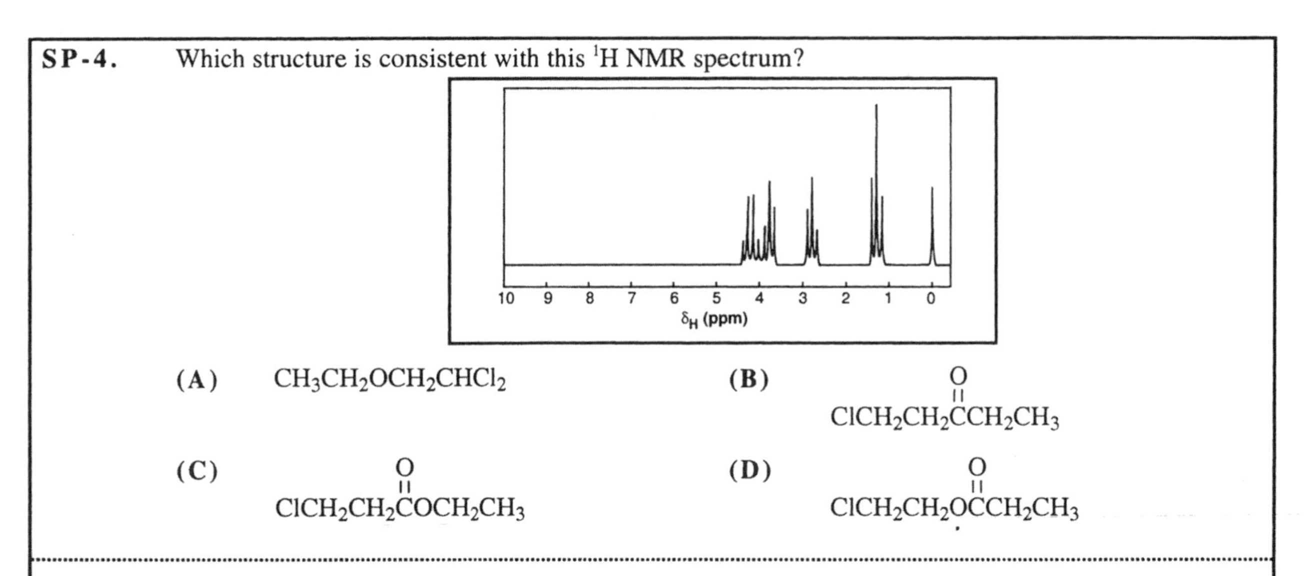

SP-4. Which structure is consistent with this ^{1}H NMR spectrum?

(A) $CH_3CH_2OCH_2CHCl_2$

(B) $ClCH_2CH_2C(=O)CH_2CH_3$

(C) $ClCH_2CH_2C(=O)OCH_2CH_3$

(D) $ClCH_2CH_2OC(=O)CH_2CH_3$

Knowledge Required: (1) Spin–spin splitting patterns. (2) Factors that affect chemical shifts. (3) Relative chemical shifts.

Thinking it Through: The best way to approach this question is to compare the expected ^{1}H NMR spectrum for each choice with the given spectrum. The given spectrum shows a quartet, two triplets of equal intensity, and a more intense triplet in that order from lowest to highest field. Only the spectra of choices **(B)**, **(C)** and **(D)** would show the same type and number of splitting patterns (mutiplets) as are seen in the given spectrum. The spectrum of choice **(A)** would show two triplets of different intensity, a doublet, and a quartet. It is clearly incorrect. Although the spectra of choices **(A)**, **(B)**, and **(D)** have the same number and type of multiplets, the order of the signals from lowest to highest field will differ. Only in choice **(C)** is the methylene of the ethyl group attached to a heteroatom so that the quartet would come at lowest field. The spectra of choices **(B)** and **(D)** would both have the two equally intense triplets at lowest field followed by the quartet and then the more intense triplet. Therefore, choice **(C)** appears to be the correct answer. Indeed, the spectrum of choice **(C)** would show the correct order of the quartet, two triplets of equal intensity, and then the more intense triplet from lowest to highest field.

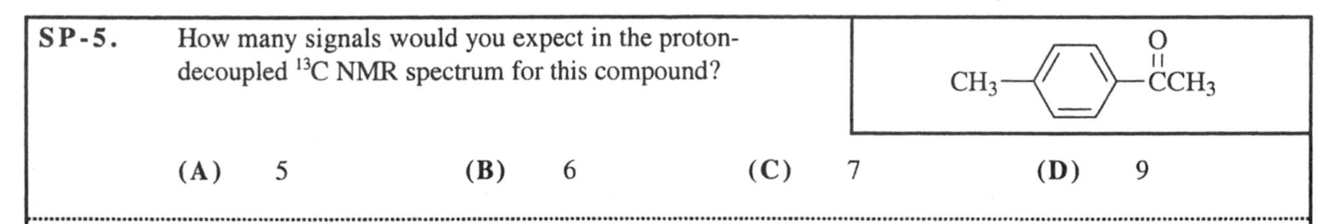

SP-5. How many signals would you expect in the proton-decoupled ^{13}C NMR spectrum for this compound?

(A) 5 (B) 6 (C) 7 (D) 9

Knowledge Required: (1) Ability to discern to whether or not various carbon atoms are chemically equivalent. (2) That symmetry causes carbon atoms to be chemically equivalent.

Thinking it Through: In a proton-decoupled ^{13}C NMR spectrum, each different type of carbon atom will give rise to a singlet signal. As a result, the number of signals is equal to the number of different types of carbon atoms in the molecule. A plane of symmetry passing through the two substituted carbon atoms in the benzene ring causes the two carbon atoms *ortho* to the acetyl group to be equivalent. This is also true for the two carbon atoms *ortho* to the methyl group. There are a total of seven different types of carbon atoms in the molecule, and choice **(C)** is the correct answer.

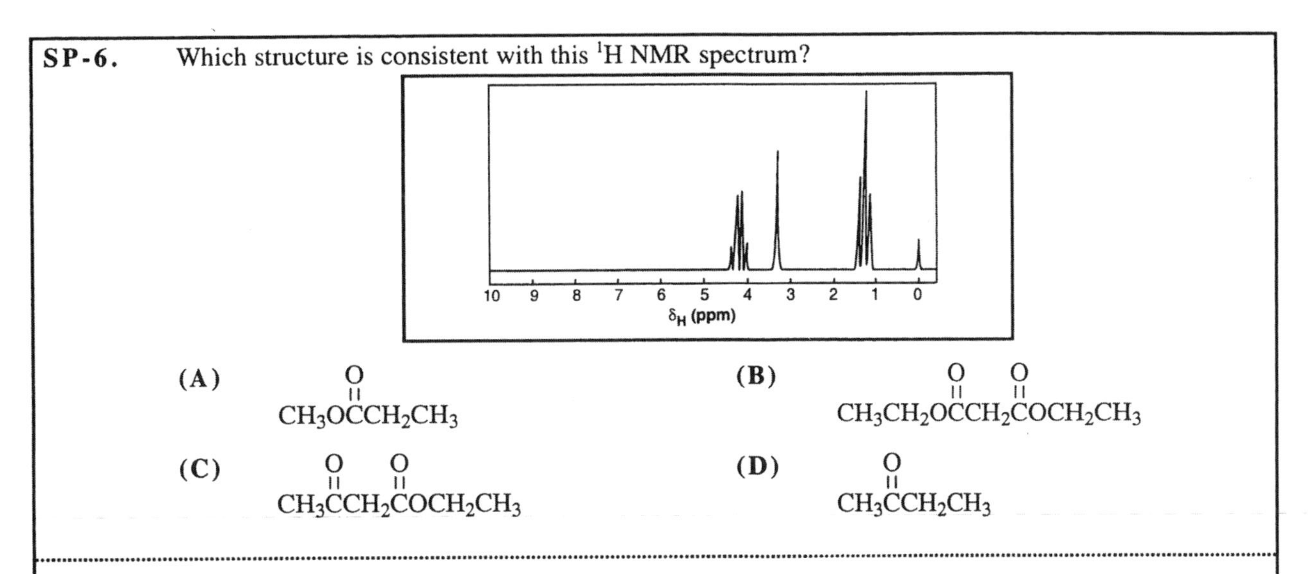

SP-6. Which structure is consistent with this ¹H NMR spectrum?

(A)

CH₃OCCH₂CH₃

(B)

O O
‖ ‖
CH₃CH₂OCCH₂COCH₂CH₃

(C)

O O
‖ ‖
CH₃CCH₂COCH₂CH₃

(D)

CH₃CCH₂CH₃

Knowledge Required: (1) Spin–spin splitting patterns. (2) Factors that affect chemical shifts. (3) Relative chemical shifts.

Thinking it Through: As was done previously, compare the expected spectrum for each choice with the given spectrum. The given spectrum shows three different signals: a quartet, a singlet, and a triplet in that order from lowest to highest field. The spectra of choices **(A)**, **(B)**, and **(D)** will all show the same type and number of multiplets as the given spectrum. However, the spectrum of choice **(C)** would show four different signals: a quartet, two singlets, and a triplet in that order from lowest to highest field. Consequently, choice **(C)** may be eliminated. The spectra for choices **(B)** and **(D)** would show the same ordering of the multiplets as the given spectrum, but the spectrum for choice **(A)** would have the singlet (CH_3 attached to an oxygen atom) at lower field than the quartet. Thus, choice **(A)** cannot be correct. Although the spectra of choices **(B)** and **(D)** exhibit the same ordering of the signals, the actual chemical shifts for the quartets are significantly different. The methylenes of the ethyl groups in choice **(B)** are attached to an electronegative oxygen atom, which causes the quartet for the methylene protons to appear near δ 4.0. But, the methylene of the ethyl group in choice **(D)** is attached to a carbonyl carbon atom, which causes the quartet for the methylene protons to appear near δ 2.0. The given spectrum is in better agreement with choice **(B)** than with choice **(D)**, and choice **(B)** is the correct answer. Note that the singlets for choices **(B)** and **(D)** would also come at different chemical shifts. The singlet for the methylene proton between the two carbonyl groups in choice **(B)** would be expected to appear near δ 3.0, whereas the singlet for the methyl group attached to the carbonyl carbon atom would be expected to appear just above the quartet near δ 2.0.

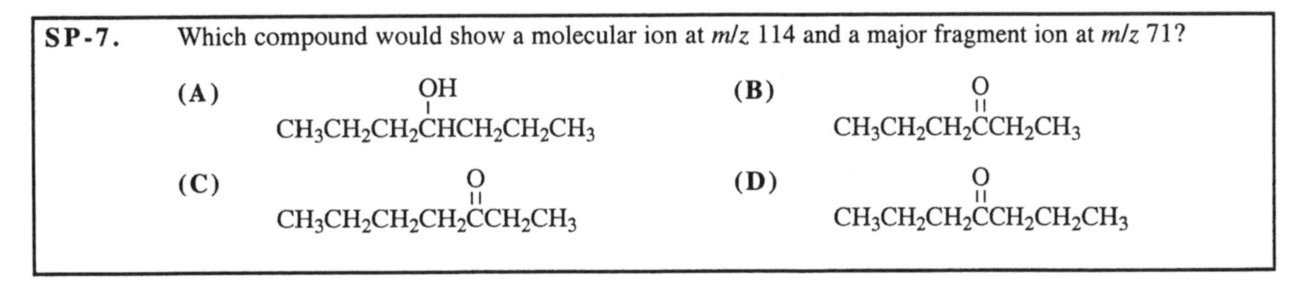

SP-7. Which compound would show a molecular ion at m/z 114 and a major fragment ion at m/z 71?
(A) CH3CH2CH2CHCH2CH2CH3
OH
(B) CH3CH2CH2CCH2CH3
(C) CH3CH2CH2CH2CCH2CH3
(D) CH3CH2CH2CCH2CH2CH3
O

Knowledge Required: (1) That the *m/z* for the molecular ion is the same as the molecular mass. (2) Factors that affect carbocation and radical stabilities. (3) That carbocation and radical stabilities determine which fragmentation processes are most likely to occur.

Thinking it Through: Choices **(C)** and **(D)** both have molecular masses equal to 114, but choice **(A)** has a molecular mass equal to 116 and choice **(B)** has a molecular mass equal to 100. Thus choices **(A)** and **(B)** cannot give rise to a molecular ion at *m/z* 114 in their mass spectra, and they may be eliminated from consideration. Since choices **(C)** and **(D)** will both give a molecular ion at *m/z* 114, we must look at the fragment ions to distinguish between them. A major fragmentation of the molecular ion of ketones is for either alkyl group attached to the carbonyl carbon atom to be lost as an alkyl radical to generate the corresponding acyl cation. For choice **(C)**, the loss of an ethyl radical (mass of 29) would produce an acyl cation at *m/z* 85 and loss of a butyl radical (mass of 57) would produce an acyl cation at *m/z* 57. For choice **(D)**, both alkyl groups are the same, propyl, and loss of a propyl radical (mass of 43) would produce an acyl cation at *m/z* 71. Choice **(D)** is the correct answer. This type of fragmentation for ketones is often called α-cleavage, because the bond between the carbonyl carbon atom and the α-carbon atom is the one that is broken. α-cleavage is a major fragmentation reaction for alcohols, amines and all types of carbonyl compounds.

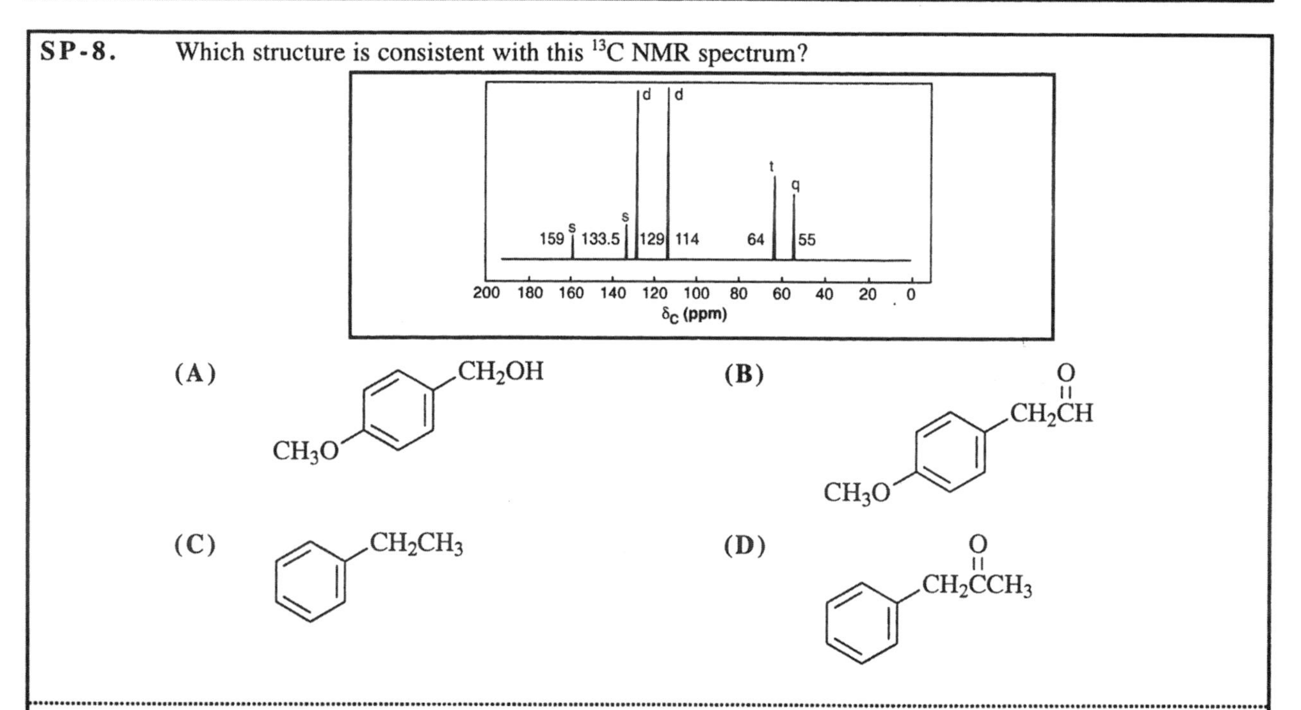

Knowledge Required: (1) Approximate chemical shifts of various types of carbon atoms. (2) The number of lines in a ^{13}C NMR spectrum is equal to the number of different types of carbon atoms in the molecule.

Thinking it Through: The spectrum shows signals for six different types of carbon atoms in the molecule. Inspection of the four choices shows that choices **(A)** and **(C)** each contain six different types of carbon atoms, but choices **(B)** and **(D)** each contain seven different types of carbon atoms. Choices **(B)** and **(D)** may be immediately eliminated. These two choices could still be eliminated even if they had contained only six types of carbon atoms because the given spectrum does not have a signal in the δ 190–200 region, where the signals for the carbonyl carbon atoms in **(B)** and **(D)** would appear. Both choices **(A)** and **(C)** have only four types of sp^2-hybridized carbon atoms and would show four signals below δ 100. However, the low intensities of the signals at δ 133.5 and δ 159 indicate that they are probably for quaternary carbon atoms. Since choice **(A)** has two quaternary carbon atoms and choice **(C)** has only one, choice **(A)** is more likely the correct answer. That choice **(A)** is the correct answer is confirmed by considering the signals at δ 55 and δ 64. Signals for sp^3-hybridized carbon atoms must be attached to heteroatoms to come at these chemical shifts. This is the case for choice **(A)**, but not for choice **(C)**. The two sp^3-hybridized carbon atoms in choice **(C)** would come above δ 30.

Practice Questions

1. Which ketone will show a carbonyl absorption at the lowest frequency (cm^{-1}) in the infrared?

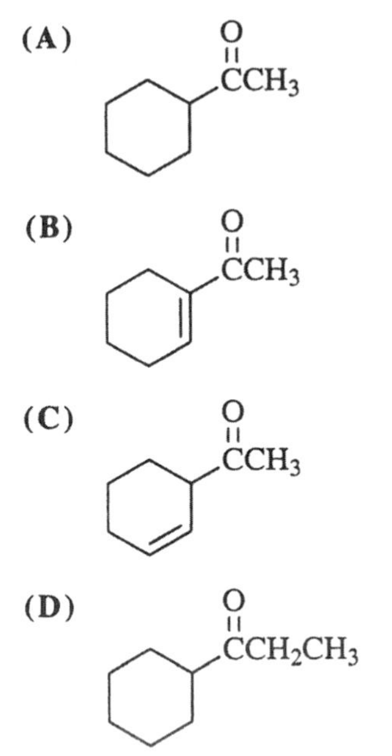

2. Which compound is consistent with this IR spectrum?

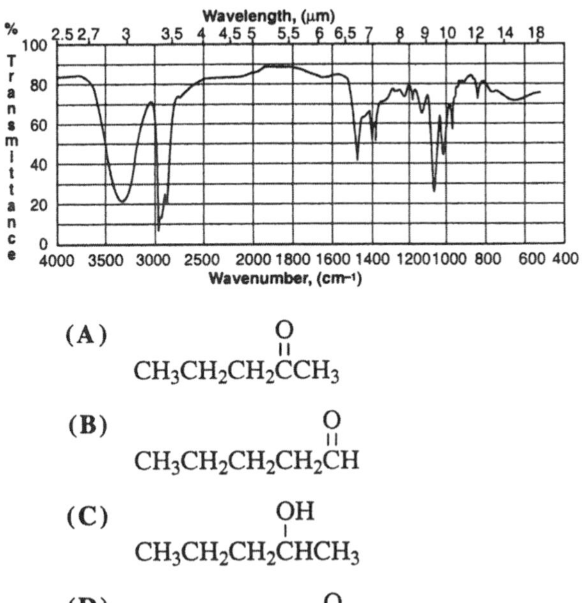

(A) $CH_3CH_2CH_2\overset{O}{\overset{\|}{C}}CH_3$

(B) $CH_3CH_2CH_2CH_2\overset{O}{\overset{\|}{C}}H$

(C) $CH_3CH_2CH_2\overset{OH}{\overset{|}{C}}HCH_3$

(D) $CH_3CH_2CH_2CH_2\overset{O}{\overset{\|}{C}}OCH_3$

3. Which is the most reasonable structure for a compound with this IR spectrum?

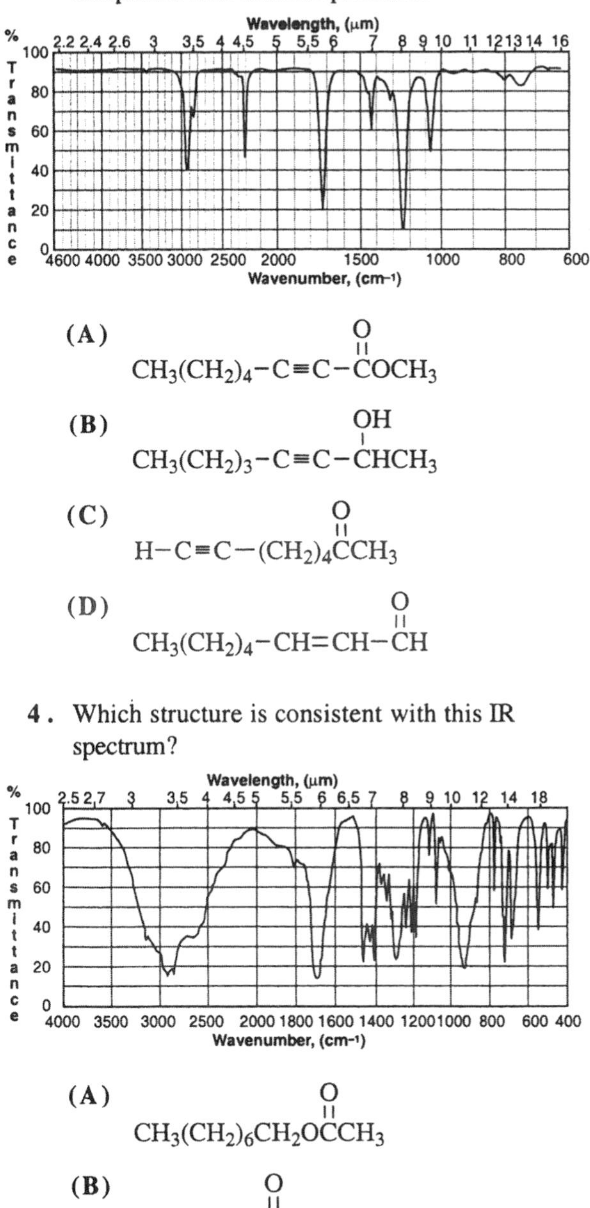

(A) $CH_3(CH_2)_4-C{\equiv}C-\overset{O}{\overset{\|}{C}}OCH_3$

(B) $CH_3(CH_2)_3-C{\equiv}C-\overset{OH}{\overset{|}{C}}HCH_3$

(C) $H-C{\equiv}C-(CH_2)_4\overset{O}{\overset{\|}{C}}CH_3$

(D) $CH_3(CH_2)_4-CH{=}CH-\overset{O}{\overset{\|}{C}}H$

4. Which structure is consistent with this IR spectrum?

(A) $CH_3(CH_2)_6CH_2O\overset{O}{\overset{\|}{C}}CH_3$

(B) $CH_3(CH_2)_{10}\overset{O}{\overset{\|}{C}}OH$

(C) $CH_3(CH_2)_8\overset{O}{\overset{\|}{C}}H$

(D) $HOCH_2(CH_2)_7\overset{O}{\overset{\|}{C}}CH_3$

5. Which structure is most consistent with this IR spectrum?

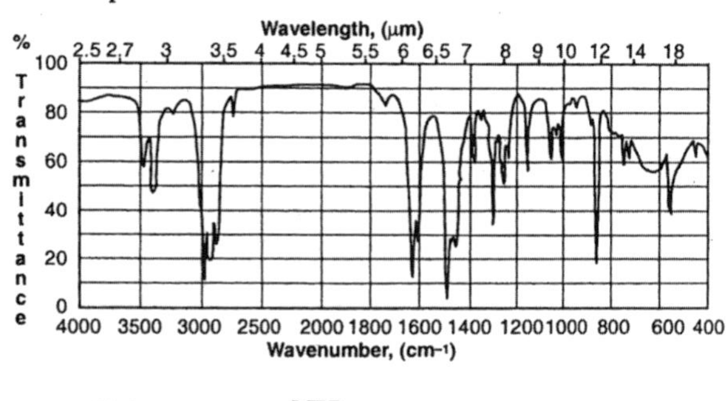

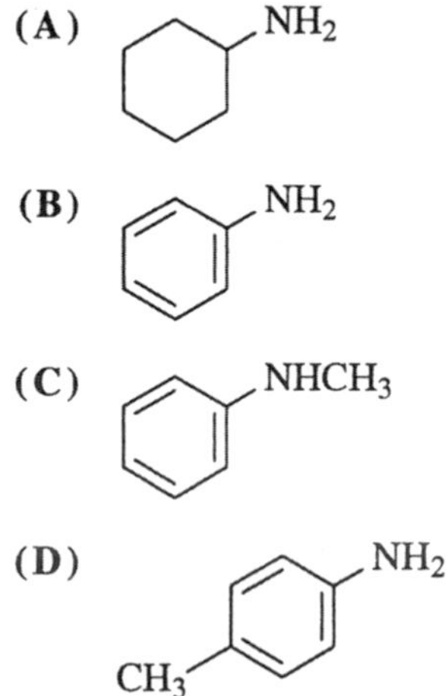

(D) NH_2, CH_3

6. Which compound corresponds to this IR spectrum?

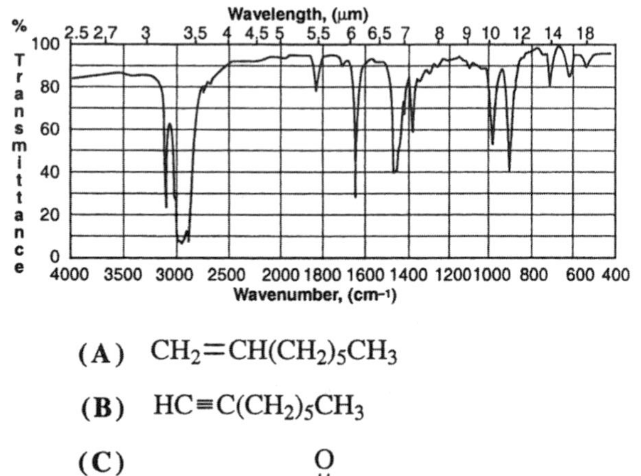

(A) $CH_2{=}CH(CH_2)_5CH_3$

(B) $HC{\equiv}C(CH_2)_5CH_3$

(C) $HC{\equiv}C(CH_2)_5\overset{O}{\overset{||}{C}}OCH_3$

(D) $CH_3(CH_2)_5\overset{O}{\overset{||}{C}}CH_2CH_3$

7. Which compound would produce this IR spectrum?

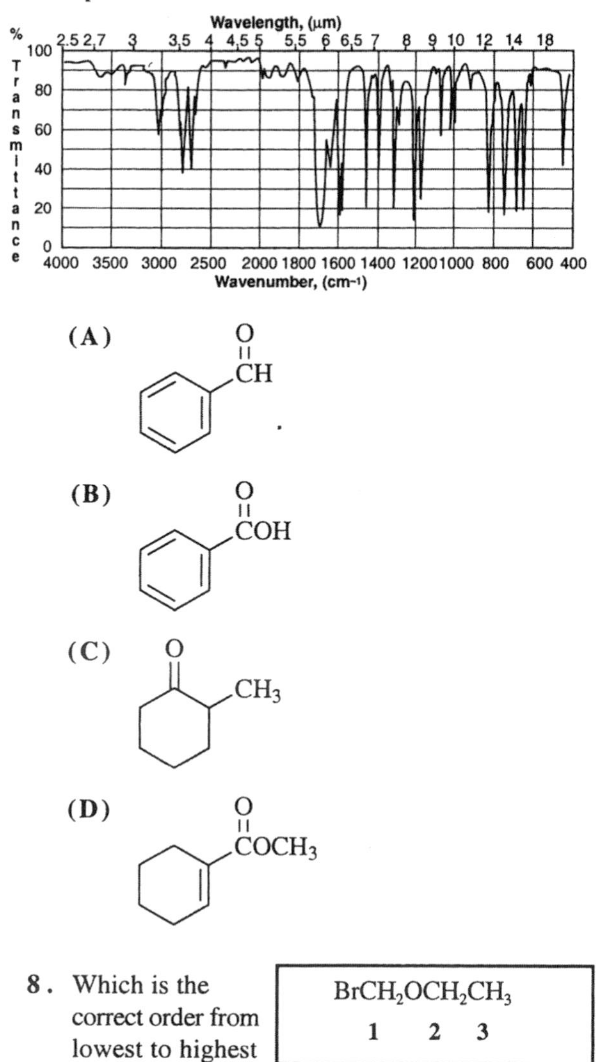

8. Which is the correct order from lowest to highest field for the chemical shifts of the numbered sets of protons in the ^{1}H NMR spectrum of this compound?

$BrCH_2OCH_2CH_3$
1 2 3

(A) 3 < 2 <1 **(B)** 2 < 1 < 3

(C) 1 < 2 < 3 **(D)** 2 < 3 < 1

9. Which of these compounds will show two triplets (among other signals) in their ^{1}H NMR spectra?

I	$ClCH_2CH_2CHCl_2$
II	$ClCH_2CH_2CH_2Cl$
III	$Cl_2CHCH_2CHCl_2$

(A) I, II, III **(B)** I only

(C) I, III only **(D)** II only

10. Which ^{1}H NMR spectrum is consistent with that of ethyl methyl sulfide? $CH_3CH_2SCH_3$

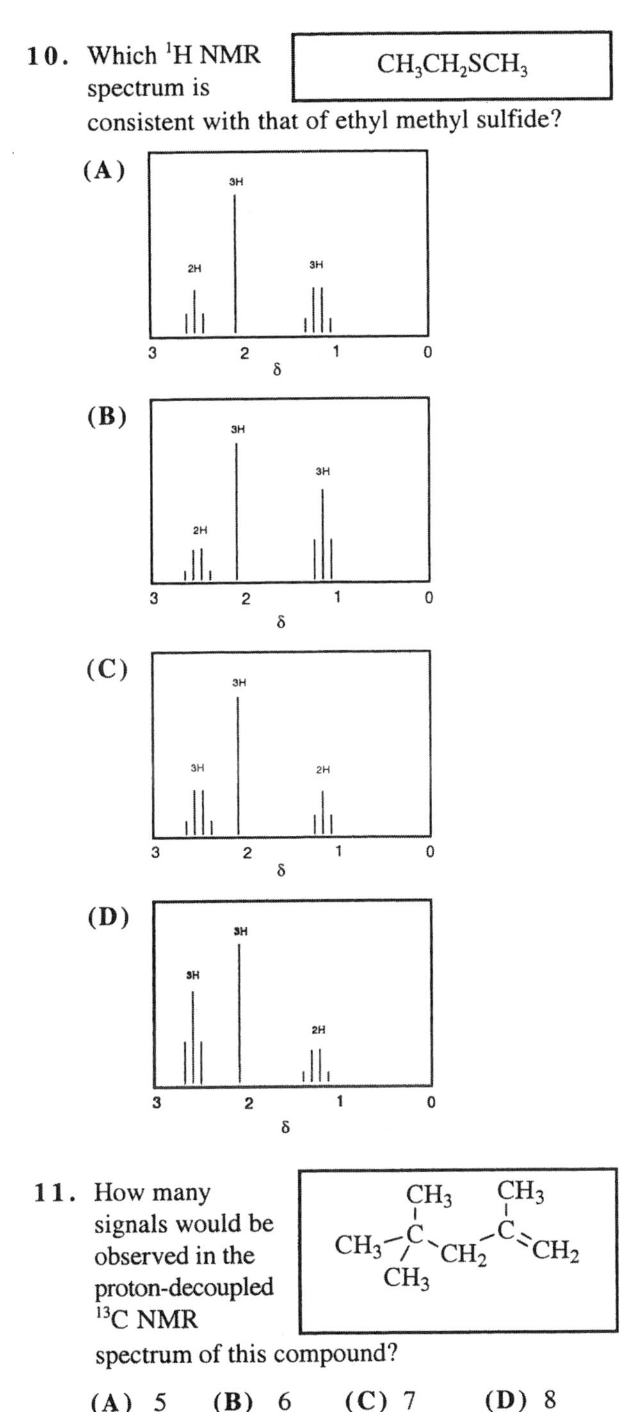

11. How many signals would be observed in the proton-decoupled ^{13}C NMR spectrum of this compound?

$(CH_3)_3C{-}CH_2{-}C(CH_3){=}CH_2$

(A) 5 **(B)** 6 **(C)** 7 **(D)** 8

12. Which structure is consistent with this ^{1}H NMR spectrum?

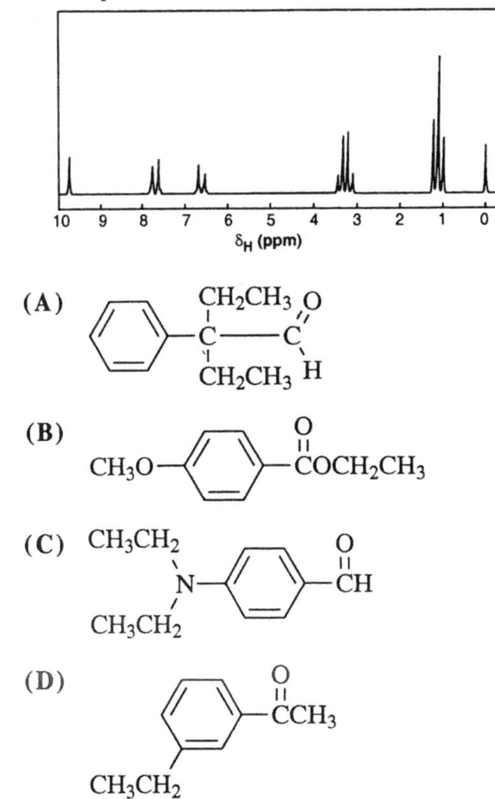

13. Which structure is consistent with this ^{1}H NMR spectrum?

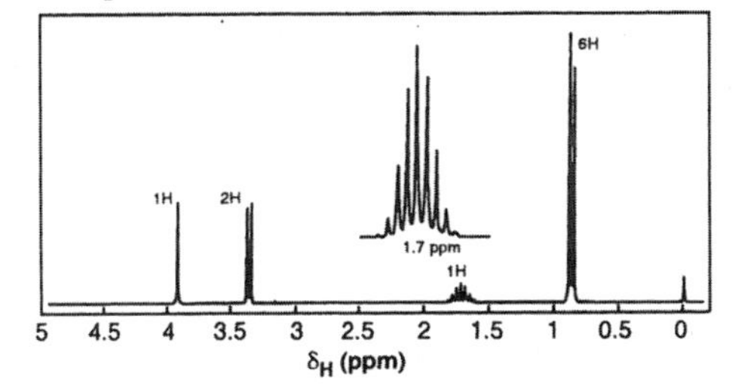

(A) $(CH_3)_3COH$

(B) $CH_3CH_2CH_2CH_2OH$

(C) $CH_3CH(CH_3)CH_2OH$

(D) $CH_3CH_2CH(OH)CH_3$

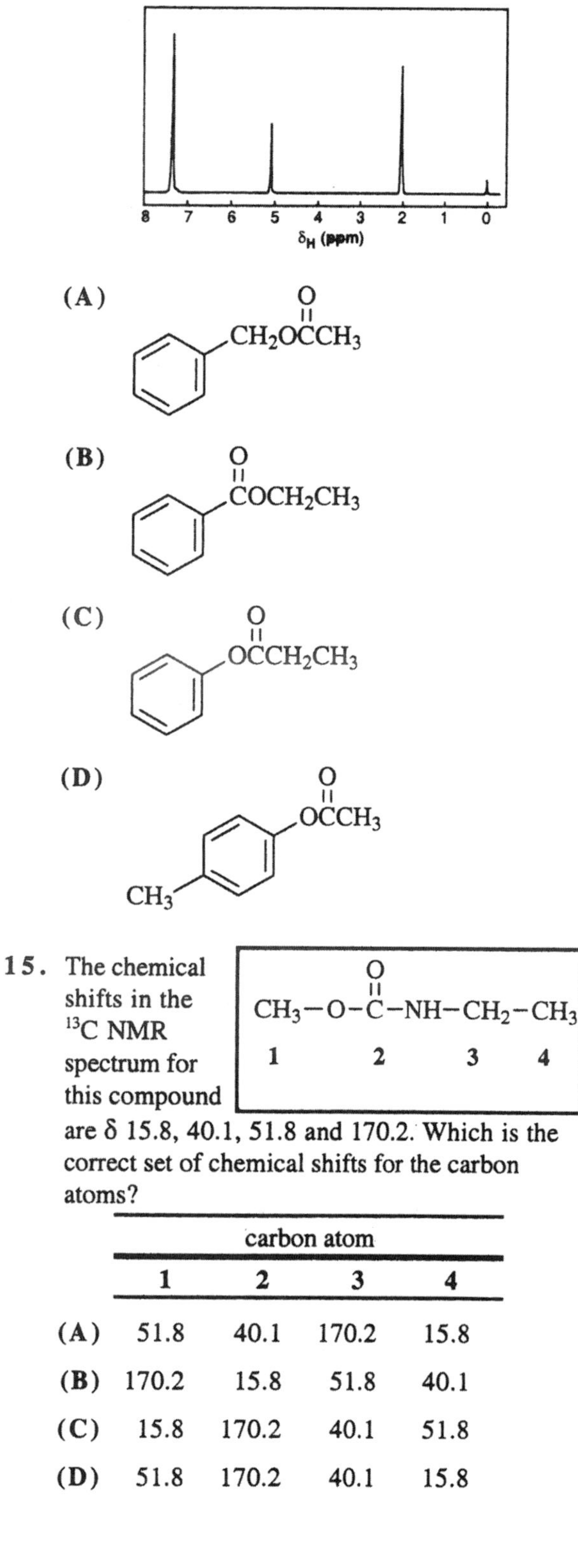

14. Which structure is consistent with this ^{1}H NMR spectrum?

(A) $CH_2O\overset{O}{C}CH_3$

(B) $\overset{O}{C}OCH_2CH_3$

(C) $O\overset{O}{C}CH_2CH_3$

(D) $O\overset{O}{C}CH_3$, CH_3

15. The chemical shifts in the ^{13}C NMR spectrum for this compound

$$\underset{1}{CH_3}-O-\underset{2}{\overset{O}{C}}-NH-\underset{3}{CH_2}-\underset{4}{CH_3}$$

are δ 15.8, 40.1, 51.8 and 170.2. Which is the correct set of chemical shifts for the carbon atoms?

	carbon atom			
	1	2	3	4
(A)	51.8	40.1	170.2	15.8
(B)	170.2	15.8	51.8	40.1
(C)	15.8	170.2	40.1	51.8
(D)	51.8	170.2	40.1	15.8

16. Which structure is most consistent with this ^{13}C NMR spectrum?

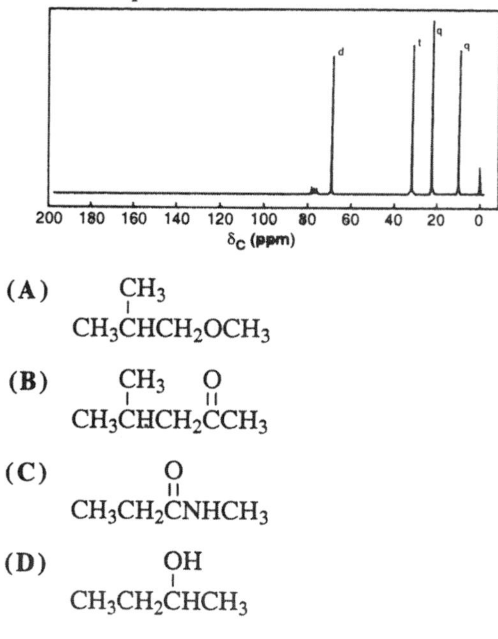

(A) $CH_3\overset{CH_3}{C}HCH_2OCH_3$

(B) $CH_3\overset{CH_3}{C}HCH_2\overset{O}{C}CH_3$

(C) $CH_3CH_2\overset{O}{C}NHCH_3$

(D) $CH_3CH_2\overset{OH}{C}HCH_3$

17. Which structure is in agreement with these ^{1}H NMR and ^{13}C NMR spectra?

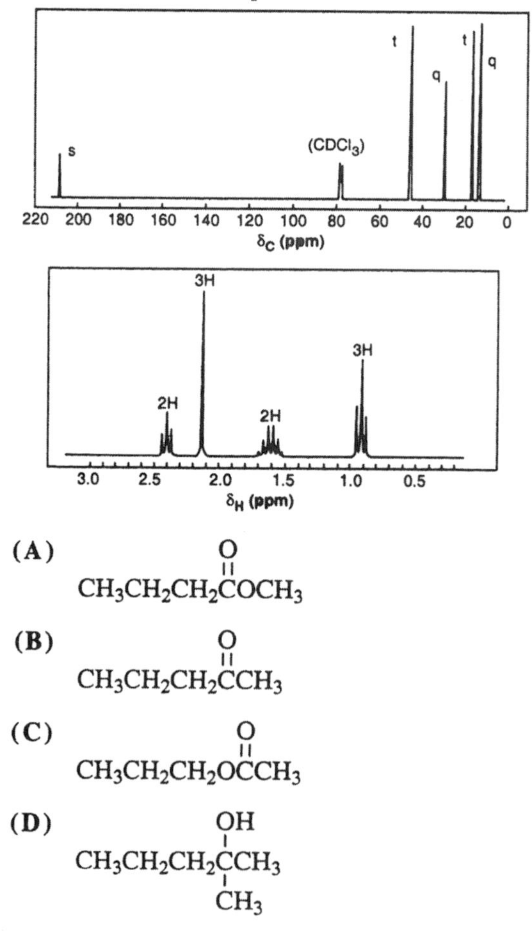

(A) $CH_3CH_2CH_2\overset{O}{C}OCH_3$

(B) $CH_3CH_2CH_2\overset{O}{C}CH_3$

(C) $CH_3CH_2CH_2O\overset{O}{C}CH_3$

(D) $CH_3CH_2CH_2\overset{OH}{\underset{CH_3}{C}}CH_3$

18. Which structure is consistent with this ^{13}C NMR spectrum?

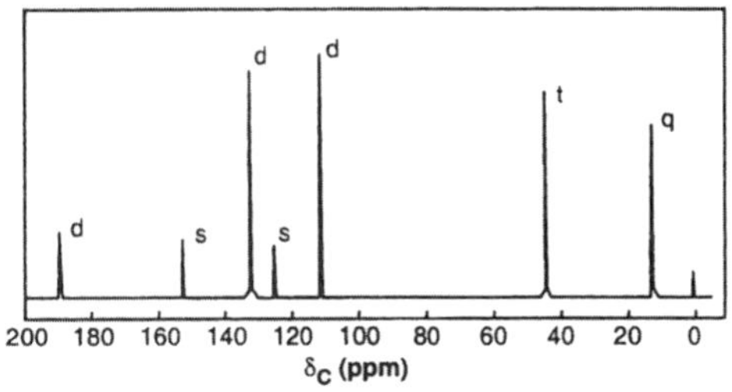

(A) 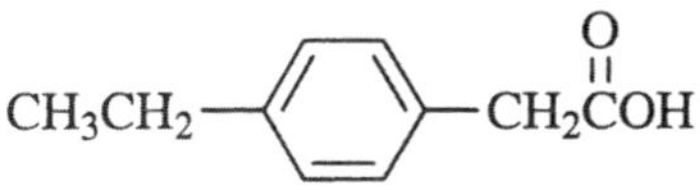

(B) 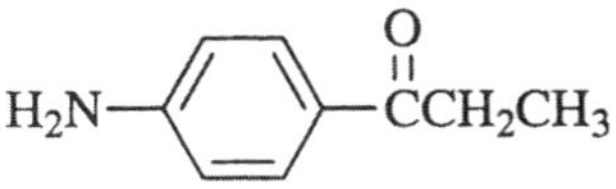

(C) 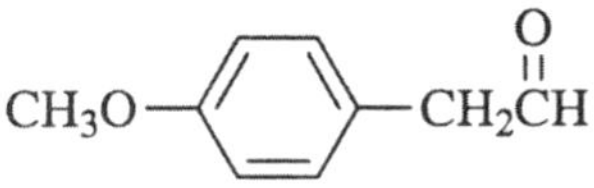

(D) 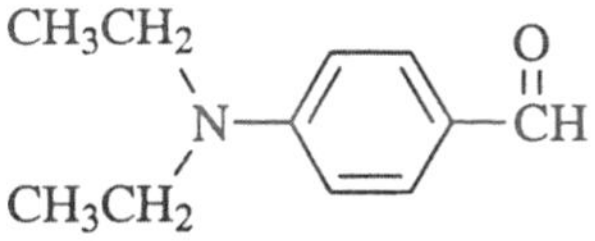

19. Which compound would show major fragment ions at *m*/*z* 73 and 59?

(A) $CH_3CH_2\overset{OH}{C}HCH_2CH_3$

(B) $CH_3CH_2CH_2\overset{OH}{C}HCH_3$

(C) $CH_3\underset{CH_3}{C}H\overset{OH}{C}HCH_3$

(D) $CH_3\underset{CH_3}{\overset{OH}{C}}CH_2CH_3$

20. Which structure for a molecule with the molecular formula $C_{11}H_{14}O_2$ is in best agreement with this ^{13}C NMR spectrum?

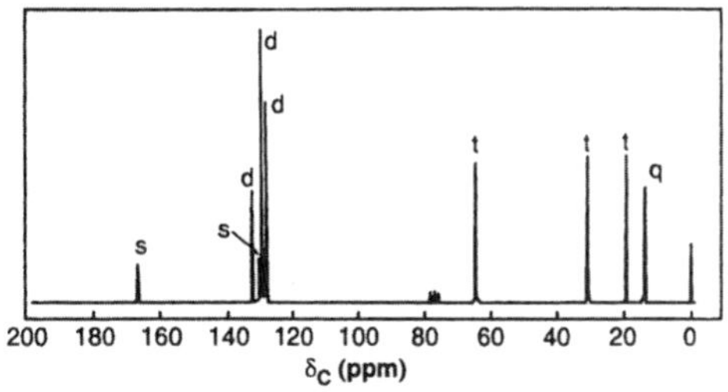

(A) 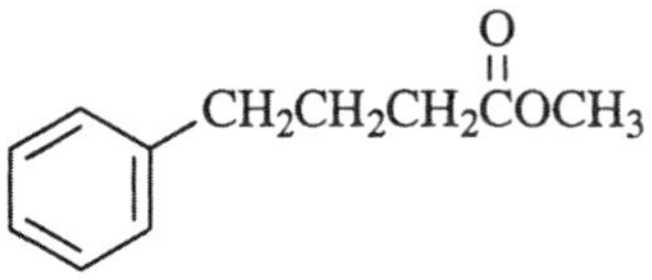

(B) 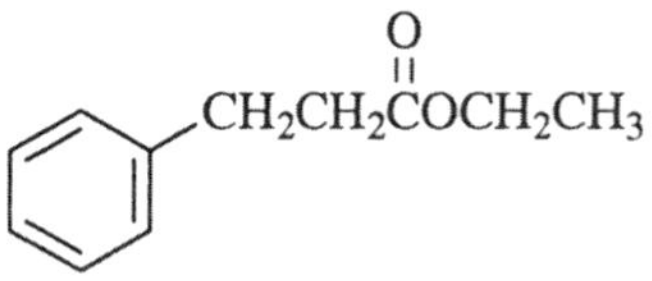

(C) 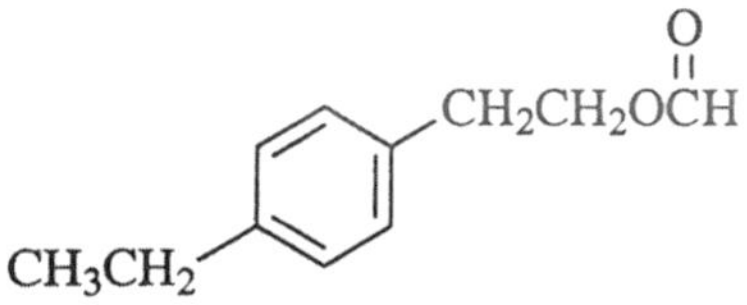

(D) 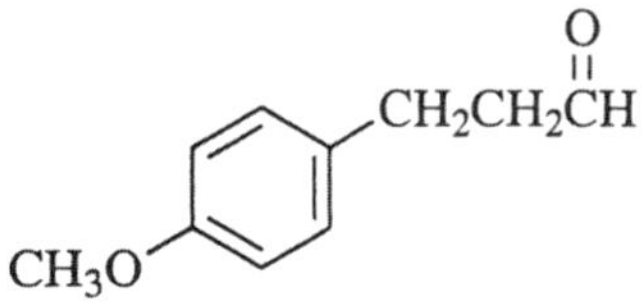

21. Which compound would show major fragment ions in its mass spectrum at *m*/*z* 85 and 57?

(A) $CH_3CH_2CH_2CH_2\overset{O}{\overset{\|}{C}}CH_2CH_3$

(B) $CH_3CH_2CH_2\overset{O}{\overset{\|}{C}}CH_2CH_2CH_3$

(C) $CH_3CH_2CH_2CH_2CH_2\overset{O}{\overset{\|}{C}}CH_3$

(D) $CH_3CH_2CH_2\overset{O}{\overset{\|}{C}}\underset{CH_3}{C}HCH_3$

22. An unknown compound was found to be unreactive towards chromic acid and gave this IR spectrum. Which is a likely structure for the unknown?

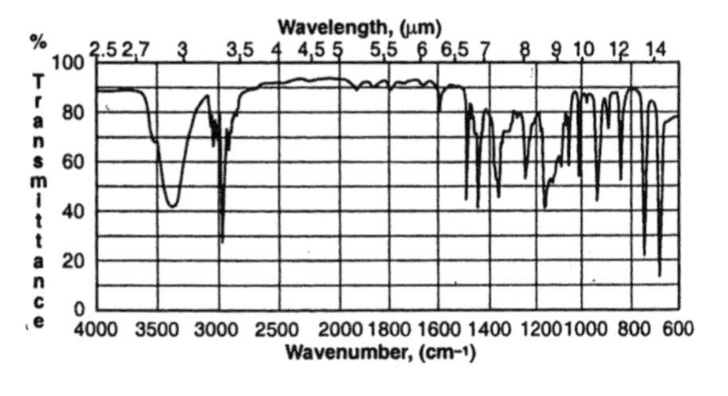

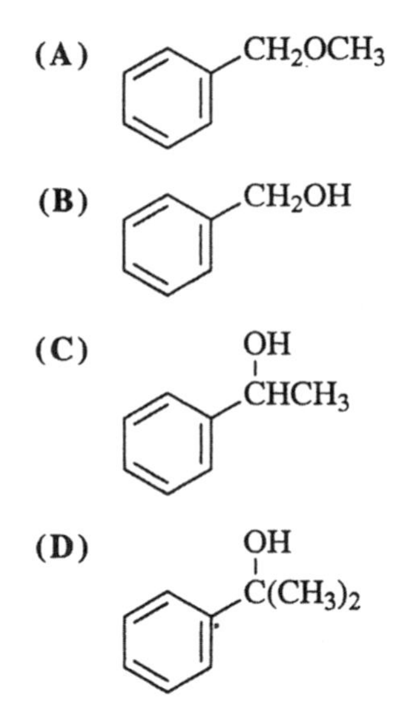

Answers to Study Questions

1. A	4. C	7. D
2. B	5. C	8. A
3. D	6. B	

Answers to Practice Questions

1. B	9. C	17. B
2. C	10. B	18. D
3. A	11. B	19. D
4. B	12. C	20. B
5. D	13. C	21. A
6. A	14. A	22. D
7. A	15. D	
8. C	16. D	

Synthesis and Qualitative Analysis

Synthesis involves the stringing together of a short series of organic reactions to prepare a desired pure compound, often with a defined stereochemistry. To accomplish this feat, it is necessary to know (1) the chemistry of the various functional groups, (2) how to choose, from among several reagents that will accomplish a desired functional group interconversion (the one that will produce the proper regiochemistry and/or stereochemistry in the best yield), and (3) how to use protecting groups to prevent one functional group from undergoing a reaction while another functional group is being transformed.

Syntheses involving a series of electrophilic aromatic substitutions require more than just knowing which reagents are necessary for the introduction of the ring substituents. The directing effects of the substituents must be taken into account, so that the reactions are performed in the correct order to obtain the desired substitution pattern.

Qualitative analysis involves using various chemical and/or spectroscopic data to identify a specific compound from among several possibilities.

Study Questions

SA-1. What is the major product of this reaction sequence?

$$C_6H_5-C{\equiv}CH \xrightarrow{NaNH_2} \xrightarrow{CH_3CH_2CH_2Br} \xrightarrow[\text{Lindlar catalyst}]{H_2}$$

(A) $C_6H_5(H)C{=}C(H)CH_2CH_2CH_3$ — C_6H_5 and $CH_2CH_2CH_3$ on the same side (H, H on the same side)

(B) $C_6H_5CH_2CH_2CH_2CH_2CH_3$

(C) $C_6H_5(H)C{=}C(H)CH_2CH_2CH_3$ — C_6H_5 and $CH_2CH_2CH_3$ on opposite sides

(D) $C_6H_5(H)C{=}C(H)CH(CH_3)_2$ — C_6H_5 and $CH(CH_3)_2$ on the same side (H, H on the same side)

Knowledge Required: (1) Formation and alkylation of anions of terminal alkynes. (2) Conditions for, and stereochemistry of, selective hydrogenation of alkynes.

Thinking it Through: Sodium amide is basic enough to quantitatively convert the terminal alkyne into its corresponding carbanion, which gives an S_N2 displacement with the propyl bromide to form an internal alkyne. Alkynes undergo catalytic hydrogenation to either alkanes or alkenes, depending upon the type of catalyst. Alkynes are reduced to alkanes with the usual catalysts (Pd or Pt), whereas they are reduced to alkenes with deactivated (poisoned) catalysts (most commonly a deactivated Pd catalyst such as Lindlar catalyst). In this question, phenylacetylene is alkylated with propyl bromide to give 1-phenyl-1-pentyne, which is then reduced to 1-phenyl-1-pentene. Therefore, choice **(B)**, the result of reduction to the corresponding alkane and choice **(D)**, which contains a rearranged alkyl group, may be eliminated. Choices **(A)** and **(C)** could both be produced by selective reduction of 1-phenyl-1-pentyne. Catalytic hydrogenation occurs by a *syn* addition of hydrogen to the alkyne and results in the formation of the *cis*, or *Z*, alkene. Consequently, choice **(A)** is the correct answer. Note that choice **(C)** would have been produced if the reducing agent had been sodium metal in liquid ammonia.

SA-2. What is the product of this reaction sequence?

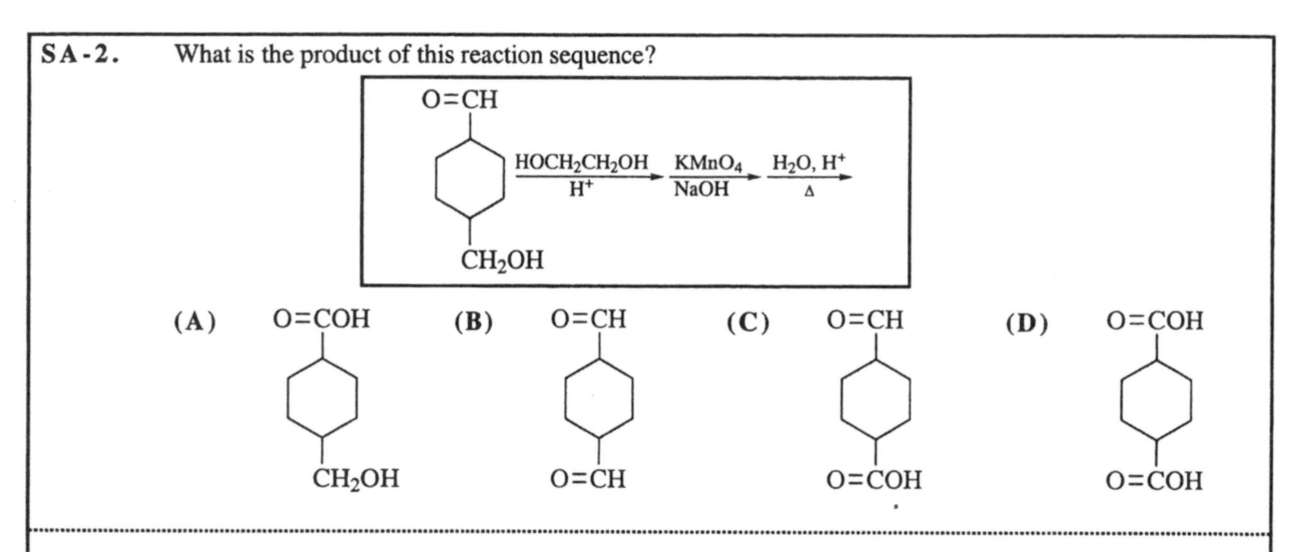

Knowledge Required: (1) Use of acetals as protecting groups. (2) Oxidation of alcohols with permanganate.

Thinking it Through: Permanganate oxidizes aldehydes and primary alcohols to carboxylic acids. The aldehyde is converted into an acetal to protect it from oxidation by the alkaline permanganate. (The acetal is stable under alkaline conditions.) After the oxidation has been completed, the acetal is hydrolyzed back to the aldehyde. Choice **(C)** is the only one that corresponds to the reaction sequence, and is the correct answer. Choice **(A)** would be the result if the acetal function oxidized instead of the alcohol. Choice **(B)** would result if the alcohol functional group were only oxidized to aldehyde. Choice **(D)** would result if both the acetal and alcohol functions were oxidized.

SA-3. Which starting material would be best suited for the preparation of this labeled tertiary alcohol using ^{13}C–labeled methylmagnesium bromide?

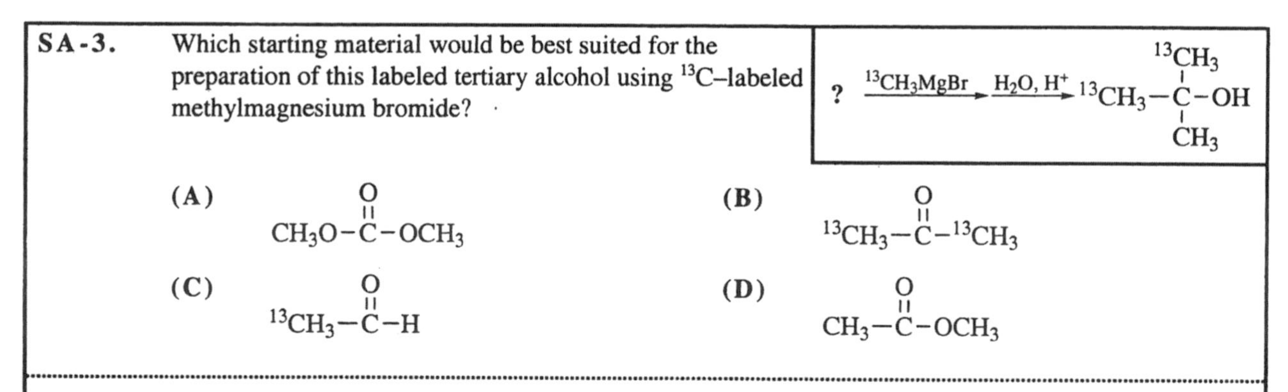

Knowledge Required: The products of the reactions of Grignard reagents with aldehydes, ketones, and esters.

Thinking it Through: Grignard reagents add to aldehydes and ketones to give secondary alcohols and tertiary alcohols, respectively. With esters, Grignard reagents give an initial substitution reaction to produce ketones, which then undergo addition of another molecule of the Grignard reagent to give tertiary alcohols as the final products. Only choices **(A)**, **(B)** and **(D)** will react with methylmagnesium bromide to yield *tert*-butyl alcohol. Thus, choice **(C)** may be eliminated. Note that only two of the three methyl groups in the *tert*-butyl alcohol are labeled with carbon-13. Since the Grignard reagent contains a carbon-13 labeled methyl group, the unlabeled methyl group must come from the carbonyl compound. Choices **(A)** and **(B)** will both produce *tert*-butyl alcohol with three carbon-13 labeled methyl groups, and both are therefore incorrect. This leaves choice **(D)**, methyl acetate, as the correct answer. This ester will react with two equivalents of the labeled Grignard reagent to give the desired product.

SA-4. Which of these reactions will give isobutyl isopropyl ether as the principal organic product?

$(CH_3)_2CHCH_2OCH(CH_3)_2$
isobutyl isopropyl ether

(A) $(CH_3)_2CHCH_2OH + CH_3CH_2CH_2OH \xrightarrow[140\ °C]{H_2SO_4}$

(B) $(CH_3)_2CHCH_2OH + CH_3CH{=}CH_2 \xrightarrow{Hg(OAc)_2} \xrightarrow[NaOH]{NaBH_4}$

(C) $CH_3C(CH_3){=}CH_2 + CH_3CH_2CH_2OH \xrightarrow{Hg(OAc)_2} \xrightarrow[NaOH]{NaBH_4}$

(D) $(CH_3)_2CHCH_2ONa + CH_3CH_2CH_2OH \longrightarrow$

Knowledge Required: (1) Methods for the preparation of ethers. (2) The regiochemistry of the oxymercuration–demercuration reactions of alkenes.

Thinking it Through: Choices **(A)**, **(B)** and **(C)** will all produce ethers as products, but choice **(D)** will not. The sodium isobutoxide in choice **(D)** would produce an ether if the other reactant were an alkyl halide or an alkyl sulfonate ester instead of an alcohol. Choice **(A)** may be eliminated because this method is useful only when the two alkyl groups of the ether are identical, or when one of them is tertiary. Choice **(A)** would give a mixture of one unsymmetrical and two symmetrical ethers, and since primary alcohols react in the presence of acid via an S_N2 process, none of the ethers would contain an isopropyl group. Choices **(B)** and **(C)** produce ethers by the same mechanism: formation of a cyclic mercurinium ion that undergoes ring opening by solvent attack on the more highly substituted carbon atom; followed by replacement of the mercury by hydrogen from sodium borohydride. Choice **(B)** would yield the desired isobutyl isopropyl ether, whereas choice **(C)** would yield *tert*-butyl propyl ether. Choice **(B)** is therefore the correct answer.

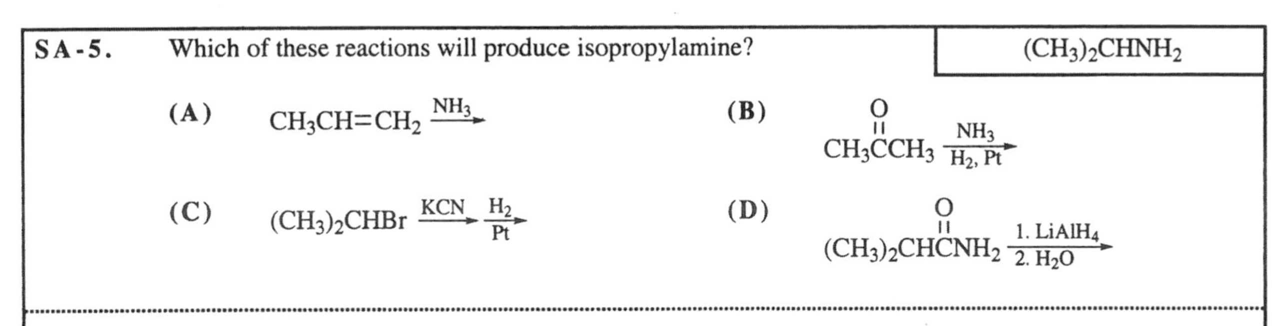

SA-5.
Which of these reactions will produce isopropylamine?
(CH3)2CHNH2
(A)
CH3CH=CH2
NH3
(B)
O
CH3CCH3
NH3
H2, Pt
(C)
(CH3)2CHBr
KCN
H2
Pt
(D)
O
(CH3)2CHCNH2
1. LiAlH4
2. H2O

Knowledge Required: (1) Methods for the preparation of amines. (2) That alkenes undergo electrophilic additions. (3) Mechanisms for reductions of nirtiles and amides. (4) Mechanism for reductive aminations of aldehydes and ketones.

Thinking it Through: Choices **(B)**, **(C)**, and **(D)** are all useful methods for preparing amines, but choice **(A)** is not. Choice **(A)** can be ruled out because it would require a nucleophilic addition to an alkene instead of the normally observed electrophilic additions. Choices **(C)** and **(D)** will produce an amine with four carbon atoms, isobutylamine, not the desired isopropylamine. This leaves choice **(B)** as the correct answer, which will indeed produce isopropylamine. The acetone reacts with the ammonia to form an imine, which is reduced to the amine faster than the acetone is reduced to isopropyl alcohol.

SA-6. What is a major product of this reaction sequence?

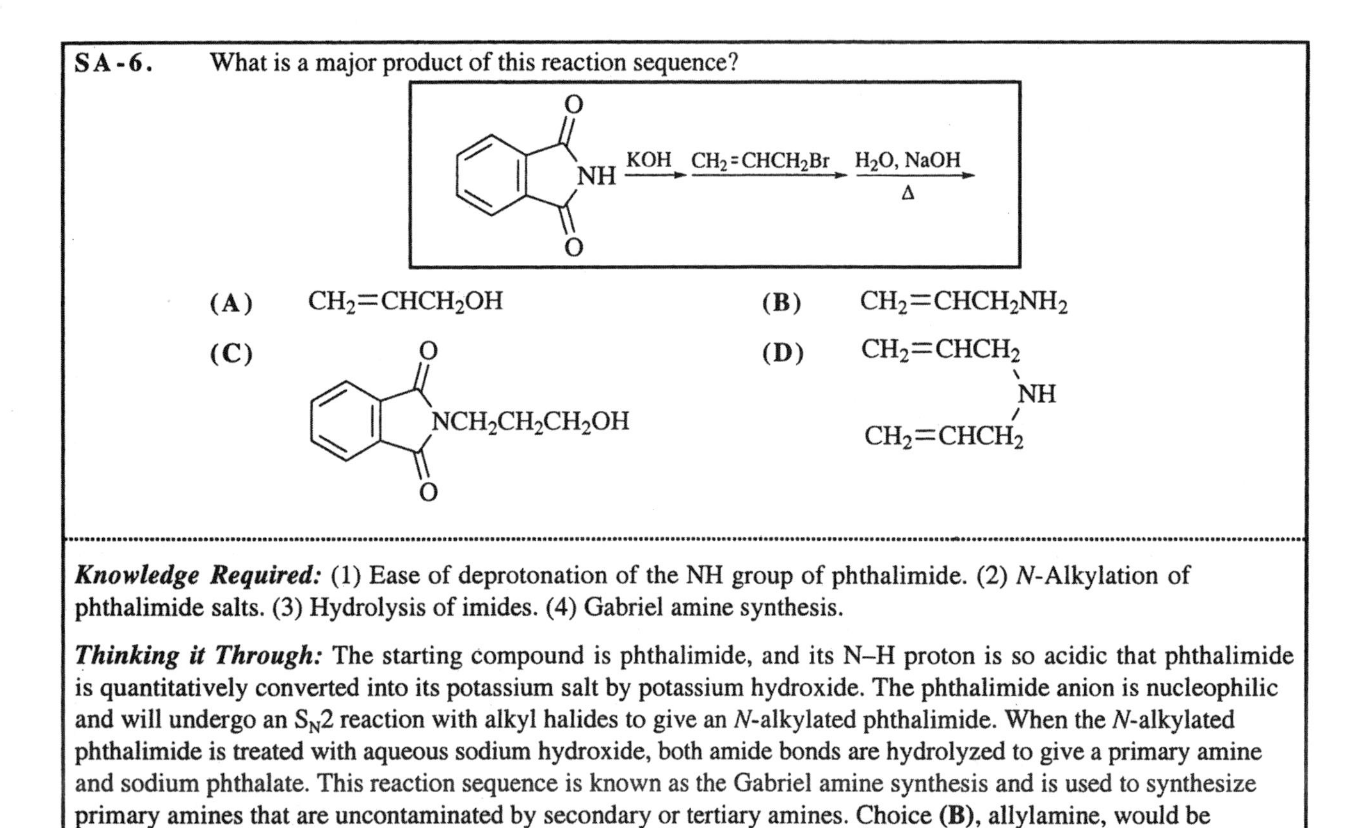

Knowledge Required: (1) Ease of deprotonation of the NH group of phthalimide. (2) *N*-Alkylation of phthalimide salts. (3) Hydrolysis of imides. (4) Gabriel amine synthesis.

Thinking it Through: The starting compound is phthalimide, and its N–H proton is so acidic that phthalimide is quantitatively converted into its potassium salt by potassium hydroxide. The phthalimide anion is nucleophilic and will undergo an S_N2 reaction with alkyl halides to give an *N*-alkylated phthalimide. When the *N*-alkylated phthalimide is treated with aqueous sodium hydroxide, both amide bonds are hydrolyzed to give a primary amine and sodium phthalate. This reaction sequence is known as the Gabriel amine synthesis and is used to synthesize primary amines that are uncontaminated by secondary or tertiary amines. Choice **(B)**, allylamine, would be produced by this sequence when allyl bromide is used as the alkyl halide in the synthesis, and choice **(B)** is the correct answer. Choice **(A)**, allyl alcohol, would be the result of an S_N2 reaction between the hydroxide and the allyl bromide, which ignores the presence of phthalimide. Choice **(C)** would require the unreasonable nucleophilic addition of the phthalimide anion to allyl bromide followed by an S_N2 displacement of bromide ion by sodium hydroxide. Even if choice **(C)** were produced, it would undergo hydrolysis to give 3-amino-1-propanol. Choice **(D)** would require dialkylation of the phthalimide followed by hydrolysis. However, a second alkylation of the monoalkylated phthalimide is not reasonable, because the electron pair on the nitrogen is stabilized by resonance with the two carbonyl groups and is non-nucleophilic. That is the reason the Gabriel synthesis produces only primary amines.

SA-7. Which reaction sequence will produce *tert*-butylamine in good yield? $(CH_3)_3CNH_2$

(A) $(CH_3)_3CBr \xrightarrow{NH_3}$

(B) $(CH_3)_3CBr \xrightarrow{KCN} \xrightarrow[Pt]{H_2}$

(C) $(CH_3)_3C\overset{O}{\overset{\|}{C}}OH \xrightarrow[\Delta]{SOCl_2} \xrightarrow{NH_3} \xrightarrow[2.\ H_2O]{1.\ LiAlH_4}$

(D) $(CH_3)_3C\overset{O}{\overset{\|}{C}}OH \xrightarrow[\Delta]{SOCl_2} \xrightarrow{NH_3} \xrightarrow[NaOH]{Br_2}$

Knowledge Required: (1) Methods for the preparation of amines. (2) Hofmann rearrangement. (3) That tertiary alkyl halides undergo eliminations, not substitutions, with basic nucleophiles.

Thinking it Through: Recall that because tertiary alkyl halides are sterically hindered to backside attack by nucleophiles, they undergo eliminations with basic nucleophiles instead of S_N2 displacements. The ammonia in choice **(A)** is basic enough to cause the *tert*-butyl bromide to undergo elimination to give isobutylene, not *tert*-butylamine. Cyanide ion is also basic enough to cause elimination of the *tert*-butyl bromide. Even if cyanide did substitute, reduction of the resulting nitrile would give neopentylamine, not *tert*-butylamine. In choice **(C)**, the carboxylic acid is converted into its amide, which is then reduced to the amine with $LiAlH_4$. Once again the amine would be neopentylamine, not *tert*-butylamine. This leaves choice **(D)** as the correct answer. The same carboxylic acid as in choice **(C)** is converted into its amide, which undergoes a Hofmann rearrangement upon treatment with an alkaline bromine solution to give *tert*-butylamine via an intermediate *tert*-butyl isocyanate.

$$(CH_3)_3C{-}\overset{O}{\overset{\|}{C}}{-}NH_2 \xrightarrow[NaOH]{Br_2} (CH_3)_3C{-}\overset{O}{\overset{\|}{C}}{-}\bar{N}{-}Br \longrightarrow (CH_3)_3C{-}N{=}C{=}O \xrightarrow[-CO_2]{H_2O}$$

SA-8. What is the expected product from this reaction sequence?

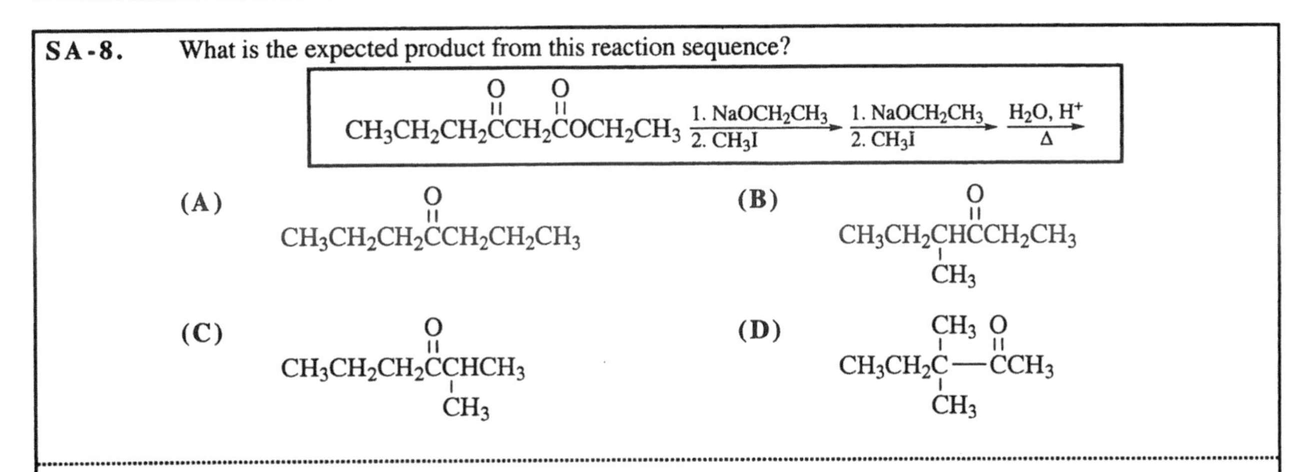

Knowledge Required: (1) The acidity of β-dicarbonyl compounds. (2) The alkylation and decarboxyklation of β-keto acids.

Thinking it Through: A hydrogen atom attached to a carbon atom between two carbonyl groups is fairly acidic ($pK_a = 11$–14) and is quantitatively abstracted by alkoxide bases to give a resonance-stabilized enolate ion. This enolate ion may be alkylated on the carbon atom with alkyl halides. If another hydrogen atom is attached to the carbon atom between the two carbonyl groups, it too may be abstracted and a second alkyl group added. If one of the carbonyl groups is an ester, hydrolysis will yield a carboxylic acid in which the carboxyl group is β to another carbonyl group. Such carboxylic acids are unstable to heat, and they decarboxylate to give a monocarbonyl compound. In this question, a β-keto ester undergoes two successive deprotonations and alkylations with methyl iodide followed by acid hydrolysis and decarboxylation, to yield an alkylated ketone. Because of the mechanism for this reaction, two methyl groups must be attached to the same carbon atom of the ketone. Thus choices **(A)** and **(B)** may be eliminated. Choice **(A)** is illogical. Choice **(B)** results from methylation of two different carbon atoms, the one that is α to both carbonyls and the one that is α to only the ketone carbonyl. Choices **(C)** and **(D)** both have two methyl groups attached to a single carbon atom. In choice **(D)** the methyl groups are attached to the carbon atom that was α to only the ketone carbonyl group, whereas in choice **(C)** the methyl groups are attached to the carbon atom that was α to both carbonyl groups. Thus, choice **(C)** is the correct answer. Had the starting material been a β-diester, the final decarboxylation product would have been a carboxylic acid instead of a ketone.

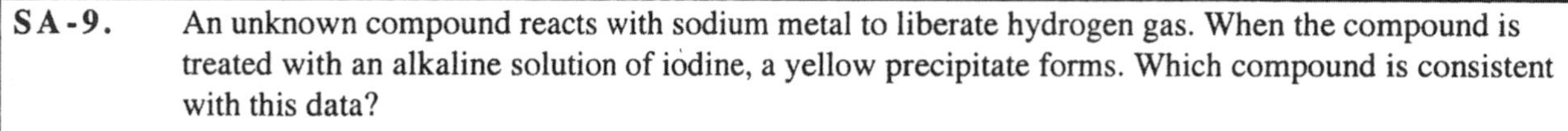

SA-9. An unknown compound reacts with sodium metal to liberate hydrogen gas. When the compound is treated with an alkaline solution of iodine, a yellow precipitate forms. Which compound is consistent with this data?

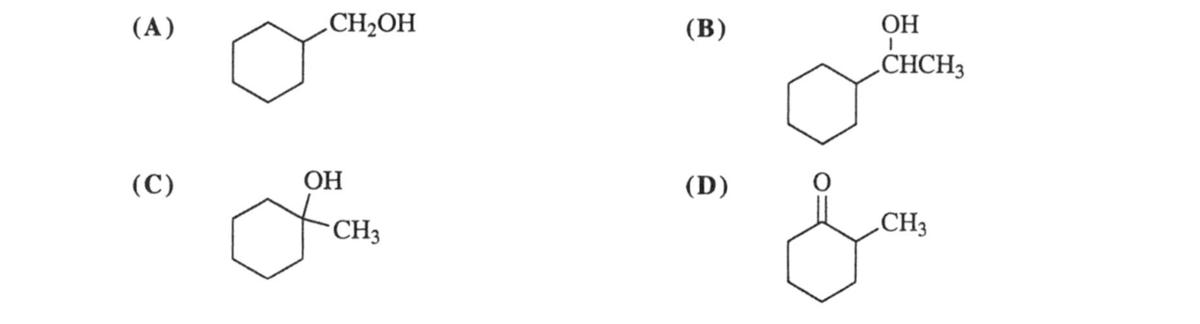

(A) CH₂OH

(B) OH / CHCH₃

(C) OH / CH₃

(D) O / CH₃

Knowledge Required: (1) The reaction of alcohols with sodium metal. (2) The classes of compounds that undergo the haloform reaction. (3) The fact that iodoform is a yellow solid.

Thinking it Through: Choices **(A)**, **(B)** and **(C)** all contain an OH group and will react with sodium metal to liberate hydrogen gas, but choice **(D)**, a ketone, will not, and may be eliminated from consideration. The cleavage of a ketone with an α methyl group, or acetaldehyde, into CHX_3 (haloform) and a carboxylate salt by the action of alkaline solution of a halogen is known as the haloform reaction. Because iodoform (CHI_3) is a yellow solid, an alkaline solution of iodine serves as the basis for a qualitative test for the presence of a methyl ketone or acetaldehyde. Alkaline solutions of halogens are oxidizing agents; therefore, any alcohol that can be oxidized to either a methyl ketone or acetaldehyde will undergo the haloform reaction, and give a positive iodoform test. Only the alcohol in choice **(B)** can be oxidized to a methyl ketone and undergo an iodoform reaction to produce the yellow CHI_3. Choice **(B)** is the correct answer.

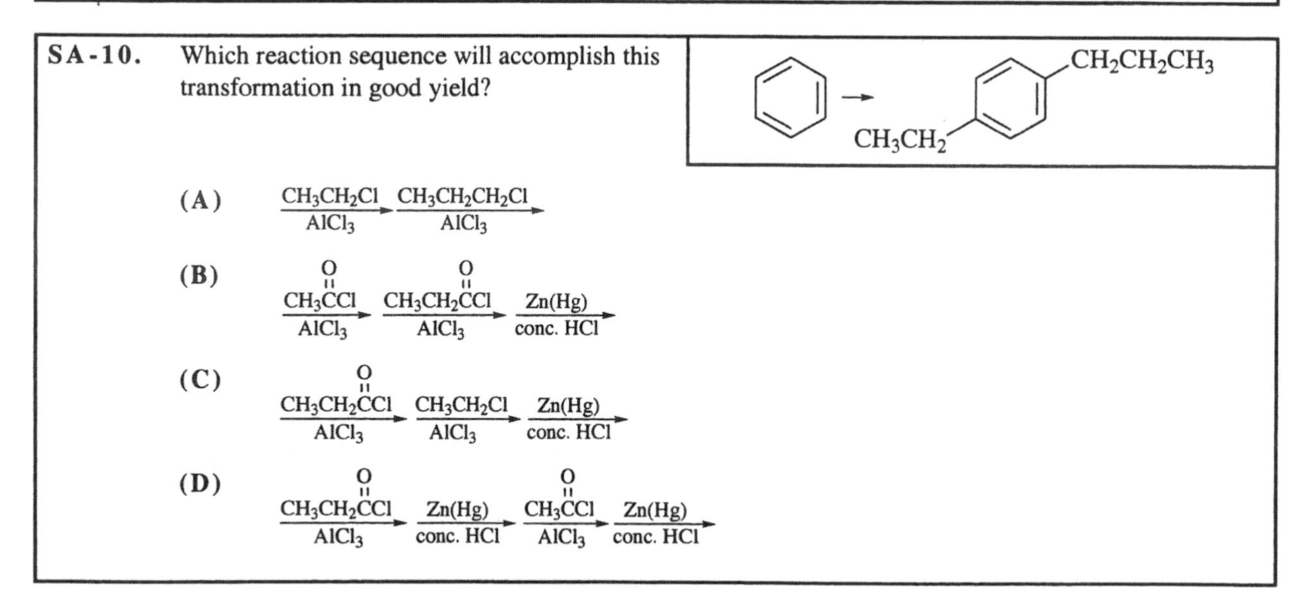

SA-10. Which reaction sequence will accomplish this transformation in good yield?

(A) $\dfrac{CH_3CH_2Cl}{AlCl_3}$ → $\dfrac{CH_3CH_2CH_2Cl}{AlCl_3}$ →

(B) $\dfrac{\overset{O}{\overset{\|}{CH_3CCl}}}{AlCl_3}$ → $\dfrac{\overset{O}{\overset{\|}{CH_3CH_2CCl}}}{AlCl_3}$ → $\dfrac{Zn(Hg)}{conc.\ HCl}$ →

(C) $\dfrac{\overset{O}{\overset{\|}{CH_3CH_2CCl}}}{AlCl_3}$ → $\dfrac{CH_3CH_2Cl}{AlCl_3}$ → $\dfrac{Zn(Hg)}{conc.\ HCl}$ →

(D) $\dfrac{\overset{O}{\overset{\|}{CH_3CH_2CCl}}}{AlCl_3}$ → $\dfrac{Zn(Hg)}{conc.\ HCl}$ → $\dfrac{\overset{O}{\overset{\|}{CH_3CCl}}}{AlCl_3}$ → $\dfrac{Zn(Hg)}{conc.\ HCl}$ →

Knowledge Required: (1) The mechanism of electrophilic aromatic substitutions. (2) Directing effects of electron-donating substituents versus electron-withdrawing substituents. (3) Friedel–Crafts reaction. (4) Clemmensen reduction of ketone carbonyl groups to methylene groups.

Thinking it Through: The product has two alkyl groups *para* to each other and could conceivably be synthesized by two successive Friedel–Crafts alkylations, as in choice **(A)**. However, recall that the products from Friedel–Crafts reactions with primary alkyl halides are primarily those derived from a rearranged alkyl group. Thus, choice **(A)** would give predominantly *p*- ethylisopropylbenzene and is incorrect. The best way to attach a primary alkyl group to a benzene ring is to perform a Friedel–Crafts acylation and then reduce the carbonyl group to a methylene group by either a Clemmensen (Zn(Hg), conc. HCl) or Wolff–Kishner (NH_2NH_2, NaOH, Δ) reduction. Choices **(B)**, **(C)** and **(D)** all involve Friedel–Crafts acylations and Clemmensen reductions, but choices **(B)** and **(C)** have fatal flaws, which allows them to be eliminated. The first flaw is that Friedel–Crafts reactions cannot be done on rings that bear substituents more deactivating than halogens. Once an acyl group is attached to the ring it prevents any further acylation or alkylation. A second flaw is that even if a further Friedel–Crafts reaction were possible, an acyl group is a *meta*-director and would give a *m*-disubstituited benzene, not a *p*-disubstituted benzene. By default, choice **(D)** must be the correct answer. In **(D)**, the first acylation followed by a Clemmensen reduction gives propylbenzene, which then undergoes a second acylation and reduction, to give the desired *p*-ethylpropylbenzene. One might have introduced the ethyl group via a Friedel–Crafts alkylation with ethyl chloride, but the propyl and ethyl groups activate the ring toward multiple alkylation. The acylation/reduction sequence ensures the formation of only a disubstituted product.

SA-11. Which of these reaction sequences will produce *m*-bromochlorobenzene?

(A) Bromobenzene $\xrightarrow[H_2SO_4]{HNO_3}$ $\xrightarrow[Pt]{H_2}$ $\xrightarrow[HCl,\ 0\ °C]{NaNO_2}$ $\xrightarrow[\Delta]{CuCl}$

(B) Nitrobenzene $\xrightarrow[Pt]{H_2}$ $\xrightarrow[FeBr_3]{Br_2}$ $\xrightarrow[HCl,\ 0\ °C]{NaNO_2}$ $\xrightarrow[\Delta]{CuCl}$

(C) Nitrobenzene $\xrightarrow[FeBr_3]{Br_2}$ $\xrightarrow[Pt]{H_2}$ $\xrightarrow[HCl,\ 0\ °C]{NaNO_2}$ $\xrightarrow[\Delta]{CuCl}$

(D) Bromobenzene $\xrightarrow[H_2SO_4]{HNO_3}$ $\xrightarrow[HCl,\ 0\ °C]{NaNO_2}$ $\xrightarrow[\Delta]{CuCl}$

Knowledge Required: (1) The mechanism of electrophilic aromatic substitutions. (2) Directing effects of electron-donating substituents versus electron-withdrawing substituents. (3) Reduction of nitro groups to amino groups. (4) Conditions for nitration and bromination of benzene rings. (5) Diazotization of anilines. (6) Displacement of molecular nitrogen from diazonium ions (Sandmeyer reaction).

Thinking it Through: In choice **(A)**, the nitration and reduction would place an amino group *ortho* or *para* to the bromine. The subsequent diazotization of the amino group and Sandmeyer reaction would produce *o*-bromochlorobenzene or *p*-bromochlorobenzene, not *m*-bromochlorobenzene. Thus choice **(A)** is incorrect. Choice **(B)** is also incorrect because the aniline produced by the initial reduction of nitrobenzene is so reactive towards bromination that it produces 2,4,6 tribromoaniline. In choice **(C)**, the bromination of nitrobenzene will give *m*-bromonitrobenzene, and the subsequent reduction, diazotization, and Sandmeyer displacement will produce the correct product, *m*-bromochlorobenzene. Choice **(D)** is incorrect because (1) the nitro group would be introduced *ortho* or *para* to the bromine and (2) no further reaction would occur as the nitro group has not been reduced to an amino group.

Practice Questions

1. Alkene derivatives of pristane (2,6,10,14-tetramethylpentadecane) can be isolated from marine zooplankton and are important in the study of the marine food chain. Which of these routes is best for the preparation of pristane?

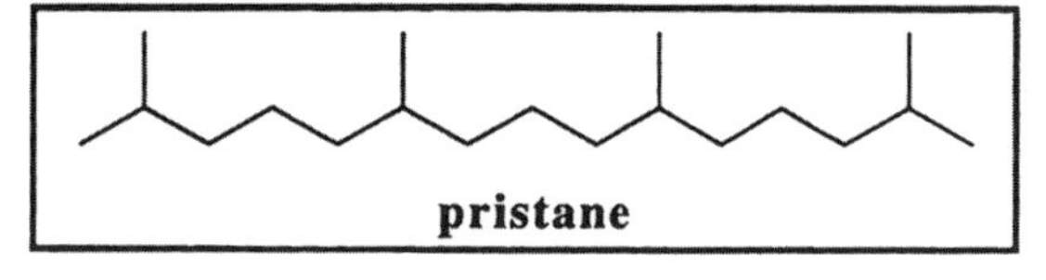

(A)

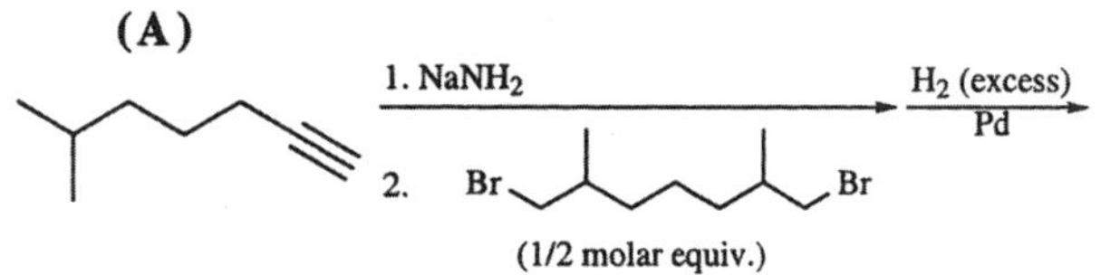

(B)

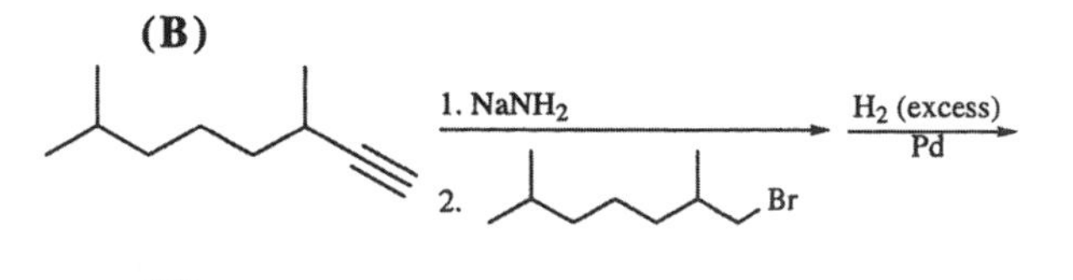

(C)

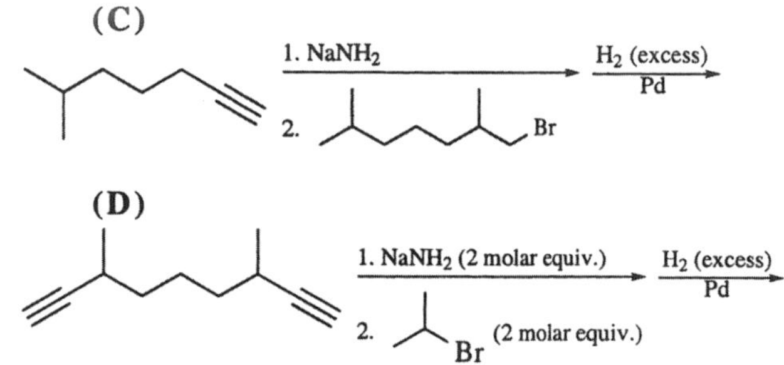

(D)

2. What is the expected product from this reaction sequence?

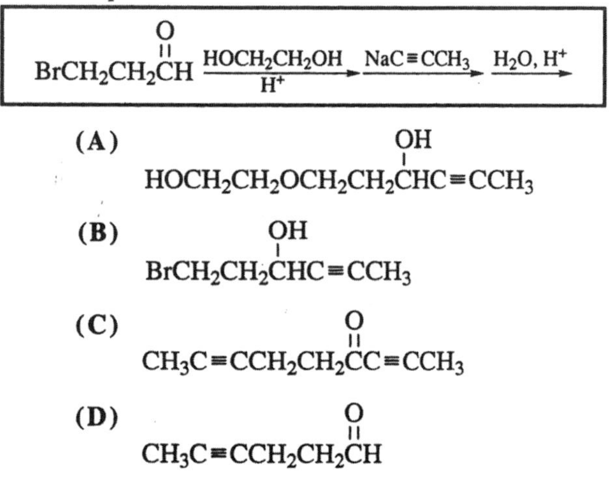

$BrCH_2CH_2CHO \xrightarrow[H^+]{HOCH_2CH_2OH} \xrightarrow{NaC\equiv CCH_3} \xrightarrow{H_2O,\ H^+}$

(A) $HOCH_2CH_2OCH_2CH_2CH(OH)C\equiv CCH_3$

(B) $BrCH_2CH_2CH(OH)C\equiv CCH_3$

(C) $CH_3C\equiv CCH_2CH_2C(=O)C\equiv CCH_3$

(D) $CH_3C\equiv CCH_2CH_2CHO$

3. What is the major product of this reaction sequence?

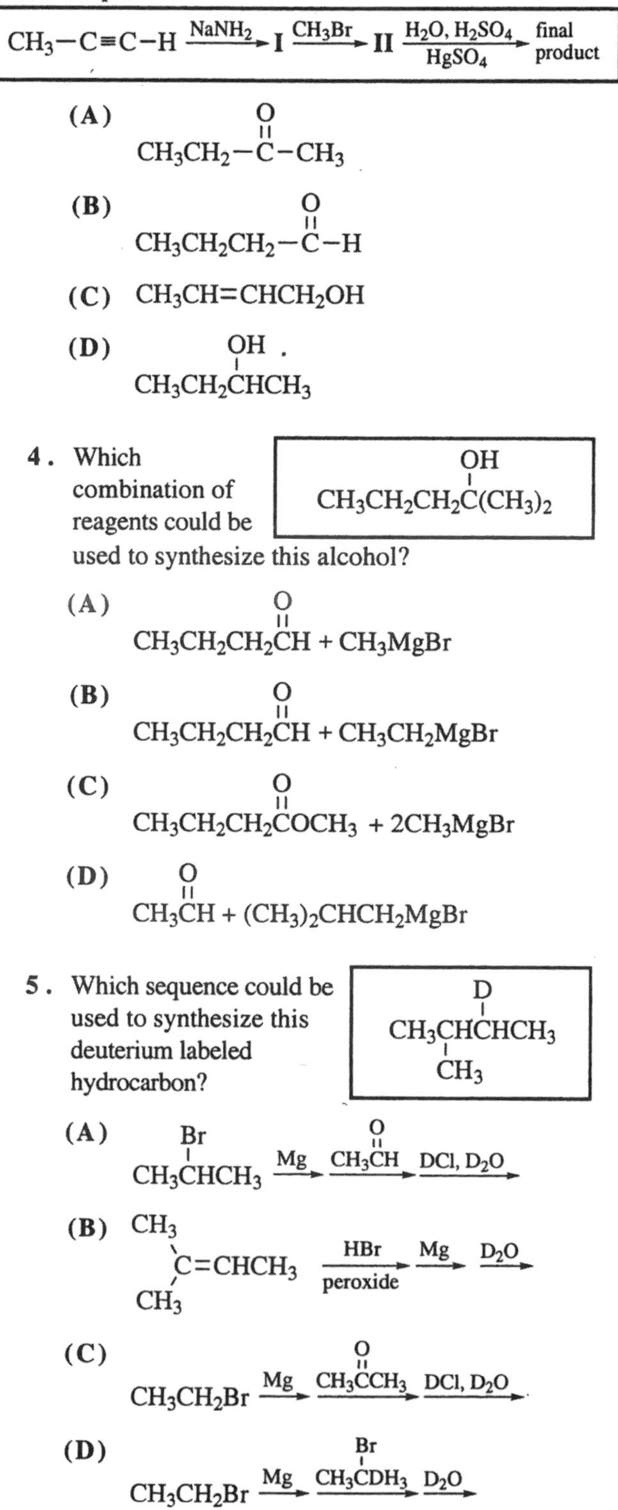

$CH_3-C\equiv C-H \xrightarrow{NaNH_2} \text{I} \xrightarrow{CH_3Br} \text{II} \xrightarrow[HgSO_4]{H_2O,\ H_2SO_4}$ final product

(A) $CH_3CH_2-C(=O)-CH_3$

(B) $CH_3CH_2CH_2-C(=O)-H$

(C) $CH_3CH=CHCH_2OH$

(D) $CH_3CH_2CH(OH)CH_3$

4. Which combination of reagents could be used to synthesize this alcohol?

$CH_3CH_2CH_2C(OH)(CH_3)_2$

(A) $CH_3CH_2CH_2CHO + CH_3MgBr$

(B) $CH_3CH_2CH_2CHO + CH_3CH_2MgBr$

(C) $CH_3CH_2CH_2C(=O)OCH_3 + 2CH_3MgBr$

(D) $CH_3CHO + (CH_3)_2CHCH_2MgBr$

5. Which sequence could be used to synthesize this deuterium labeled hydrocarbon?

$CH_3CH(D)CH(CH_3)CH_3$

(A) $CH_3CH(Br)CH_3 \xrightarrow{Mg} \xrightarrow{CH_3CHO} \xrightarrow{DCl,\ D_2O}$

(B) $(CH_3)_2C=CHCH_3 \xrightarrow[peroxide]{HBr} \xrightarrow{Mg} \xrightarrow{D_2O}$

(C) $CH_3CH_2Br \xrightarrow{Mg} \xrightarrow{CH_3C(=O)CH_3} \xrightarrow{DCl,\ D_2O}$

(D) $CH_3CH_2Br \xrightarrow{Mg} \xrightarrow{CH_3CBrDH_3} \xrightarrow{D_2O}$

6. Which reaction sequence would accomplish the synthesis of this alcohol?

$$CH_3CH_2CH(OH)CH_3$$

(A) $CH_3CH_2CH_2OH \xrightarrow{HBr} \xrightarrow[\text{ether}]{Mg} \xrightarrow[2.\ H_2O,\ H^+]{1.\ HCHO}$

(B) $CH_3CH_2CH_2OH \xrightarrow[2.\ H_2O,\ H^+]{1.\ CH_3MgBr} \xrightarrow[2.\ H_2O,\ H^+]{1.\ LiAlH_4}$

(C) $CH_3CH_2CH_2OH \xrightarrow[180\ ^\circ C]{\text{conc. } H_2SO_4} \xrightarrow[H^+]{H_2O}$

(D) $CH_3CH_2CH_2OH \xrightarrow{C_5H_5NH^+\ CrO_3Cl^-\ (PCC)} \xrightarrow[2.\ H_2O,\ H^+]{1.\ CH_3MgBr}$

7. Which of these methods will produce 1-pentanol?

I $CH_3CH_2CH_2MgBr \xrightarrow[2.\ H_2O,\ H^+]{1.\ \text{ethylene oxide}}$

II $CH_3CH_2CH_2CH{=}CH_2 \xrightarrow[2.\ H_2O_2,\ NaOH]{1.\ B_2H_6,\ THF}$

III $CH_3CH_2CH_2CH_2CH_2Br \xrightarrow{KOC(CH_3)_3} \xrightarrow[H^+]{H_2O}$

IV $CH_3CH_2CH_2CH_2MgBr \xrightarrow[2.\ H_2O,\ H^+]{1.\ HCHO}$

(A) I, II, IV (B) I, II, III

(C) I, III, IV (D) II, III, IV

8. Which would be the best method to prepare propylamine uncontaminated with dipropylamine?

(A) $CH_3CH_2CH_2Br \xrightarrow{NH_3}$

(B)

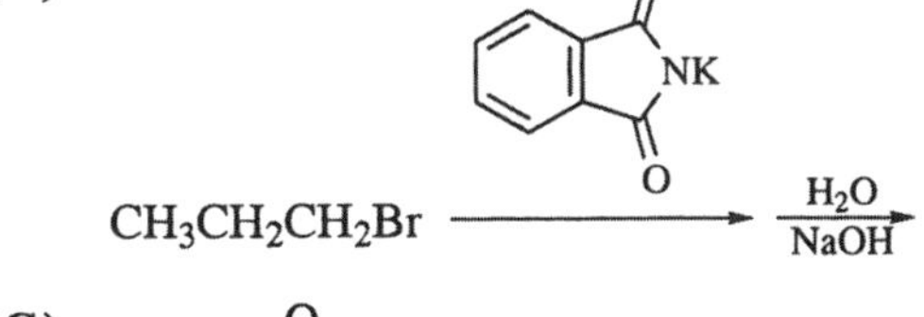

$CH_3CH_2CH_2Br \longrightarrow \xrightarrow[NaOH]{H_2O}$

(C) $CH_3CH_2C(O)NH_2 \xrightarrow[NaOH]{Br_2}$

(D) $CH_3CH_2CH_2Br \xrightarrow{KCN} \xrightarrow[Pt]{H_2}$

9. What intermediate, leading to the major product, is formed in this reaction?

$$(CH_3)_2CHCH_2C(O)NH_2 \xrightarrow[NaOH]{Br_2} (CH_3)_2CHCH_2NH_2$$

(A) $(CH_3)_2CH\bar{C}HC(O)NH_2$

(B) $(CH_3)_2CHCH_2C(O^-)(OH)NH_2$

(C) $(CH_3)_2CHCH_2N{=}C{=}O$

(D) $(CH_3)_2CHCH_2C(O)O^-$

10. What is the product of this reaction?

$$C_6H_5CH_2C(O)NH_2 \xrightarrow[NaOCH_3,\ CH_3OH]{Br_2}$$

(A) $C_6H_5CH_2NH_2$

(B) $C_6H_5CH_2C(O)OCH_3$

(C) $C_6H_5CH_2NHC(O)OCH_3$

(D) $C_6H_5CH(Br)C(O)NH_2$

11. Which sequence of reagents would best effect this transformation?

$$CH_3C(O)CH_2C(O)OCH_3 \rightarrow CH_3C(O)CH_2CH_2C_6H_5$$

(A) $\xrightarrow{NaOCH_3} \xrightarrow{C_6H_5CH_2Br} \xrightarrow[\Delta]{H_2O,\ H^+}$

(B) $\xrightarrow{NaOH} \xrightarrow{C_6H_5CH_2Br} \xrightarrow[\Delta]{H_2O,\ H^+}$

(C) $\xrightarrow{LiAlH_4} \xrightarrow{HBr} \xrightarrow{C_6H_5Li}$

(D) $\xrightarrow{C_6H_5CH_2Li} \xrightarrow{NaBH_4} \xrightarrow{H_2O,\ H^+}$

12. Which would be a suitable solvent for the preparation of ethylmagnesium bromide from ethyl bromide and magnesium?

(A) $CH_3CO_2CH_2CH_3$

(B) CH_3CH_2OH

(C) $CH_3OCH_2CH_2OCH_3$

(D) CH_3CO_2H

13. What is the product of this reaction sequence?

$CH_2(CO_2CH_3)_2 \xrightarrow[]{NaOCH_3} \xrightarrow[(1/2\ \text{molar equiv.})]{BrCH_2CH_2Br} \xrightarrow[\Delta]{H_2O,\ H^+}$

(A) ▷—CO_2H

(B) $HO_2CCH_2CH_2CH_2CH_2CO_2H$

(C) $BrCH_2CH_2CH_2CO_2H$

(D) $HO_2CCH_2CH_2CO_2H$

14. Which reaction sequence might be used to synthesize this compound?

CO_2H

NO_2

Br

(A)

CH_3

$\xrightarrow{Br_2}{FeBr_3}$ $\xrightarrow{\text{1. }KMnO_4,\ NaOH,\ \Delta}{\text{2. }H_2O,\ H^+}$ $\xrightarrow{HNO_3}{H_2SO_4}$

(B)

CH_3

$\xrightarrow{\text{1. }KMnO_4,\ NaOH,\ \Delta}{\text{2. }H_2O,\ H^+}$ $\xrightarrow{Br_2}{FeBr_3}$ $\xrightarrow{HNO_3}{H_2SO_4}$

(C)

NO_2

$\xrightarrow{Br_2}{FeBr_3}$ $\xrightarrow{CH_3Cl}{AlCl_3}$ $\xrightarrow{\text{1. }KMnO_4,\ NaOH,\ \Delta}{\text{2. }H_2O,\ H^+}$

NO_2

$\xrightarrow{CH_3Cl}{AlCl_3}$ $\xrightarrow{\text{1. }KMnO_4,\ NaOH,\ \Delta}{\text{2. }H_2O,\ H^+}$ $\xrightarrow{Br_2}{FeBr_3}$

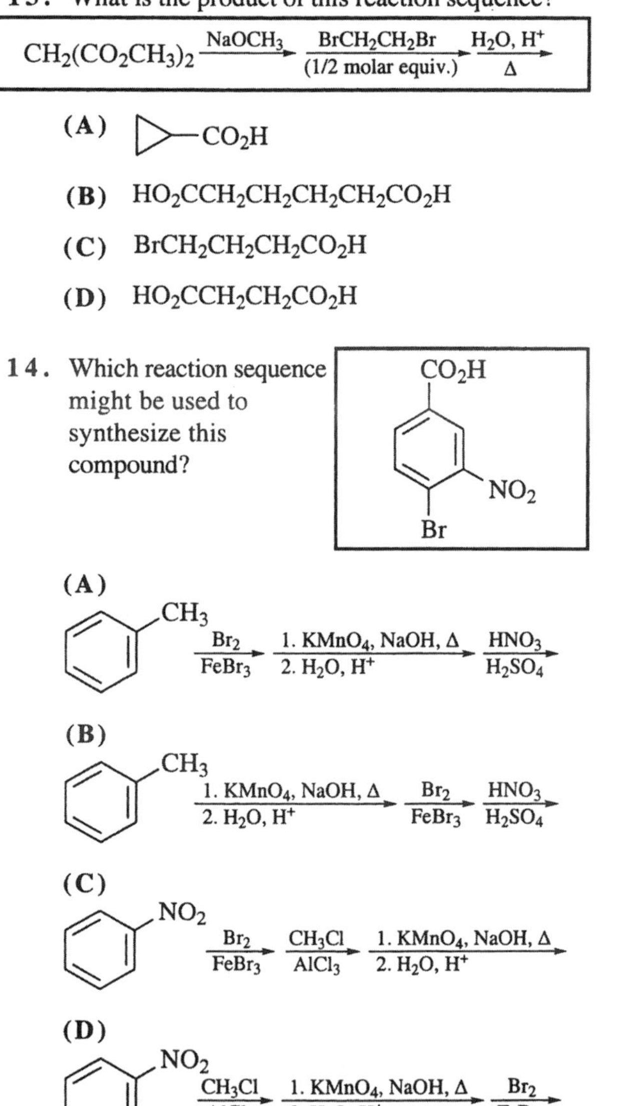

15. Which sequence of reagents would effect this conversion in highest yield?

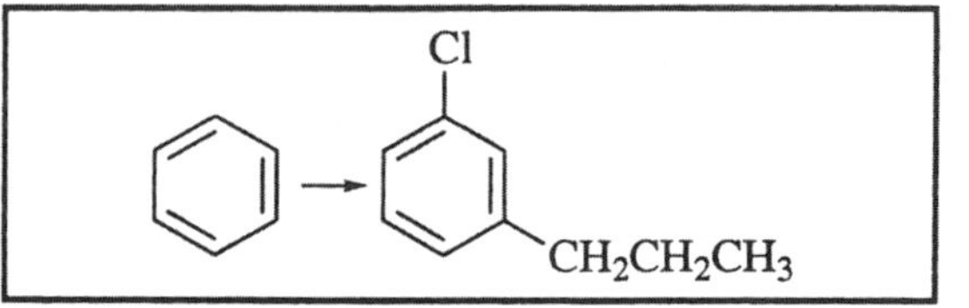

(A) $\xrightarrow[AlCl_3]{CH_3CH_2CH_2Cl} \xrightarrow[FeCl_3]{Cl_2}$

(B) $\xrightarrow[FeCl_3]{Cl_2} \xrightarrow[AlCl_3]{CH_3CH_2CH_2Cl}$

(C) $\xrightarrow[AlCl_3]{CH_3CH_2C(=O)Cl} \xrightarrow[H_2O,\ H^+]{Zn(Hg)} \xrightarrow[FeCl_3]{Cl_2}$

(D) $\xrightarrow[AlCl_3]{CH_3CH_2C(=O)Cl} \xrightarrow[FeCl_3]{Cl_2} \xrightarrow[H_2O,\ H^+]{Zn(Hg)}$

16. What is the product of this reaction sequence?

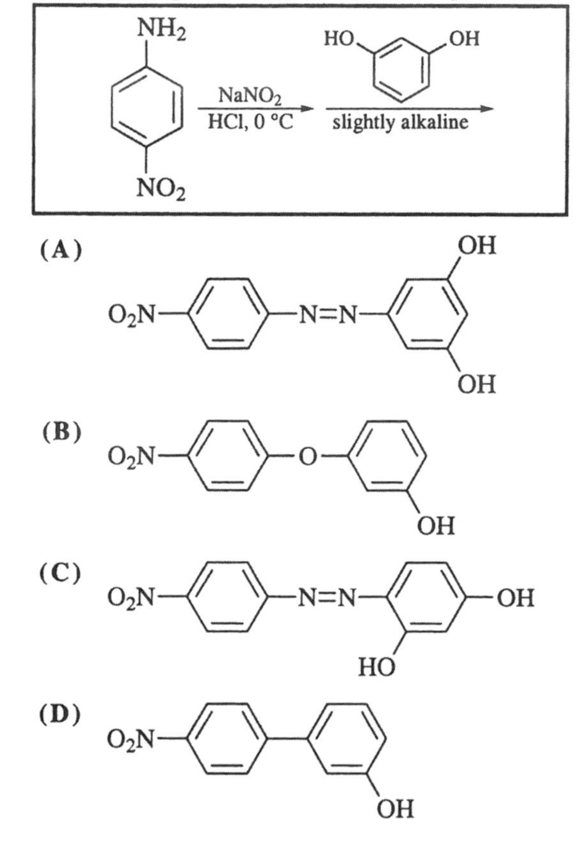

17. Which sequence of reagents will accomplish this transformation?

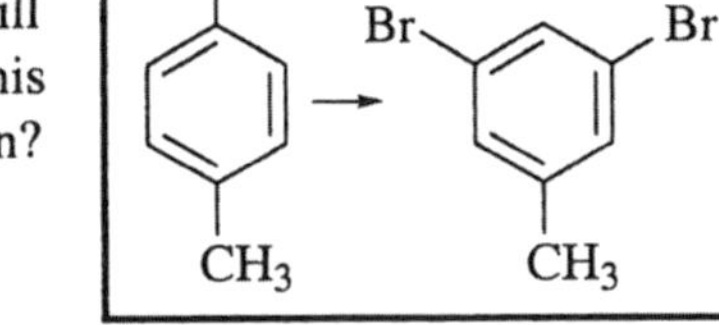

(A) $\xrightarrow[\text{HCl, 0 °C}]{NaNO_2}$ $\xrightarrow[\Delta]{CuBr}$

(B) $\xrightarrow[\text{HCl, 0 °C}]{NaNO_2}$ $\xrightarrow{H_3PO_2}$ $\xrightarrow{Br_2}$

(C) $\xrightarrow{Br_2}$ $\xrightarrow{LiAlH_4}$

(D) $\xrightarrow{Br_2}$ $\xrightarrow[\text{HCl, 0 °C}]{NaNO_2}$ $\xrightarrow{H_3PO_2}$

18. Predict the major product of this reaction sequence.

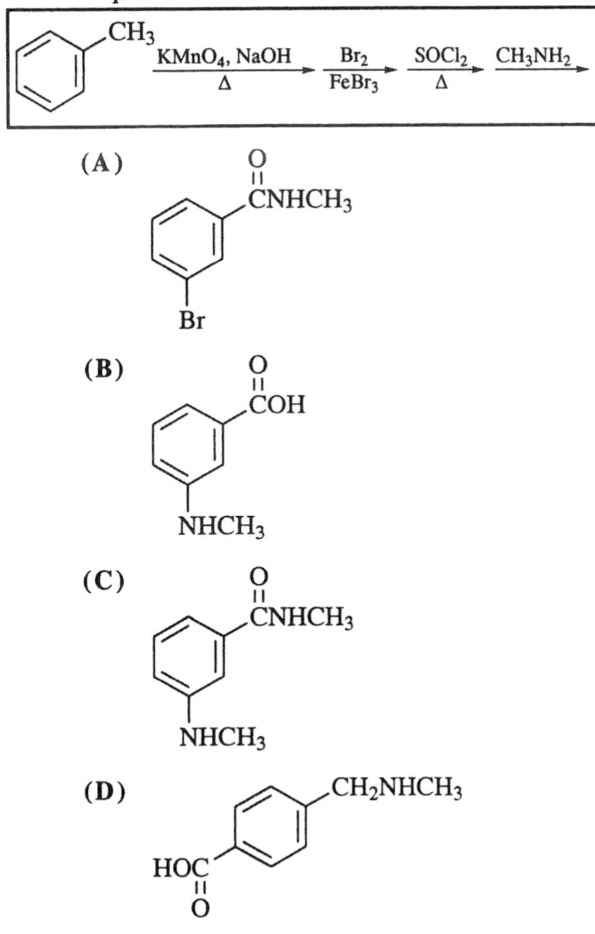

19. Which sequence of reagents will effect this transformation?

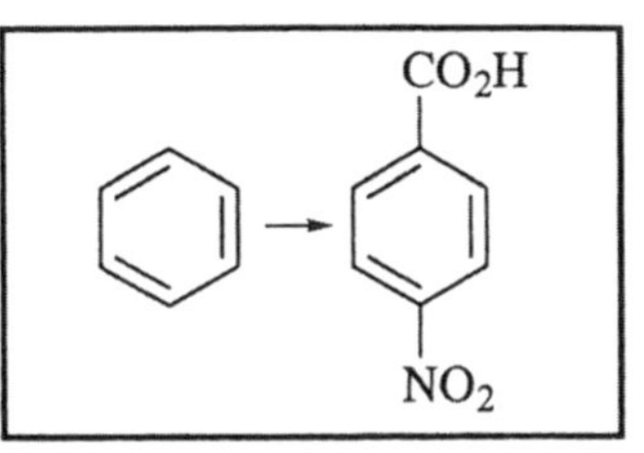

(A) $\xrightarrow[AlCl_3]{CH_3CH_2Cl}$ $\xrightarrow[H_2SO_4]{HNO_3}$ $\xrightarrow[\Delta]{\text{KMnO}_4\text{, NaOH}}$

(B) $\xrightarrow[H_2SO_4]{HNO_3}$ $\xrightarrow[FeBr_3]{Br_2}$ $\xrightarrow[\text{ether}]{Mg}$ $\xrightarrow[\text{2. } H_2O, H^+]{\text{1. } CO_2}$

(C) $\xrightarrow[H_2SO_4]{HNO_3}$ $\xrightarrow[AlCl_3]{CH_3CH_2Cl}$ $\xrightarrow[\Delta]{\text{KMnO}_4\text{, NaOH}}$

(D) $\xrightarrow[AlCl_3]{CH_3CH_2Cl}$ $\xrightarrow[\Delta]{\text{KMnO}_4\text{, NaOH}}$ $\xrightarrow[H_2SO_4]{HNO_3}$

20. A compound yields an orange precipitate when treated with 2,4-dinitrophenylhydrazine in an acidic solution. The compound also reacts with $Ag(NH_3)_2^+$ ^-OH (Tollens reagent). Which compound is consistent with these observations?

(A) 2-pentanone (B) pentanal

(C) 2-pentanol (D) methyl pentanoate

21. An unknown alkene gives propanal as the only product upon ozonolysis (reductive workup). When the unknown is treated with bromine in carbon tetrachloride, an optically inactive dibromide is formed, which cannot be resolved into enantiomers. Which of these alkenes is consistent with this data?

(A) CH_3, CH_2CH_3 / C=C / H, H

(B) H, CH_2CH_3 / C=C / CH_3, H

(C) H, CH_2CH_3 / C=C / CH_3CH_2, H

(D) CH_3CH_2, CH_2CH_3 / C=C / H, H

22. A tetrapeptide, containing 2 moles of glycine (Gly) and 1 mole each of leucine (Leu) and alanine (Ala) of unknown sequence, gives a dinitrophenyl derivative of Gly after treatment with 1-fluoro-2,4-dinitrobenzene and complete hydrolysis. Partial hydrolysis of the tetrapeptide gives Gly–Ala and Gly–Gly. What is the amino acid sequence in the original tetrapeptide?

(A) Ala–Gly–Gly–Leu

(B) Gly–Ala–Gly–Leu

(C) Leu–Gly–Gly–Ala

(D) Gly–Gly–Ala–Leu

23. An unknown compound is ***not*** oxidized by chromic acid solution. When dissolved in Lucas reagent (conc. $HCl/ZnCl_2$), the solution immediately becomes cloudy with the separation of a second liquid phase. Which compound is consistent with this data?

(A) $CH_3CH_2CH_2CH(=O)$

(B) $CH_3CH_2C(OH)(CH_3)CH_3$

(C) $CH_3CH_2CH(OH)CH_3$

(D) $CH_3CH_2C(=O)CH_3$

24. An optically active alcohol with the molecular formula of C_5H_8O undergoes hydrogenation over platinum to give an optically inactive alcohol. Which of these alcohols is consistent with this data?

(A) $HC{\equiv}CCH(OH)CH_2CH_3$

(B) $HC{\equiv}CCH_2CH(OH)CH_3$

(C) 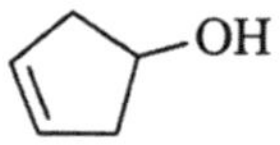 cyclopentenol (ring with –OH)

(D) 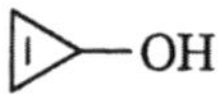 cyclopropene–OH

25. Which two of these reaction sequences will effect this transformation?

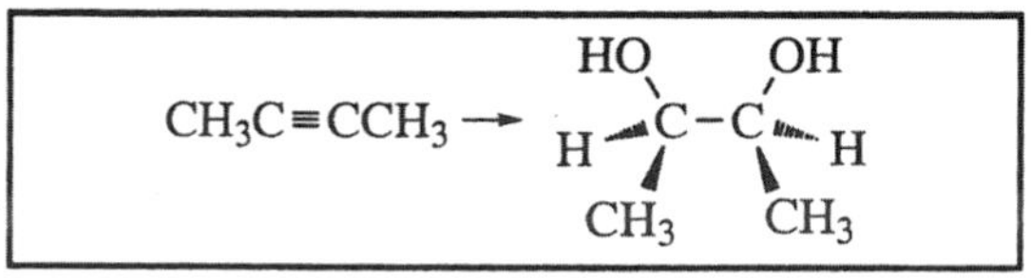

$CH_3C{\equiv}CCH_3 \rightarrow$ H–C(HO)(CH₃)–C(OH)(CH₃)–H

I $\xrightarrow[\text{deactivated Pd}]{H_2}$ $\xrightarrow[H_2O_2\text{ (excess)}]{OsO_4\text{ (cat. amt.)}}$

II $\xrightarrow[\text{liq. }NH_3]{Na}$ $\xrightarrow[H_2O_2\text{ (excess)}]{OsO_4\text{ (cat. amt.)}}$

III $\xrightarrow[\text{deactivated Pd}]{H_2}$ $\xrightarrow{CH_3CO_3H}$ $\xrightarrow[NaOH]{H_2O}$

IV $\xrightarrow[\text{liq. }NH_3]{Na}$ $\xrightarrow{CH_3CO_3H}$ $\xrightarrow[NaOH]{H_2O}$

(A) II and III **(B)** I and III

(C) I and IV **(D)** II and IV

Answers to Study Questions

1. A
2. C
3. D
4. B
5. B
6. B
7. D
8. C
9. B
10. D
11. C

Answers to Practice Questions

1. B
2. D
3. A
4. C
5. B
6. D
7. A
8. B
9. C
10. C
11. A
12. C
13. A
14. A
15. D
16. C
17. D
18. A
19. A
20. B
21. C
22. D
23. B
24. A
25. C